FAUNA ENTOMOLOGICA SCANDINAVICA

Volume 3   1975

The Tachydromiinae
(Dipt. Empididae)
of Fennoscandia and Denmark

by

M. Chvála

SCANDINAVIAN SCIENCE PRESS LTD.

Klampenborg . Denmark

Copyright for the World
Scandinavian Science Press Ltd.

Edited by
Societas Entomologica Scandinavica

Editor of this volume
Leif Lyneborg

World list abbreviation
Fauna ent. scand.

Printed by
Vinderup Bogtrykkeri A/S,
7830 Vinderup, Denmark

ISBN 87-87491-04-4

Contents

Introduction

The subfamily Tachydromiinae (Diptera, Empididae) is a group of small flies, 0.7 to 5.5 mm in length, that is generally highly adapted for predaceous activity. The adults may be found in practically every kind of biotope and in varying conditions, often in large numbers in almost any month of the year. The Palaearctic fauna is very inadequately known, but the subfamily certainly contains several hundred species in this region. Except for Drapetis and perhaps Crossopalpus, the genera are primarily Holarctic in distribution and their origin should be sought in the cold and temperate zones of the northern hemisphere.

An exact knowledge of the North European fauna is very important, not only because of its presumed origin in these regions but also because the first taxonomic works to include descriptions of new taxa were based solely on material from this region. The pioneers of Scandinavian and world entomology tended to overlook the tiny tachydromiines: for instance, Linné (1761) described only 1 species and Fabricius (1775, 1794) 2 species, of which only one is valid, but shortly afterwards Fallén (1815, 1816) described 12 and Zetterstedt (1819 - 1859) even 21 currently valid species. By the middle of the 19th century, the latter two dipterists had thus distinguished more than one quarter of the currently known Fennoscandian fauna. During the era of Fallén and Zetterstedt, however, there were several famous Dipterists in Germany, France and England, such as Meigen, Panzer, Macquart, Haliday, Walker and von Roser, who were at the same time also describing new taxa, usually without any close contact. The situation became still more complicated in the second half of the 19th century, during the period of the greatest activity of Loew, Mik, Becker and particularly Strobl, who mostly based their descriptions of new taxa on comparisons with inadequate older literary records and without seeing any type-material. Important taxonomic papers by Frey (1907 - 1943) and Lundbeck (1910) should also be mentioned here. The first step when studying the Fennoscandian and the entire European fauna of the Tachydromiinae was to bring some order into the taxonomy of this very complicated group, which involved the precise study of all types with the identification of holotypes or the designation of lectotypes when necessary. The first serious work of this kind was done by Collin (1926, 1961); this author revised the most important collections, such as those of Fallén, Meigen, Macquart, and of course those of Haliday and Walker, but he did not see some parts of Zetterstedt's Collection or the collections of Strobl and Frey.

These revisional studies have been a primary task of the present author, and as a result of intensive study over several years the individual taxa have been defined and delimited.

Altogether 128 species of Tachydromiinae in 9 genera have been found in Denmark and Fennoscandia; a few further species may be expected to occur there but the number will certainly not increase very much. All the species included here are distinguished on the basis of the male genitalia in addition to the external characters, and the more important diagnostic features including the male genitalia are illustrated, most of them for the first time. Each species, is very briefly described or differentiated from the most closely related species, and, so far as the limited scope of the Fauna allows, brief records on the distribution in Fennoscandia, the entire range of distribution, the flight-period and notes on the biology are given. Species which may also be found in Fennoscandia are mentioned with a short differential diagnosis, and are often included in the key, where they appear without numbering.

It is hoped that this Fauna will help all students with the identification of the Fennoscandian species of Tachydromiinae, and that it will also help students in other parts of Europe. It should be borne in mind that limitation of size have made it impossible to present the more detailed descriptions that are so necessary for comparison with closely related species from temperate and warmer zones of Europe, and in these cases the references cited to full descriptions where these are available must be consulted or the illustrations of the male genitalia compared.

Acknowledgements

The author is very grateful to the following colleagues for the loan of material, and for very valuable information on collections, localities or literature: Mr. H. Andersson (Lund), Dr. C.E. Dyte (Slough), Prof. W. Hackman (Helsinki), Mr. L. Hedström (Uppsala), Prof. W. Hennig (Ludwigsburg), Prof. H. Kauri (Bergen), Dr. L. V. Knutson (Washington), Dr. F. Kühlhorn (Munich), Dr. A. Lillehammer (Oslo), Mr. L. Lyneborg (Copenhagen), Dr. A. Løken (Bergen), Mr. P. I. Persson (Stockholm), Mr. A. C. Pont (London), Dr. T. Saigusa (Fukuoka), Mr. K. G. V. Smith (London) and Dr. J. R. Vockeroth (Ottawa).

I am very indebted for loans of type-material to Mr. D. M. Ackland of Oxford (Collin, Bigot), Dr. P. Arnaud Jr. of San Francisco (Melander), Dr. A. Kaltenbach of Vienna (Wiedemann, Meigen, Mik), Prof. E. Lindner of

Ludwigsburg by Stuttgart (von Roser), Dr. habil. G. Morge of Eberswalde by Berlin (Strobl, Oldenberg), Dr. H. Schumann of Berlin (Loew, Becker, Duda) and Dr. G. C. Steyskal of Washington (Coquillett). Dr. V. G. Kovalev, Moscow, was very kind in providing me with information about the fauna of the adjacent European part of the USSR and with manuscript copies of his own papers in advance of publication.

I am very much indebted to Professors S. L. Tuxen (Copenhagen), Carl H. Lindroth (Lund) and W. Hackman (Helsinki) for their kind hospitality during my stay in their institutes, and my colleague Leif Lyneborg, Chief Editor of the Fauna, is thanked for his enthusiastic and very valuable help and for many suggestions offered during the preparation of the manuscript.

I am very grateful to the artist Mrs. Grete Lyneborg for her careful execution of the drawings; the outlines for the detailed drawings were prepared by the author. Mr. Adrian C. Pont of the British Museum (Nat.Hist.) has kindly checked the English of the manuscript and I am very indebted to him for many valuable suggestions.

The Danish State Research Foundation supported my one-year stay in Copenhagen, and the Board of the Entomological Society of Helsinki offered financial support during my stay in Helsinki.

Last but not least I would like to thank my wife for her enthusiastic acceptance of my study of Diptera as a normal part of our family life.

Material

The present monograph is based on the examination of extensive material in many collections from Denmark and Fennoscandia, including material collected by the author in Denmark in 1968 and 1969. Altogether over 18.000 specimens of Fennoscandian Tachydromiinae (exactly 18.065) have been examined by the author during the past seven years, and a much larger number of specimens including type-material has been seen from other parts of the Holarctic region and particularly from Europe.

The basic Fennoscandian collections are those of Frey, deposited in Helsinki (including the Finnish Collection), and of Zetterstedt, deposited at Lund. The latter contains several parts, viz., Diptera Scandinaviae, Insecta Lapponica, Göteborg, and Diptera Exotica Collections, and some of Zetterstedt's syntypes are also located in the collections of Roth and Wallengren; the results of a study of these collections have been published separately (Chvála, 1971). In addition to the new recently collected material, taken mainly by

9

Hugo Andersson, there is also the fine Ringdahl Collection at Lund. In Stockholm, besides the Fallén Collection (revised by Collin), there is the small but very important Boheman Collection which also includes Wahlberg's material.

The basis of the material from Denmark consists of the Stæger and Lundbeck Collections in Copenhagen, but the Danish collection has been substantially increased during the past few years thanks to the activity of Leif Lyneborg and of several collecting expeditions to Jutland and Bornholm. Unfortunately less representative material is available from Norway; in Oslo there is the rather small but well-preserved Siebke Collection (published by Siebke in 1877), and a more extensive collection is located at Bergen, mainly consisting of the recently collected alcohol material obtained during the Hardangervidda investigations. The scanty records from Northern Norway are based on the material collected by Zetterstedt during his Norwegian trips and on several expeditions undertaken by Hugo Andersson of Lund.

Practically all the published Fennoscandian material was revised by the author, except for the Fallén Collection which was revised by Collin, and the material of Bonsdorff which was revised in the earlier papers of Frey and is now incorporated with other Finnish material.

To save space, the results of these revisions originally listed in chronological order in the "Synonymy" under the heading of each species, are deposited and available for further study in the Universitetets Zoologiske Museum, Copenhagen. Other literary records from non-Scandinavian countries are not included, except for the two fundamental monographs of Engel (1938 - 1939) in Lindner, Die Fliegen der Palaearktischen Region, and Collin (1961) in British Flies; these are listed together with the regular synonyms.

Life history

So far as is known, the adults of all Tachydromiinae are predaceous in both sexes and their predaceous activity is not restricted to copulatory activity, as in the Empidinae. On the contrary, epigamic behaviour in Tachydromiinae generally takes place without simultaneous feeding. Some species of Platypalpus and Chersodromia are highly predaceous, attacking all insects and frequently also each other, even in captivity. The prey consists of small insects or Collembola, but above all small Diptera, and the size of the prey is often larger than that of the predator; generally the prey is the commonest insect available. When resting Tachydromiinae suck the prey whilst holding it in the front legs, but when moving they hold it impaled on the proboscis, except for

Platypalpus species with thickened raptorial mid legs which hold their prey with the mid pair beneath the body and use only the front and hind legs for running. Some species of Platypalpus and Drapetis also visit flowers accidentally. Detailed observations of these problems with Finnish Empididae were published by Tuomikoski (1952).

The adults may be found under various conditions: on ground-vegetation, in meadows and fields, on bushes, trees and large leaves or on heaps of vegetable matter, on tree-trunks, stones, rocks, palings, guard-stones, or in sandy biotopes; some of them are typical coastal species. The behaviour and ecology is well-defined and characteristic for some genera, such as for instance Symballophthalmus (vegetation), Tachypeza (tree-trunks), Chersodromia (coasts) or Stilpon (terrestrial habitats), but in other genera the behaviour is characteristic for each species, as for instance in Tachydromia, Platypalpus, Drapetis or Crossopalpus; further details are given under each genus or species in the Systematic part.

Species of Symballophthalmus and some primitive Platypalpus species fly slowly over ground-vegetation rather like ocydromiine empidids, but all others take flight only rarely, usually when disturbed, and fly very quickly for short distances. The only good fliers seem to be Drapetis species, but unlike the primitive Tachydromiini they fly very quickly. Some brachypterous species cannot fly or use their wings for characteristic jumps, as for instance Chersodromia arenaria.

The flight-period of most of the species is confined to the summer months, although some of them are typical early spring species, such as for instance Platypalpus pallidicoxa, agilis, pseudorapidus, cothurnatus, laticinctus, Tachydromia connexa; and some are typical representatives of late summer or autumn, such as for instance Platypalpus ciliaris, nigritarsis, or Crossopalpus curvinervis. The phenology of the Finnish species was studied by Tuomikoski (1938). Several species hibernate as adults, viz. Platypalpus maculipes, Stilpon graminum and at least some Crossopalpus species, and may be found in any month of the year. The adults hibernate in the same places as they live, and are usually fully active during winter.

The flight-period, of course, differs slightly in northern and southern populations; in species with a long flight-period, both extreme dates are usually found in the southern populations in Denmark and southern Sweden, owing to the short summer in the north. However, species with a short flight-period, particularly spring or early summer species, occur first in the south but at the end of the flight-period may be found almost a month later in the north (Lapland). The flight-period given under each species shows the approximate situation in Denmark and Fennoscandia, and the differences between the south

11

and the north are not shown; this corresponds roughly with the flight-period known from Great Britain. If scanty records are available from Fennoscandia, the dates from Great Britain or from the adjacent part of the USSR are interpolated; the precise dates known from Fennoscandia are given in Table 1. The different, and generally earlier, occurrence in central and southern parts of Europe is recorded in the text in brackets or is otherwise discussed.

The larvae of the Tachydromiinae still remain unknown, except for several inadequate old literary records, although they have been collected and reared several times. They are very probably predaceous, and have been found several times in soil or under moss (Platypalpus) or in rotten wood and in debris of hollow trees (Tachypeza, Drapetis).

Geographical or obligatory parthenogenesis is found in some Platypalpus species and is fully discussed under this genus.

Economic importance

The economic importance of Tachydromiinae as predators of important agricultural pests cannot be overlooked. Their significance is shown by two factors, (1) by their great abundance and often mass-occurrence in small areas - for instance, there are often as many as twenty specimens of Platypalpus per square metre in meadows and fields, and (2) in the niche they occupy, preying on small insects, they are practically without competitors. There is evidence that some noxious species would be far more abundant were it not for natural predation already taking effect (Smith, 1969).

The Tachydromiinae are undoubtedly very important predators of phytophagous insects and mainly of Diptera. The most important feature for biological control is the mass-occurrence of some Tachydromiinae in fields of wheat, above all of Platypalpus pallidicornis and Drapetis incompleta. Species of Platypalpus and Drapetis as destroyers of agromyzid flies have also been mentioned by Kovalev (1966). The only literary record known to me of deliberate biological control is that of Whitfield (1925) who used Platypalpus minutus to destroy the agromyzid leaf-miner Phytomyza aconiti Hendel in a glasshouse, and this seems to be one of the possible ways of using the remarkable predatory activity of Tachydromiinae.

Tachydromiinae and other Empididae have been recorded in the literature several times as predators of both adults and larvae of Simuliidae (Peterson, 1960; Peterson and Davies, 1960; Chillcott, 1961 or Sommerman, 1962), of Culicidae (Service, 1969) and Ceratopogonidae (Becker, 1958). I myself have

several times observed species of Tachydromia, Tachypeza and sometimes
also Platypalpus preying on psyllids on tree-trunks of fruit trees, and Fleschner and Ricker (1953) referred to Drapetis micropyga Melander as an important predator of citrus red mite Panonychus citri (Mc Gregor) in the U.S.A.

Zoogeography

Most of the 9 genera of Tachydromiinae recorded from Fennoscandia are mainly Holarctic in distribution, with the exception of Platypalpus, Drapetis and Crossopalpus; Drapetis subgenus Elaphropeza is best represented in tropical and subtropical regions. On the other hand only Platypalpus unguiculatus, Tachypeza winthemi, and perhaps also Drapetis (s.str.) assimilis are known to be Holarctic in distribution. Although the Tachydromiinae are typical representatives of the cold and temperate zones in Europe, and the fauna of colder regions is quite well known at present, it seems likely that the fauna of central and probably also of southern Europe is richer in species. Unfortunately very little is known of the Siberian fauna, and practically nothing of that of Asia.

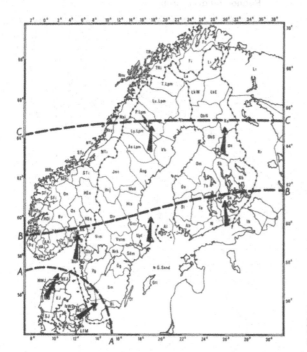

Fig. 1. Distribution-pattern of Fennoscandian Tachydromiinae. - Groups A-C.

13

The 128 recorded Fennoscandian species may be divided into seven groups (A - G) of different distribution-patterns, which also demonstrates the supposed origin of the species. The groups may be characterised as follows (cf. Figs. 1 and 2):

A - Atlantic elements, extending from England to Denmark and the extreme south of Sweden (Scania , Hallandia). About 12% of the Fennoscandian fauna belongs here.

B - Species of southern distribution within a larger area covering Denmark, the southern parts of Norway, Sweden, and the southern part of Finland along the Baltic coast including NW of European USSR, reaching approximately 62°N. The largest group of species, including about 32 % of the fauna.

C- Species with the same distribution-pattern as under B, but extending further northwards into central and northern parts of Fennoscandia, to approximately $65 - 66^{\circ}$N; absent in the extreme north and in the Kola Peninsula. A large group of species, including about 20 % of the fauna.

D - Widely distributed species, extending from the south (Denmark, S. Sweden) up to and including the extreme north (Lapland) and also the Kola Peninsula. About 8 % of the Fennoscandian fauna.

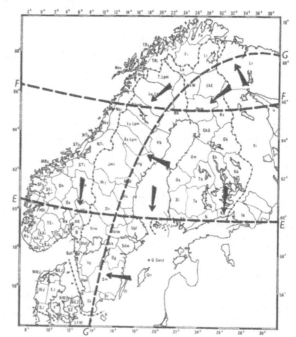

Fig. 2. Distribution-pattern of Fennoscandian Tachydromiinae. - Groups E-G.

E - Northern species, covering a large area including the extreme northern
and central parts of Fennoscandia, absent in the south of Norway, Sweden
and in Denmark, but often reaching the Baltic coast of Finland. The smal-
lest group of species, including about 5 % of the fauna.

F - Extremely northern species, distributed within a smaller area covering
Norwegian, Swedish and Finnish Lapland including the Kola Peninsula,
usually extending to or from Siberia. About 9 % of the Fennoscandian species.

G - Continental Eastern European species, extending through NW parts of
European USSR to the eastern parts of Fennoscandia, and sometimes probably along the Gulf of Bothnia as far as southern Sweden. Absent in western Fennoscandia, Denmark and Great Britain. A larger group of species,
including about 14 % of the fauna.

A survey of all the species is given in Table 1, with the symbols A - G to
show the type of distribution and further data on the known occurrence in Europe. The supposed pattern of distribution is given in brackets if only single
specimens from Fennoscandia are available; if the known occurrence in Fennoscandia shows a different type of distribution from what would be expected
on the basis of other records from adjacent countries, the probable correct
distribution-symbol is given in brackets, for instance A(G) which means that
the G type of distribution is probably correct.

It is clear from the Table 1 that only two-thirds of the 128 known Fennoscandian species are common to both Great Britain (87 species) and Central
Europe (88 species), and only 30 species have been found in Southern Europe;
95 species have also been found in the NW parts of European USSR (Carelia,
Leningrad region, Estonia) and only 32 species in the north of the USSR (mainly the Kola Peninsula). Four Platypalpus species (albisetoides, longicornioides,
tuomikoskii and subbrevis) are still known only from Fennoscandia.

Of the 88 Fennoscandian species also found in Central Europe, at least
four (Platypalpus confiformis, commutatus, nigricoxa and brevicornis) are
very probably boreoalpine in distribution, and at least a further seven (Platypalpus stigmatellus, maculus, boreoalpinus, alpinus, difficilis, Tachypeza
truncorum, heeri), if not of the same origin, are restricted to the mountainous
areas of Central Europe. However, the Central European fauna includes many
more species, certainly almost two hundred, but many of them still await description, as do many from Southern Europe. Engel (1938-1939) erroneously
recorded many Scandinavian species from the whole of Europe, including southern areas, as was clearly shown in the genus Chersodromia (Chvála, 1970a).

Table 1

	type of distribution	Great Britain		European USSR			Europe			Asia	Flight period in Denmark and Fenno-scandia
		England	Scotland	northwest	north	central	central	south	southwest	Asia	
Symballophthalmus											
1. dissimilis	C-D		+	+		+					V-VII
2. fuscitarsis	B-C	+		+		+	+				V-VII
3. pictipes	(B)	+	+				+				VII
Platypalpus											
4. ciliaris	C	+	+	+		+	+				VI-X
5. confiformis	E				+		+				VII-VIII
6. confinis	E		+	+	+						VI-VIII
7. stigmatellus	E		+	+	+		+				VI-IX
8. pectoralis	C	+	+	+		+	+	+			V-X
9. mikii	A(G)		+				+	+			VII
10. nonstriatus	G		+				+				?
11. maculus	D	+	+	+	+		+				V-VIII
12. pallipes	C	+	+	+							V-X
13. albiseta	B	+		+		+	+	+			VI-IX
14. albisetoides	B										?
15. albocapillatus	C-D	+	+	+		?					VI-IX
16. niveiseta	A	+					+		+		?
17. unguiculatus	D		+	+	+						VI-VIII
18. zetterstedti	F			+							VI-VII
19. alter	F		+	+							VI
20. laestadianorum	F			+							VII-VIII
21. lapponicus	F			+							VII
22. sahlbergi	F			+							V-VII
23. boreoalpinus	E		+	+			+				VI-VIII
24. alpinus	G						+				V-VII
25. commutatus	F		+		+		+				?
26. longicornis	D	+	+	+	+	+	+	+			V-X
27. longicornioides	(A)										VI
28. brunneitibia	G		+				+	+	+		VI-VIII
29. difficilis	B-C	+	+	+		+	+				VI-IX
30. scandinavicus	G		+								VII-IX
31. tuomikoskii	(G)										?
32. exilis	C	+	+	+		+	+	+			V-VIII
33. pulicarius	B	+		+			+				VI-IX
34. nigricoxa	F			+			+				VI-VII
35. luteus	C	+	+	+		+	+	+			V-IX
36. nigritarsis	D	+	+	+	+	+	+	?			V-XI
37. excisus	C	?	?	+			+	+			IX
38. sylvicola	B	+	+	+							VII
39. fuscicornis	B			+		+	+				VI-VII
40. ruficornis	A	+					+				VII-VIII
41. fenestella	G			+		+					VIII
42. ater	F-G			+	+						VII
43. minutus	C-D	+	+	+		+	+	?			V-X
44. niger	A(B)	+		+		+	+				VI-VIII
45. aeneus	B(A)	+					+		+		VI-VII

	type of distribution	Great Britain		European USSR			Europe				Flight period in Denmark and Fenno-scandia
		England	Scotland	northwest	north	central	central	south	southwest	Asia	
46. maculipes	C	+	+	+		+	+	+	+		throughout
47. rapidus	G(B)	+					+		+		IV-VII
48. pallidicoxa	B		+	+			+				V-VII
49. agilis	B	+	+	+		+	+				V-VII
50. pseudorapidus	G			+			+				V-VII
51. nigrosetosus	B(G)			+			+				VI-VIII
52. cothurnatus	B	+		+		+	+	+	+		V-VII
53. cryptospina	B	+		+		+	+	?			VI-VII
54. optivus	A	+					+				VII
55. annulatus	C	+	+	+		+	+	+	?	+	V-IX
56. melancholicus	G(C)	+	+	+			+				V-VII
57. notatus	C	+	+	+			+				V-X
58. strigifrons	B	+	+	+							V-X
59. infectus	B	+	+	+		+	+				V-IX
60. interstinctus	C	+	+	+		+	+				V-IX
61. coarctatus	B	+	+	+		+	+				VI-IX
62. clarandus	A	+	+								VI-VII
63. articulatus	C	+		+		+	+		+		V-IX
64. articulatoides	C			+	+		+				V-VII
65. annulipes	B-C	+	+	+			+				V-IX
66. ecalceatus	D		+	+		+	+				V-VIII
67. calceatus	C	+	+	+		+	+				VI-IX
68. stabilis	B	+		+							VI-VIII
69. pallidiventris	C	+	+	+		+	+	+			V-IX
70. longiseta	C	+	+	+		+	+	+			V-IX
71. laticinctus	A	+	+				+				V-VII
72. albicornis	B	+		+		+	+				V-VII
73. flavicornis	B	+		+		+	+	+			VI-IX
74. pallidicornis	B	+		+		+	+				III-VIII
75. major	C	+	+	+		+	+	+			V-VIII
76. analis	A	+					+				VII-VIII
77. candicans	C-D	+	+	+		+	+	+			V-VIII
78. cursitans	D	+	+	+	+	+	+	?			V-VIII
79. verralli	C	+	+	+		+	+				VI-VII
80. brevicornis	G-D			+			+				IV-VIII
81. sordidus	G-D			+	+						VII-VIII
82. subbrevis	F										?
83. hackmani	G-D			+	+						VI-VII
Dysaletria											
84. atriceps	G			+			+				VI-VII
Tachypeza											
85. nubila	C	+	+	+		+	+	+	+		V-XI
86. truncorum	D		+	+	+	+	+				VI-IX
87. fuscipennis	C	+		+		+	+				V-VIII
88. heeri	E		+	+	+		+			+	VI-IX
89. fennica	G-D			+	+	+	+	+		+	V-VIII

17

	type of distribution	England	Scotland	northwest	north	central	central	south	southwest	Asia	Flight period in Denmark and Fenno-scandia
		Great Britain		European USSR			Europe				
90. winthemi	E				+						VI-VIII
91. sericeipalpis	F				+						?
Tachydromia											
92. terricola	B			+		+	+				V-VII
93. sabulosa	B			+			+				VI-VII
94. connexa	B	+		+		+	+	+	+		VI
95. morio	D	+	+	+	+		?				VI-VIII
96. lundstroemi	G			+							VII-IX
97. arrogans	B-C	+		+			+	+			III-X
98. aemula	B-C	+	+	+			+	+	+		VI-X
99. punctifera	F				+						VII
100. incompleta	F				+					+	VI-VIII
101. umbrarum	D	+	+	+	+	+	+				V-IX
102. woodi	A(B)	+		+		+	+				?
Drapetis											
103. (E.)ephippiata	B	+		+		+	+	?			VI-VIII
104. (D.)assimilis	G(B)	+		+		+	+				V-IX
105. (D.)ingrica	G			+							VII-VIII
106. (D.)arcuata	B	+					+				VI-VIII
107. (D.)simulans	B	+		+		+	+				VI-IX
108. (D.)pusilla	B	+		+			+	+			VI-X
109. (D.)exilis	B	+		+		+	+	+	+		V-IX
110. (D.)infitialis	G(B)	+		+		+	+	+			VII-VIII
111. (D.)parilis	B	+	+	+		+	+				VI-VII
112. (D.)incompleta	A-B					+	+				V-VIII
Crossopalpus											
113. setiger	B	+		+		+	+	+	+		V-IX
114. curvipes	D	+	+	+	+						V-IX
115. nigritellus	B	+		+		+	+	+	+		throughout
116. humilis	C	+	+	+	+	+	+	+			throughout
117. curvinervis	G			+		+	+			+	V-IX
118. abditus	G			+		+					?
Chersodromia											
119. hirta	A	+	+								V-VIII
120. cursitans	C	+		+		+	+	+			V-VIII
121. difficilis	A	+	+								VI-X
122. arenaria	D	+	+		+						V-IX
123. speculifera	A	+	+								V-VII
124. beckeri	B			+							VI
125. incana	A	+									V-VIII
Stilpon											
126. graminum	B	+		+		+	+	+			throughout
127. nubila	A-B	+		+				+			VI-X
128. lunata	A	+	+	+						+	VI

The British fauna is undoubtedly the best known in Europe, thanks to Collin (1961) who recognised 110 British species of Tachydromiinae. Since that time a further six species (Platypalpus pygialis, longimanus, analis, Tachydromia acklandi, Drapetis (Drapetis) infitialis and Chersodromia cursitans) have been found, so that the total number of British species has now increased to 116. There are 29 species in the British fauna which are not known from Denmark and Fennoscandia, and most of them are southern elements penetrating from SW or C parts of Europe along the Atlantic coast north to England. On the other hand the Fennoscandian fauna is richer by 41 species which belong to a group formed equally by eastern species, extending to Fennoscandia from the Continent through European USSR and Finland (G type of distribution), and by northern species, often probably of Siberian origin, occupying the north of Fennoscandia (F and E types of distribution).

Adult morphology

HEAD (Figs. 3-12).

The head is closely set upon thorax in Drapetini, almost globular with rounded eyes in Symballophthalmus (Fig. 3) and primitive Platypalpus-species, with a tendency to be distinctly deeper than long or broad in all other genera with eyes more or less reniform; sometimes distinctly produced below eyes because of deep jowls (Crossopalpus (Fig. 10) and some Chersodromia (Fig. 11)). Eyes bare (Tachydromiini) or microscopically pubescent (Drapetini), with facets equal in size or slightly enlarged at middle or below antennae (Symballophthalmus, Stilpon), often contiguous on face or on frons (without sexual distinctions) which provides a good generic or specific (Platypalpus) character. Usually 1 or 2 pairs of ocellar and vertical bristles developed, sometimes very conspicuously so, or entirely absent. Vertex with 3 distinct ocelli which are sometimes (Tachypeza) placed lower on frons.

ANTENNAE (Figs. 13-21).

The antennae are placed at about middle of head in profile, sometimes directed upwards (Drapetis s.str.), paraantennal excisions distinct (small in Symballophthalmus). Basal segment (scape) always very small, often entirely concealed; segment 2 (pedicel) cup-shaped, generally uniform in shape and smaller than segment 3, distinctly enlarged in Stilpon (Fig. 21); apically with a circlet of short bristly-hairs, or with longer bristles beneath at tip in Drape-

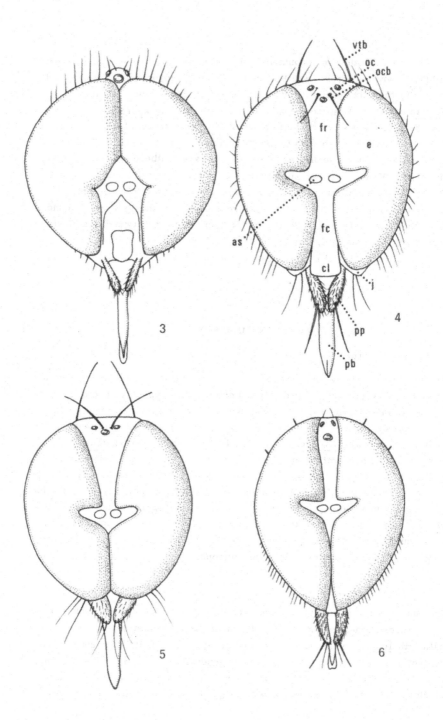

tis (Fig. 18) and particularly Crossopalpus (Fig. 19). The rest of antenna (flagellum) is formed by segment 3 and arista; segment 3 is usually pointed apically and more or less elongate, or short-ovate to almost spherical, very small and deeper than long in Stilpon (Fig. 21), generally longer and more densely pubescent in more primitive Tachydromiini; the shape of segment 3 is practically constant in all genera except for Platypalpus, where it varies conspicuously both in shape and length and is of great importance for the classification at specific level. The arista is distinctly two-segmented with basal segment very small, in the form of a short hook-like projection at tip or in dorsal part of segment 3, rest of arista usually very long and bristle-like, bare or microscopically pubescent; its length and thickness vary considerably

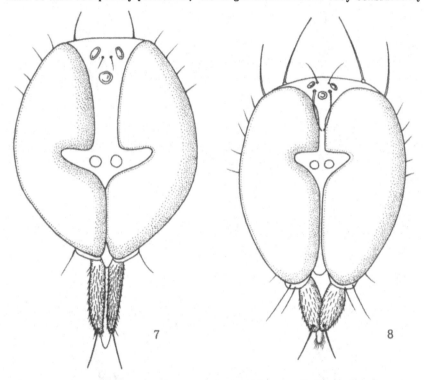

Figs. 7-8. Heads in frontal view of Tachydromiinae. - 7: Tachydromia arrogans (L.); 8: Drapetis (Elaphropeza) ephippiata (Fall.).

Figs. 3-6. Heads in frontal view of Tachydromiinae. - 3: Symballophthalmus pictipes (Beck.) ; 4: Platypalpus strigifrons (Zett.); 5: Dysaletria atriceps (Boh.); 6: Tachypeza nubila (Meig.); as - antennal socket, cl - clypeus , e - eye, fc - face, fr - frons, j - jowl, oc - ocellus, ocb - ocellar bristle, pb - proboscis, pp - palpus, vtb - vertical bristle.

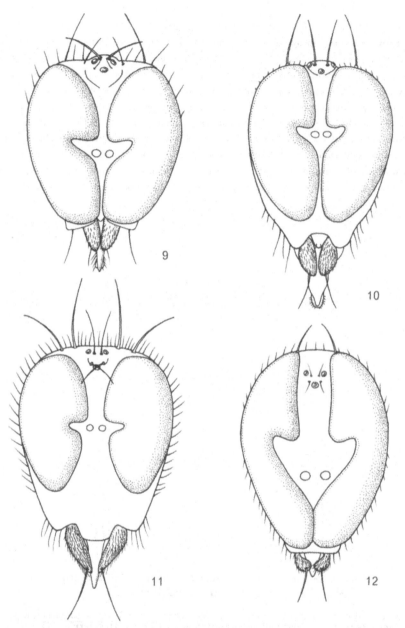

Figs. 9-12. Heads in frontal view of Tachydromiinae. - 9: Drapetis (Drapetis) assimilis (Fall.); 10: Crossopalpus nigritellus (Zett.); 11: Chersodromia hirta (Walk.); 12: Stilpon graminum (Fall.).

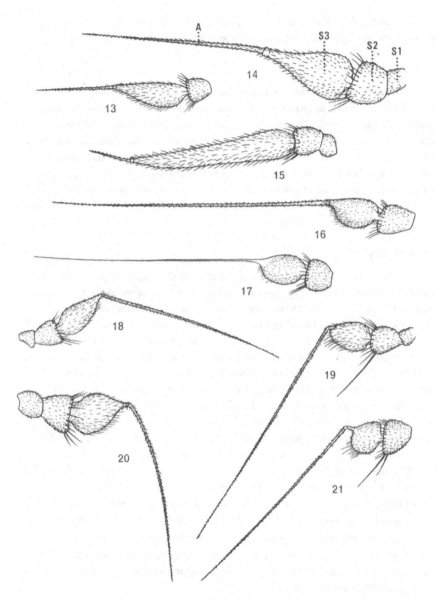

Figs. 13-21. Antennae of Tachydromiinae. - 13: Symballophthalmus fuscitarsis (Zett.); 14: Platypalpus major (Zett.); 15: Platypalpus longicornis (Meig.); 16: Tachypeza nubila (Meig.); 17: Tachydromia umbrarum Hal.; 18: Drapetis (Drapetis) incompleta Coll.; 19: Crossopalpus nigritellus (Zett.); 20: Chersodromia hirta (Walk.); 21: Stilpon nubila Coll.

only in Platypalpus. The arista is generally terminal, only rarely (Stilpon and some Chersodromia-species) obviously supra-terminal to almost dorsal.

MOUTH PARTS (Fig. 4).

These are formed by a rather short but strong proboscis, at most as long as head is high, pointed downwards but sometimes distinctly recurved (Crosso-palpus, Chersodromia). The proboscis consists of a sclerotised labrum, tubular hypopharynx and stout labium ending in small labella (labial palpi). Maxillae absent but maxillary palpi very distinct, one-segmented and connected with special sclerites, the so-called palpifers (Krystoph, 1961). Palpi usually flattened, ovate, and attached to base of proboscis, narrower in Tachypeza and Tachydromia, apically often armed with one or several bristles. Clypeus distinct, membraneous laterally.

THORAX (Fig. 22).

The thorax is rather short and robust, often with slightly convex dorsum, distinctly elongated only in Tachypeza and Tachydromia. Most of the thorax consists of the mesothorax with well separated pleural sclerites, metapleura always devoid of distinct hairs or bristles. Humeri well-developed in Tachydromiini, conspicuously long in Tachypeza and Tachydromia, but not differentiated in Drapetini; the same applies to postalar calli; scutellum always well-developed. Prothorax very small, often partly membraneous in Drapetini and better developed in Tachydromiini. Prothoracic notum present as a small "prothoracic collar" above neck, prothoracic episternum developed in the form of a narrower sclerite between prothoracic collar and fore coxae and often with distinct setae (Drapetis, Crossopalpus). Prothoracic sternum (or prosternum) a small usually V-shaped sclerite between the bases of fore coxae. The metathorax is better developed in Drapetini, where it is conspicuously enlarged, and simulates the basal abdominal segment which is virtually hidden beneath it. Acrostichal (acr) and dorsocentral (dc) bristles (cf. Fig. 22), if present, generally small and hair-like, either separated or evenly distributed over mesonotum. Large thoracic bristles often distinct, usually 1 or more humeral, 1 - 3 notopleural, a postalar, 1 or 2 pairs of apical scutellars, sometimes also a posthumeral or even an intrahumeral, and several bristle-like prescutellar dorsocentrals.

LEGS.

The legs vary considerably in form and shape in individual genera, and one of the anterior two pairs is often modified for catching and/or holding prey

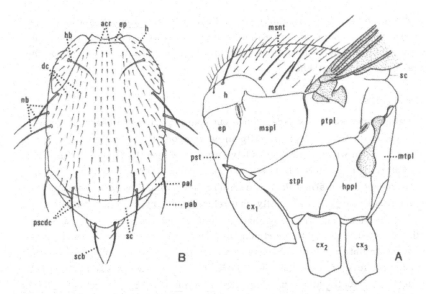

Fig. 22. Thorax in lateral view (A) and dorsal view (B) of <u>Platypalpus analis</u> (Meig.), pattern omitted; acr - acrostichal bristles, cx - coxa, dc - dorso-central bristles, ep - prothoracic episternum, h - humerus, hb - humeral bristle, hppl - hypopleuron, msnt - mesonotum, mspl - mesopleuron, mtnt - metanotum, nb - notopleural bristles, pab - postalar bristle, pal - postalar callus, pscdc - prescutellar dorsocentral bristles, pst - prothoracic sternum, ptpl - pteropleuron, sc - scutellum, scb - scutellar bristle, stpl - sternopleuron.

and bears special raptorial modifications such as spines, spurs, tubercles and excavations; specialised parts of the legs are often lengthened or very thickened, probably to accomodate strong muscles. With the exception of the most primitive <u>Symballophthalmus</u> and some <u>Platypalpus</u>-species with quite simple and rather long legs, the anterior femora are more or less thickened (or hind femora in some <u>Crossopalpus</u>), and the most specialised legs are the mid pair in most <u>Platypalpus</u>; the mid femora are very thickened and armed with a double row of short ventral spines which are directed in closed legs against the ventral row of short bristly-hairs on mid tibia, posteroventrally behind the spines there is usually a row of long probably sensory bristles; the tibia bears apically a strong spur-like projection (spur) and is often slightly curved; the whole mid leg of some <u>Platypalpus</u> forms a highly modified rap-torial structure analogous to the raptorial fore legs of Hemerodromiinae or <u>Mantis</u>. The fore tibiae are often more or less spindle-shaped dilated and al-ways have the distinct but small opening of a tubular gland anteriorly or ante-roventrally near base; the opening is clearly visible owing to its dense pubes-cence. The legs are often covered with dense hairs (<u>Platypalpus</u>, <u>Crossopalpus</u>)

25

or distinct bristles, particularly on hind pair (Crossopalpus, Chersodromia). Tarsi usually slender but sometimes apical segments very shortened and dilated, often in species inhabiting sandy biotopes and coastal areas, or on the other hand very lengthened.

WINGS (Figs. 23-31).

The wings are usually well-developed but some species are brachypterous (Tachydromia, Chersodromia, Stilpon) or apterous (Chersodromia), with extreme cases in which the halteres have also been lost (Pieltainia Arias, 1919 and Apterodromia Oldroyd, 1949). Wings usually clear or slightly clouded, sometimes with a distinct dark pattern (Tachydromia (Fig. 27), Tachypeza, Stilpon), very large, with usually undeveloped axillary angle and complete venation including anal cell (Symballophthalmus and primitive Platypalpus), with a gradual tendency to shortening of wings and anterior veins, and consequent incomplete venation (absence of anal cell). All veins without forks except for the fork of veins R2+3 and R4+5, discal cell always absent, anal cell present in Symballophthalmus (Fig. 23), Platypalpus (Fig. 24) and Dysaletria (Fig. 25), otherwise vein A absent (or in the form of an indistinct fold in Drapetis, Crossopalpus and Chersodromia (Fig. 30) and the lower branch of vein Cu (vein Cu 2 in Fig. 24) closing anal cell still present in most species of Tachypeza (Fig. 26).

Vein C (costa or costal vein) ending at wing-tip, at end of vein M, covered with short bristly-hairs on the whole length and often with a strong bristle near base (costal bristle). Vein Sc (mediastinal or auxillary vein) very indistinct, attached close to vein R1 and fading before reaching costa. Vein R1 (subcostal or first longitudinal vein) ending in costa at or before middle of wing (except for Chersodromia), but very short in Stilpon. Vein R2+3 (radial or second longitudinal vein) usually long, ending in costa near wing-tip, but much shorter in Stilpon (Fig. 31). Veins R4+5 (cubital or third longitudinal vein) and M (discal or fourth longitudinal vein) ending at wing-tip, usually more or less parallel or conspicuously bowed or undulating. Upper branch of vein Cu (postical or fifth longitudinal vein; vein Cu 1 in Fig. 24) ending in the membraneous posterior wing-margin, usually almost straight or more or less abruptly bent downwards (Tachydromia) or abbreviated (Stilpon). Vein A (anal or sixth longitudinal vein) always fine, complete in Symballophthalmus, abbreviated in Platypalpus and Dysaletria, otherwise absent.

The nomenclature of cells is given in Fig. 24; both basal cells, R (first or upper basal cell) and M (second or middle basal cell), are always distinct even if variable in length and size; practically equal in length in Tachydromiini

26

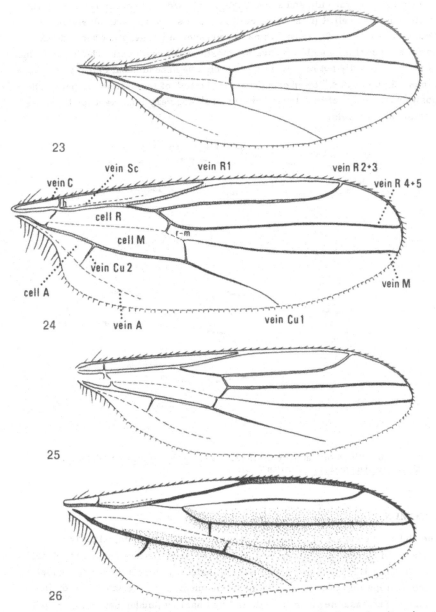

Figs. 23-26. Wings of Tachydromiinae. - 23: Symballophthalmus fuscitarsis (Zett.); 24: Platypalpus strigifrons (Zett.); 25: Dysaletria atriceps (Boh.); 26: Tachypeza nubila (Meig.). For explanation see opposite page.

(except for Tachydromia and some Tachypeza) and Chersodromia of Drapetini, in other genera cell R (first basal cell) is much shorter, particularly small in Drapetis (Fig. 28). Two crossveins are present: h (humeral), often indistinct, and r-m (discal or middle crossvein), closing first basal cell. The lower branch of vein M (actually not crossvein m-cu), closing second basal cell, is for simplicity named as "lower crossvein" in the text of the systematic part. The position and shape of both these "crossveins" are very important for the classification at all levels.

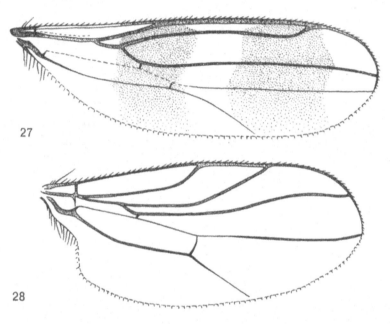

Figs. 27-28. Wings of Tachydromiinae. - 27: Tachydromia umbrarum Hal.; 28: Drapetis (Drapetis) assimilis (Fall.).

ABDOMEN.

The abdomen consists of 8 usually fully sclerotised segments, slightly mem-braneous in Symballophthalmus and Stilpon, or very distinctly so on basal seg-ment in Drapetis; more or less conical in males, stouter and bluntly conical in most Drapetini, telescopic with apical segments forming an ovipositor in females. The basal segment in male is often narrow (hidden beneath enlarged metathorax in Drapetis), segment 8 often ring-like and hidden beneath seg-ment 7. Segment 4 in Drapetis is very large and, like the narrow segment 5

28

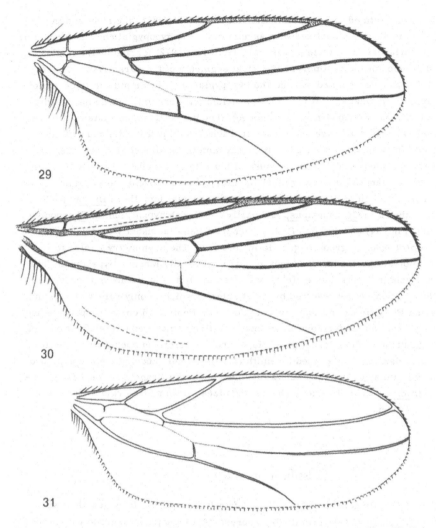

Figs. 29-31. Wings of Tachydromiinae. - 29: *Crossopalpus nigritellus* (Zett.);
30: *Chersodromia hirta* (Walk.); 31: *Stilpon graminum* (Fall.).

(partly hidden beneath segment 4), is armed with curious short curved bristles at sides; the shape of the bristles provides an important diagnostic feature.

GENITALIA (Figs. 32-34).

The male genitalia were studied in detail in the Empididae by Bährmann (1960).

They are rotated in Tachydromiinae through 90 - 180° to the right and the rotation is always associated with asymmetry. The hypopygial shell is formed by the periandrium, a term proposed by Griffiths (1972) for a sclerite originating in a fusion of the basimeres, the basal segments of the parameres; Ulrich (1972) has also pointed out that the hypopygial shell in Empididae (side-lamellae of Empidinae) is not the true epandrium (9th tergum), and he named this sclerite as "Gonopoden" (= gonopodes). The periandrium consists of two lamellae, originally fused near base in Tachydromiini (Fig. 32) and quite separated in Drapetini (Figs. 33, 34); right lamella is almost always larger, often very conspicuously so, and because of the rotation is placed beneath or on the left; left lamella is usually much smaller, placed above or on the right, much smaller and quite separated from right periandrial lamella in Drapetis (Fig. 33), absent in Crossopalpus, and partly fused with hypandrium in Chersodromia (Fig. 34) and Stilpon. Both periandrial lamellae usually bear one or several processes of great diagnostic importance. The asymmetry is also to be seen on the cerci, and the right cercus is generally larger. Hypandrium (9th sternum or "ventral sclerite"), placed on the left or quite above because of the rotation, is represented by a flat sclerite which is only rarely convex in some Platypalpus and Tachydromia species. Penis with co-axial apodeme and very complicated terminal filaments, and firmly attached to the distal part of hypandrium. The internal genitalia, without sclerotised spermathecae in female, have been discussed by Smith (1969). Female abdomen telescopic, produced into an ovipositor, but rather blunt-ended in Drapetini except for Chersodromia; cerci of various shapes and sizes, from short-ovate to very long and slender.

Classification and Phylogeny

It is generally considered that the subfamily Tachydromiinae, together with the Hybotinae and Ocydromiinae, represents a monophyletic group with a series of apomorphic characters. Tuomikoski (1966) listed eight such characters and Hennig (1970) eleven, some of which were based on the very important studies of Bährmann (1960) and Krystoph (1961). In the ocydromiine group of subfamilies (subfamily group Ocydromioinea sensu Hennig), the male hypopygium is quite different from that of other Empididae, because of its enlarged periandrium (sensu Griffiths) which covers the greater part of the genital segment. Bährmann (1960) interpreted this as a plesiomorphic condition but Griffiths (1972) as an apomorphic character derived from the fusion of the basime-

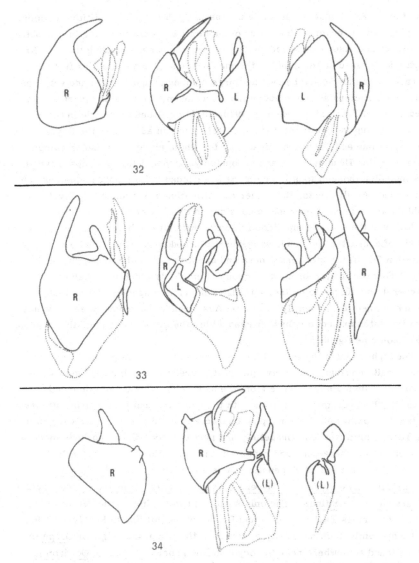

Figs. 32-34. Male genitalia of Tachydromiinae (schematically). - 32: Platypalpus analis (Meig.); 33: Drapetis (Drapetis) simulans Coll.; 34: Chersodromia
speculifera Walk. R - right periandrial lamella; L - left periandrial lamella;
cerci (above) and hypandrium (below) marked with dotted line.

res of other Empidinae in the same way as in Cyclorrhapha. Griffiths' opinion, implying that the ocydromiine group of subfamilies is apomorphic, is undoubtedly correct and also corresponds with other conclusions, including those of Krystoph (1961) based on the study of the mouth-parts. When evaluating further characters, as discussed below, the large number of apomorphic and very specialised characters show the subfamily Tachydromiinae to be the most specialised group of Empididae. Hennig (1970, 1971) proposed to include in the ocydromiine group the Microphorinae and Atelestinae in addition to the Ocydromiinae, Hybotinae and Tachydromiinae, but the whole situation is rather complicated (see also Hennig, 1971) and undoubtedly further detailed comparative studies including those of fossil specimens are needed before any conclusions can be reached. At all events, the higher classification at the subfamily level is a problem which I do not propose to consider in this chapter.

Although the family Empididae is a very difficult one because of the great variety of characters on head, wing-venation, abdomen, etc., and no simple definition for all the different genera and subfamilies can be found, the subfamily Tachydromiinae represents a group of empidids which is rather uniform in general and simple to define, although very rich in species. Since 1862, when Schiner defined the subfamily in "Fauna Austriaca", there have been no doubts as to its status and nothing has changed within the group apart from the description of some new genera.

The higher classification and the arrangement of genera proposed in this work is quite unusual, but in my opinion it expresses best the phylogeny of the subfamily. The main difference from the usual classification of other authors is that the Tachydromiini, starting with the genus Symballophthalmus, possess the largest number of plesiomorphic characters and represent the most primitive Tachydromiinae. The Drapetini, and particularly Stilpon, possess most of the apomorphic characters and represent the specialised forms of Tachydromiinae. The ascendant phylogenetic row is as follows: Symballophthalmus - Platypalpus - Dysaletria - Tachypeza - Tachydromia - Drapetis - Crossopalpus - Chersodromia - Stilpon. The only serious problem lies in the relationship between the tribes Tachydromiini and Drapetini. According to Kovalev (1969a) the Tachydromiinae is not really a monophyletic group: the origin of Drapetis may be found somewhere near the ocydromiine genera Trichina - Ocydromia - Leptopeza - Bicellaria, and the origin of Symballophthalmus near the genera Anthalia - Oedalea. Detailed comparative studies of the Hybotinae and Ocydromiinae need to be made to solve this problem but, at all events, the apomorphic characters within the subfamily Tachydromiinae are concentrated in the Drapetini.

The classification of the Tachydromiinae is based mainly on the wing-venation and the structure of male genitalia, and the trends of the development are obvious from the section on adult morphology. The primitive complete venation of the subfamily is present in Symballophthalmus, Platypalpus and Dysaletria, combined with the wing-shape in Symballophthalmus and primitive Platypalpus-species (large wing with undeveloped axillary lobe, broadened towards tip), with gradual loss of the anal cell (Tachypeza and Tachydromia), to the highly specialised wings of the Drapetini up to the apomorphic condition in Stilpon. Parallel specialisation of the male genitalia has also taken place, from the primitive structure in Tachydromiini, through gradual separation of the periandrial lamellae, to reduction or fusion of the periandrial (left lamella) and hypandrial parts in Drapetini. The structure of the thorax, including the shape and insertion of the head, and all adaptations to the predatory habit in combination with a preference for rapid movement on legs instead of on wings, show the tribe Drapetini to be a derived and highly specialised group, unlike the phylogenetically older and more primitive Tachydromiini. The predatory adaptations on the mid legs of some Platypalpus species (tibial spur, as well as femoral and tibial spines) seem to be a parallel side-development of the predatory activity of the more primitive species (ciliaris-, pallipes- and albiseta-groups) and, by the way, are also present in Tachydromia species. Similar developments may be found on the hind legs of many species of Hybotinae and Ocydromiinae.

The dendrogram (Fig. 35) shows the supposed phylogeny of the Palaearctic Tachydromiinae; the genera Pieltainia and Ariasella were not examined, and Pseudostilpon should have the same position as Stilpon.

Symballophthalmus is a small homogeneous genus with the largest number of plesiomorphic characters: (1) Head almost globular, not closely-set upon thorax; (2) vertical bristles not differentiated; (3) antennal segment 3 long-pointed, with terminal arista; (4) eyes bare, rounded; (5) metathorax weakly developed; (6) legs all equally slender, without raptorial adaptations; (7) wings large, narrow at base and broadened apically, venation complete including anal cell; (8) male genitalia with periandrial lamellae connected near base, cerci long and slender.

Platypalpus is a large and very heterogeneous genus. The more primitive groups (groups 1 - 3) have the most plesiomorphic characters as in Symballophthalmus, but group 11 (hackmani-group) is more closely allied to Dysaletria than to most other Platypalpus species.

Dysaletria is a monotypic genus that has separated phylogenetically from Platypalpus only rather recently, and its close relationship with some Platypalpus species makes its generic value a little problematic. Perhaps a transfer of Platypalpus group 11 to Dysaletria, because of the thickened fore femora and other characters, could be made.

Tachypeza and Tachydromia are two clearly distinct genera the origin of which may be found somewhere in the common ancestor of Platypalpus groups 4 or 5. They are undoubtedly highly modified on account of their prolonged adaptation to special ecological conditions.

The systematic position of Drapetis, Elaphropeza and Crossopalpus is the only contraversial feature within the Tachydromiinae. For instance, Engel (1939) differentiated 2 genera, Elaphropeza and Drapetis, Crossopalpus being included in Drapetis; Collin (1961) gave generic status to all three groups, whilst Smith (1969) gave them subgeneric status within the genus Drapetis. However, I accept the division recently used by Kovalev (1972) who separated the genus Crossopalpus as distinct from Drapetis, the latter including the subgenera Drapetis s.str. and Elaphropeza. Kovalev (1968) based this subdivision of "Drapetis s.lat." on the use of a numerical taxonomy, comparing phenetic relations after the Smirnov school, but the evaluation of phyletic characters also fully justifies this division: (1) jowls below eyes linear in Elaphropeza and Drapetis, very deep in Crossopalpus; and (2) left periandrial lamella in the form of a distinct even if small sclerite separated from the right lamella in Elaphropeza and Drapetis, but absent in Crossopalpus.

Drapetis subgenus Elaphropeza possesses some plesiomorphic characters, and Drapetis s.str. should be derived from Elaphropeza; the antennae in Elaphropeza have a longer and conical segment 3 (shorter and rather ovate, at least lower margin convex, in Drapetis s.str.); the left periandrial lamella is larger in Elaphropeza, but smaller and more reduced in Drapetis s.str.

Chersodromia is a rather heterogeneous genus but, unlike Drapetis and Crossopalpus, the left periandrial lamella is partly fused with hypandrium (the same situation as in Stilpon), basal cells on wing equal in length, and no specialised terga on male abdomen; in addition to these characters, jowls are deep below eyes in some species but almost linear in others, and antennal arista has a tendency to be supra-apical or almost dorsal. The origin of Chersodromia may be found in the common ancestor of the Chersodromia hirta-group and Crossopalpus.

Stilpon is a small homogeneous genus with a larger number of apomorphic characters: (1) head very deep (although without distinct jowls below eyes) and closely-set upon thorax; (2) vertical bristles differentiated; (3) antennal segment 3 very small, ovate to somewhat kidney-shaped, with arista supra-apical; (4) eyes microscopically pubescent, vertically elongate; (5) metathorax strongly developed; (6) legs short and thickened; (7) wings small and rather narrow, venation incomplete with veins R1 and R2+3 considerably shortened and anal

34

cell absent; (8) genitalia very asymmetrical, left periandrial lamella very small and quite separated from right lamella (partly fused with hypandrium), cerci conspicuous and with a very complicated armature. Its origin may be found somewhere in the common ancestor of Chersodromia incana-group.

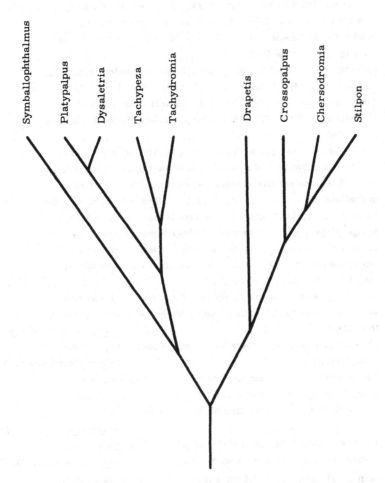

Fig. 35. Phylogenetic relationships within the Palaearctic Tachydromiinae.

Key to genera and subgenera of Tachydromiinae

1   Humeri more or less differentiated, often very sharply so (cf. Fig. 53). Eyes bare; head with distinct neck, not closely set upon front of thorax (cf. Figs. 36, 51) (tribe Tachydromiini) 2

-     Humeri not differentiated. Eyes microscopically pubescent; head without a distinct neck, closely set upon front of thorax (cf. Fig. 583) (tribe Drapetini)..................................... 6

2(1) Anal cell present (cf. Figs. 23-25). Humeri rounded or not very long (cf. Fig. 53). Wings clear or very indistinctly clouded 3

-     Anal cell absent (cf. Figs. 26, 27). Humeri conspicuously elongated (cf. Figs. 519, 540). Wings usually distinctly clouded or banded 5

3(2) Eyes (Fig. 3) touching on frons. Wings (Fig. 23) large, narrow at base, broadened apically and blunt-tipped. Legs (Fig. 40) with all femora equally slender, without special raptorial modifications .... Symballophthalmus Beck. (p. 37)

-     Eyes (Figs. 4, 5) at least very narrowly separated on frons. Wings (Figs. 24, 25) with more or less developed axillary lobe, not conspicuously broader at tip. Legs usually more or less thickened on anterior four femora and often with special raptorial modifications 4

4(3) Eyes (Fig. 4) at least narrowly separated on facial part below antennae. Humeri not conspicuously large. Mid femora usually thickened .... Platypalpus Macq. (p. 45)

-     Eyes (Fig. 5) touching below antennae. Humeri very large. Fore femora the stoutest .... Dysaletria Loew (p. 210)

5(2) The lower branch of vein Cu (usually closing anal cell) present (Fig. 26); wings usually uniformly dark clouded. Occiput below neck with conspicuous whitish bristle-like setae................ .... Tachypeza Meig. (p. 212)

-     The lower branch of vein Cu absent (Fig. 27); wings usually banded. Occiput below neck with usual hairs.. Tachydromia Meig. (p. 225)

6(1) Both basal cells equal in length (Fig. 30). Face usually broad and often with deep jowls below eyes (Fig. 11); coastal species .... Chersodromia Walk. (p. 276)

-     First basal cell distinctly shorter than second basal cell (cf. Figs. 28, 29, 31). Face linear or eyes practically touching below antennae (cf. Figs. 8-10, 12). 7

7(6) Veins R1 and R2+3 very short, latter ending in costa at about
      middle of wing; wings short and often narrowed (Fig. 31).
      Frons rather broad, parallel-sided (Fig. 12)....... Stilpon Loew (p.290)

-     Veins R1 and R2+3 longer, latter ending in costa beyond middle
      of wing; wings not conspicuously shortened or narrowed (cf.
      Figs. 28, 29). Frons narrow in front, more or less widening
      towards vertex (cf. Figs. 9, 10)..................................... 8

8(7) Jowls linear, anterior pair of ocellar bristles present (Figs.
      8, 9). The common stem of veins R2+3 and R4+5 rather long
      (Fig. 28) .... Drapetis Meig. 9

-     Jowls deep below eyes (Fig. 10). Anterior pair of ocellar brist-
      les absent, posterior pair very long. The common stem of veins
      R2+3 and R4+5 very short (Fig. 29) .... Crossopalpus Bigot (p.264)

9(8) Hind tibiae with distinct long anterodorsal bristles. Yellow
      species .... Elaphropeza Macq. (p.250)

-     Hind tibiae without distinct long bristles. Black to blackish-
      brown species.................................. Drapetis s.str. (p.252)

TRIBE TACHYDROMIINI

Genus *Symballophthalmus* Becker, 1889

Macroptera Becker, 1889, Wien.ent.Ztg, 8: 80 (nec Macroptera Lioy,
    1863: 224).

Symballophthalmus Becker, 1889, Wien.ent.Ztg, 8: 285 (nom.n.for Macrop-
    tera Becker, nec Lioy).

Type-species: Macroptera pictipes Becker, 1889 (mon.).

Medium-sized, rather shining black or rarely yellowish, slender species.
Head rather small and almost rounded in profile, distinctly convex both in
front and on occiput. Eyes bare, broadly contiguous on frons in both sexes,
widely separated on face below antennae, and with conspicuously enlarged
facets anteriorly. 3 ocelli on a small tubercle placed just on the vertex and
1 pair of rather long but fine ocellar bristles almost as long as numerous similar hairs on the upper part of occiput; no vertical bristles differentiated.
Antennae inserted slightly below middle of head in profile; basal segment indistinct, segment 3 laterally compressed, elongate and tapering, with upper
margin straight but convex below, almost equal in length to microscopically

pubescent arista. Palpi very small and narrow, covered with several long
bristly-hairs, generally resembling those of <u>Tachypeza</u>. Proboscis half as
long as head is high, sometimes directed slightly forwards. Thorax rather

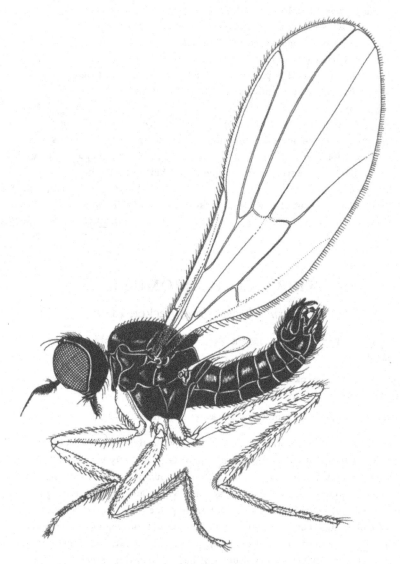

Fig. 36. Male of <u>Symballophthalmus fuscitarsis</u> (Zett.).
Total length: 2.0 - 2.3 mm. Redrawn after Collin (1961).

long and narrow, slightly arched above, with distinct but only very small humeri. Acr and dc bristles very fine and hair-like, latter not stouter posteriorly, and only 1 or 2 notopleural bristles and 2 pairs of scutellars long and bristle-like, even if fine. Legs long and slender without any special raptorial modifications, mid femora not thickened and only with weak bristles in two rows beneath. Wings clear, long and conspicuously broadened apically, narrow at base, anal lobe not developed. Vein R1 ending at middle of wing, vein R2+3 not far from tip of wing; veins R4+5 and M very slightly bowed, almost parallel, vein C extending practically right around wing. Both basal cells long, subequal, anal cell conspicuously small with the vein closing it slightly recurrent; vein A fine. Abdomen consisting of eight rather weakly sclerotised segments (particularly so in ♀), basal and last segment reduced in ♂. Male genitalia rather large, usually turned upwards, both periandrial lamellae well-developed but unlike Platypalpus small, and the elongated cerci and dorsal process of right lamella overlapping the lamellae - resembling in this way some species of the Platypalpus albiseta-group. Female abdomen telescopic, long and pointed, with long slender cerci.

The adults are to be found on the leaves of bushes and on ground-vegetation; they fly slowly like ocydromiine empidids or the more primitive Platypalpus species, and are all predaceous.

The genus is Holarctic in distribution, with only a few known species. 3 species are known with certainity from Europe, and they all occur in Fennoscandia. S. cyanophthalmus (Strobl, 1880) may well be identical both with fuscitarsis Zett. (=scapularis Coll.) and pictipes Beck. (=pollinosus Coll.) considering that "Fühler gelbbraun, das dritte Glied an den Rändern etwas dunkler bis ganz schwarzbraun". 1 species is known from North America (Chillcott, 1958) and 1 species from Japan (Saigusa, 1963).

Key to species of Symballophthalmus

1   Mesonotum entirely very thinly grey pilose, not shining.
    Pleura with similar pile, leaving only lower part of me-
    sopleura, whole of sternopleura and hind part of hypopleu-
    ra shining. Anterodorsal row of black spines on fore tibiae
    very distinct in both sexes. Antenna (Fig. 39) uniformly
    brownish to dark brown, basal segments rarely paler..........
    ....3. pictipes (Beck.)

-    Mesonotum polished black (Fig. 36), not in the least thin-
    ly dulled by greyish pile. Pleura extensively shining. The

small black spines in anterodorsal row on fore tibiae less
distinct in female... 2

2(1) Antenna (Fig. 37) unicolourous brownish, or at least segment 3 not obviously darker. Whole of thorax including prothorax and humeri polished black............... 1. dissimilis (Fall.)

- Basal antennal segments (Fig. 38) yellow, segment 3 brownish, distinctly darker. Prothorax, humeri and upper margin of mesopleura covered with fine grey pile, not shining........
....2. fuscitarsis (Zett.)

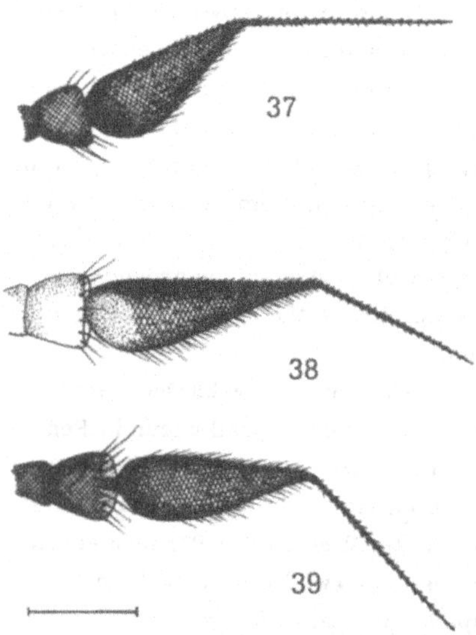

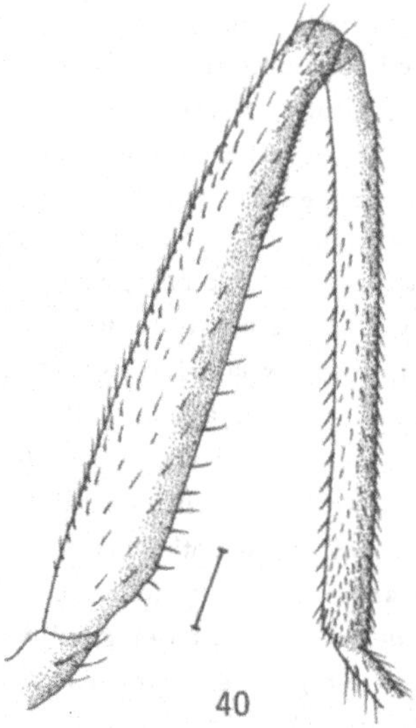

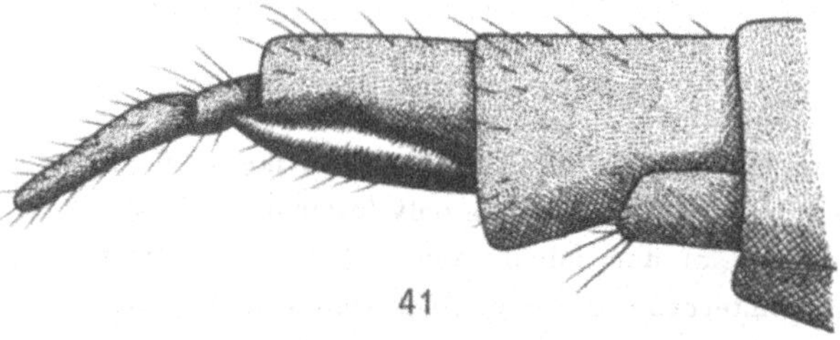

Figs. 37-39. Antennae of Symballophthalmus. - 37: dissimilis (Fall.), ♂; 38:
fuscitarsis (Zett.), ♂; 39: pictipes (Beck.), ♂. Scale: 0.1 mm.
Fig. 40. Mid leg in posterior view of Symballophthalmus fuscitarsis (Zett.), ♂.
Scale: 0.1 mm.
Fig. 41. Terminal segments of female abdomen in Symballophthalmus pictipes
(Beck.). Scale 0.1 mm.

1. SYMBALLOPHTHALMUS DISSIMILIS (Fallén, 1815)

Figs. 37, 42-44, 668.

Tachydromia dissimilis Fallén, 1815: 9; Collin, 1961: 217 (as Symballophthalmus).

Mesonotum including humeri and whole of prothorax polished black. Antennae unicolourous rather light brown.

♂. Head polished black on vertex, occiput above neck slightly grey pilose; face polished black, very broad and almost parallel. A pair of long but fine ocellar bristles and similar forwardly directed hairs on vertex and upper part of occiput equal in length, all brownish. Antennae brownish, segment 2 at tip and base of segment 3 often translucent yellowish; segment 3 about 3 times as long as deep, arista subequal. Palpi very small, dark, with several long pale apical hairs as long as or longer than palpus. Proboscis blackish.

Thorax polished black but scutellum, postalar calli, a narrow stripe between scutellum and humeri, and narrow margins of pleural sclerites very thinly grey pilose. Acr and dc pale, very minute; humeri with several small hairs, but 2 notopleural bristles and 2 pairs of apical scutellars long and bristle-like, although fine.

Legs yellow with blackish tarsi, hind femora towards tip and sometimes also mid femora and four anterior tibiae more or less brownish. Mid femora slender, ventrally with a double row of very weak bristles becoming longer,

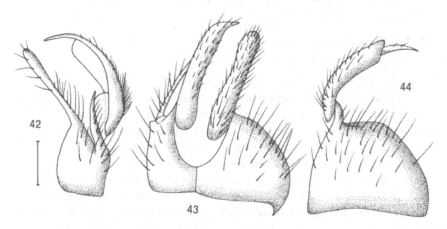

Figs. 42-44. Male genitalia of Symballophthalmus dissimilis (Fall.). - 42: right periandrial lamella with cerci; 43: periandrium with cerci; 44: left periandrial lamella with cerci. Scale: 0.1 mm.

41

darker and more bristle-like towards base; mid tibiae beneath with the short bristles in one row, more distinct towards tip. Fore tibiae slightly thickened, anterodorsally with a row of short black, conspicuously adpressed spine-like bristles; fore metatarsus anteroventrally with very long pale hairs, becoming shorter towards tip.

Wings clear and considerably iridescent, yellowish at extreme base; veins brown. Squamae and halteres pale yellow.

Abdomen brown, somewhat yellowish at base and on venter, very thinly grey dusted and covered with fine pale hairs. Genitalia large, dark in contrast with the rather pale abdomen, subshining blackish-brown.

Length: body 2.0 - 2.5 mm, wing 2.9 - 3.2 mm.

♀. Resembling male but fore metatarsus with short hairs anteroventrally, and all bristles on mid legs only weak and pale. Abdomen paler with respect to a very slight sclerotisation, cerci dull grey.

Length: body 2.3 - 2.7 mm, wing 3.0 - 3.2 mm.

Commoner in SE Fennoscandia incl. Russian Carelia, in Sweden N to Nb., in Finland to Le; rare in Denmark and in Norway N to NTi. - Scotland, NW and C parts of European USSR. The records from C. and S. Europe are very probably mistakes. A northern species in distribution. - May-July.

2. SYMBALLOPHTHALMUS FUSCITARSIS (Zetterstedt, 1859)
Figs. 36, 38, 40, 45-47, 669.

Tachydromia fuscitarsis Zetterstedt, 1859: 4990.
Symballophthalmus scapularis Collin, 1961: 218.

Resembling dissimilis but basal antennal segments yellow, much paler in contrast to dark segment 3, and prothorax and humeri grey pilose.

♂. Head black with occiput thinly grey dusted, the bristly-hairs on vertex and occiput as in dissimilis. Antennae with the club-shaped segment 2, and sometimes also with the extreme base of segment 3, pale yellow, segment 3 darker brown. Proboscis brownish.

Thorax distinctly polished black on mesonotum but humeri, whole of prothorax and adjacent upper parts of mesopleura slightly dulled by greyish pile.

Legs as in dissimilis, with slender mid femora and slightly thickened fore tibiae, but anterior four tibiae and more than basal half of corresponding metatarsi pale yellow; apical half of hind femora and all of hind tibiae often brownish, tarsi very darkened. Fore metatarsi with the same long pale an-

42

teroventral hairs but mid femora with thinner ventral bristles in anterior row towards base, the bristles not so black and spine-like as in dissimilis.

Wings often faintly tinged light brown on costal margin towards tip. Abdomen usually more yellowish with perhaps longer pale pubescence. Genitalia large, similar to those of dissimilis, but lamellae more greyish dusted and there are slight differences in the cerci.

Length: body 2.0 - 2.3 mm, wing 3.0 - 3.3 mm.

♀. Tarsi including all fore tarsi deeper black, front metatarsus without long anteroventral hairs, fore tibiae with less distinct dark anterodorsal bristly-hairs, and the ventral bristly-hairs on mid femora rather hair-like. Abdomen with last sternite polished black towards tip, cerci dull.

Length: body 2.2 - 2.9 mm, wing 3.0 - 3.2 mm.

Rather common in Denmark and southern Fennoscandia, in Sweden N to Sdm. (a pair taken in Jmt.), in Finland only on the Baltic coast; no records from Norway available. - Great Britain, NW and C parts of European USSR, C. Europe. - May-July.

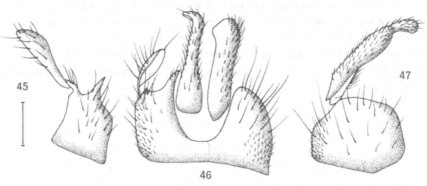

Figs. 45-47. Male genitalia of Symballophthalmus fuscitarsis (Zett.). - 45: right periandrial lamella; 46: periandrium with cerci; 47: left periandrial lamella with cerci. Scale: 0.1 mm.

3. SYMBALLOPHTHALMUS PICTIPES (Becker, 1889)
   Figs. 41, 48-50, 670.

Macroptera pictipes Becker, 1889: 80.
Symballophthalmus pollinosus Collin, 1961: 218, syn.n.

Mesonotum thinly covered with greyish pile, not polished. Antennae unicolourous brownish or basal segments slightly paler.

43

♂. Head as in <u>dissimilis,</u> antennae unicolourous brownish or rarely basal
segments yellowish and slightly paler than the brown segment 3. Palpi conspicuously small, dark, with very long pale hairs at tip.

Thorax thinly covered with greyish microscopic pile on mesonotum, not
polished; similar grey pile present also on prothorax and most of pleura,
leaving only lower part of mesopleura, practically whole of sternopleura and
hind part of hypopleura polished black. Biserial acr, uniserial dc, and a few
fine bristly-hairs as in <u>dissimilis.</u>

Legs with the same colour as in <u>fuscitarsis</u> with fore metatarsi mostly
pale, but the black bristly-hairs in anterior row on mid femora beneath rather
stouter, as in <u>dissimilis.</u> The long pale anteroventral hairs on fore metatarsi numerous and conspicuously long, at least twice as long as metatarsus is
deep.

Wings indistinctly and faintly brownish clouded along costal margin. Abdomen yellowish at base, blackish towards tip, with large black genitalia,
lamellae thinly grey dusted and covered with sparse pale hairs.

Length: body 2.0 - 2.3 mm, wing 2.9 - 3.2 mm.

♀. Fore metatarsi practically all black and anteroventrally with only short
hairs, but the anterodorsal short black bristly-hairs distinct on the whole
length of fore tibiae as in the male. Mid femora ventrally with only hair-like
paler bristles towards base. Abdomen more yellowish, very narrowly pointed and thinly dull grey dusted, but last sternite almost entirely polished black;
cerci dull grey.

Length: body 2.4 - 2.6 mm, wing 3.1 - 3.4 mm.

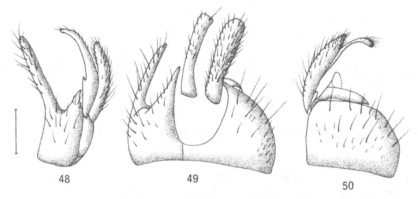

Figs. 48-50. Male genitalia of <u>Symballophthalmus pictipes</u> (Beck.). - 48: right
periandrial lamella with cerci; 49: periandrium with cerci; 50: left periandrial
lamella with cerci. Scale: 0.1 mm.

Rare. Norway: HOi, Eidfjord, Hjølmodalen, ♂♀ (T. Nielsen). - Great Britain, C. Europe. - May-August.

Note. Becker (1889) described this species from two females from Zermatt and Partenkirchen: the former is preserved in coll. Becker, Berlin, and is herewith designated as lectotype of S. pictipes (Beck.) The recently described S. pollinosus Coll. is identical with pictipes and becomes a junior synonym of the latter. Becker (1889) erroneously described the mesonotum of pictipes as without acrostichals.

Genus *Platypalpus* Macquart, 1827

Coryneta auct., nec Meigen.

Tachydromia auct., nec Meigen.

Platypalpus Macquart, 1827, Ins. Dipt. Nord Fr., 3: 92.

Type-species: Musca cursitans Fabricius, 1775 (design. by Westwood, 1840: 132).

Very small to large species, ranging from 1 to over 5 mm in body-length, also including the largest representatives of the subfamily; shining black or yellow with all gradations to densely grey dusted species. Head (cf. Fig. 4) rather deep, semicircular in front and somewhat concave behind below neck, in more primitive species somewhat rounded in profile. Eyes large, bare, facets equal in size. Both frons and face vary from linear to very broad, but eyes always at least very narrowly separated in both sexes. 2 pairs of ocellar bristles, varying from very minute hairs to large bristles; 1 or 2 pairs of vertical bristles, usually erect and bristle-like, but sometimes not differentiated from other adpressed hairs on vertex and occiput. Antennae (cf. Fig. 14) with segment 3 more or less conical, rarely very elongate, terminal arista of various lengths. Palpi flat, ovate, sometimes very broadly so, outer side with microscopic pile and some longer hairs or bristles towards tip. Proboscis strong but usually small, directed downwards.

Thorax (cf. Fig. 22) slightly elongated and only very little arched above, humeri well differentiated but not very long, sometimes very small (group 3) or conspicuously prominent (group 11). Acr and dc varying in size and number, latter with a tendency to be more bristle-like at least posteriorly. Large thoracic bristles, if present (sometimes only hair-like), as follows: 1-3 humeral, 1 posthumeral (sometimes replaced by third anterior notopleural), 1-3 notopleural, 1 postalar, and usually 2 pairs of scutellars. Pleura when dusted usually with a bare polished patch on sternopleura.

Anterior four femora more or less thickened but mid femora almost always the stoutest, and mid tibiae often arcuate. Mid femora with a double row of short spine-like bristles beneath which vary in length and thickness, behind them often with a row of long posteroventral bristles. Mid tibiae ventrally with a row of small bristly-hairs (opposite the femoral spines), and a more or less developed apical ventral spur which is rarely entirely absent. Hind legs always slender and with only fine hairs.

Wings (cf. Fig. 24) clear or faintly tinged, vein Sc indistinct, vein R1 long, ending beyond middle of wing, vein R2+3 ending not far from tip of wing; veins R4+5 and M ending at tip of wing, often more or less bowed. Both basal cells long, usually subequal (with crossveins contiguous) or 2nd basal cell slightly longer. Anal cell present, vein A often inconspicuous, the vein Cu2 closing anal cell very recurrent to almost at right-angles to vein Cu1.

Abdomen consisting of eight fully sclerotised segments. Male genitalia (cf. Fig. 32) of various sizes, twisted round to the right (left lamella placed dorsally, cerci on the right), both periandrial lamellae well separated, only narrowly connected at base, left lamella usually smaller and simple; female abdomen very telescopic with slender, more or less elongate cerci.

The adults are found running amongst ground-vegetation or over the surfaces of leaves on bushes and trees, but only rarely on tree-trunks, where they search for small insects, especially other Diptera. They are very predaceous, even in captivity (can only be compared with Chersodromia species), attacking other insects and also quite frequently each other. They usually carry their prey held in the thickened mid legs. The immature stages are unknown but larvae have been found in soil or under moss and are obviously predaceous.

The genus Platypalpus is world-wide in distribution but is best represented in the northern temperate and montane regions. Over 200 species are known from Europe, but many species still await description, particularly in the central and southern areas, and several hundred species will certainly be found to occur in the Palaearctic region. 80 species are now known from Fennoscandia and Denmark, but even this figure is not final; additional species which are likely to be found there are also included in the key. 107 species have recently been found in the Nearctic region (Stone et al., 1965) but so far only P.unguiculatus is known to be Holarctic in distribution.

It is a strange fact that in some common species the male sex is very rare or is still unknown. Geographical parthenogenesis has been found in P.candicans and P.cursitans. The former is apparently parthenogenetic in the north of Europe, but in southern areas of Central Europe and in the south males occur even if only rarely. On the other hand P.cursitans is parthenogenetic in

46

Fig. 51. Male of <u>Platypalpus agilis</u> (Meig.).
Total length: 2.7 - 3.3 mm. Redrawn after Collin (1961).

the southern parts of its range, including Central Europe, but in North and North-West Europe males are sometimes found. However, the very common P.major is most probably a parthenogenetic species throughout its entire range.

In general, considering the world-wide distribution, the presence of many diverse forms, and the very wide differences between extreme groups of species, which has no analogy in other rather homogeneous tachydromiine genera the genus is undoubtedly very ancient phylogenetically. On the other hand, the large number of very similar and closely related species (group 9) in the northern temperate regions indicates a recent explosion of species for which the only parallel is Drapetis subgenus Elaphropeza in tropical and subtropical regions.

The large numbers of species urgently need to be subdivided into species-groups, and several attempts have already been made. The genus is very often divided into 2 subgenera, Cleptodromia and Phoroxypha, based on the presence of 1 or 2 pairs of vertical bristles respectively, as for instance by Engel (1939) or Collin (1961). However, this division seems to be quite artificial since there are several pairs of very closely related species with 1 or 2 pairs of vertical bristles (longicornioides - longicornis, laestadianorum - lapponicus, unguiculatus - alter, etc.), which undoubtedly belong to the same natural group and cannot be separated subgenerically; similarly, phylogenetically close species with different numbers of vertical bristles are also present among the species of Chersodromia or Drapetis. Frey (1943) and Chillcott (1962) did not use this subgeneric division, quite justifiably; the former author divided the Palaearctic species into 5 groups, into species with yellow ground-colour, polished black species with or without apical tibial spur, and grey dusted species with black or pale antennae. Although this division is quite artificial, it is very practical and I have partly used it in the differential key to species.

Collin (1926, 1961), although separating the species with 2 pairs of vertical bristles, made the first serious attempt to find truly natural groups, and his groups 1 - 3 seem to be well defined. I have attempted to divide the Fennoscandian species into eleven species-groups that should be natural and should also fit the other non-Scandinavian species. I am aware that this division is far from being definitive, and some changes and additional groups will certainly have to be made. However, groups 1 - 4, 10 and 11 (the first three groups correspond with those of Collin) perhaps represent well-defined natural groups; group 9 (partly group 8 of Collin) is the most complex one and further subdivision seems likely.

In the systematic part, the species are arranged under species-groups. A short diagnosis for each group is given in the following table.

1(4)   Vt bristles not differentiated, vertex and upper part of oc-

48

ciput covered with numerous equally long hairs; in doubtful cases, the species with thoracic hairs evenly distributed over mesonotum belong here. No humeral bristle. Mid tibia without or with only a very small apical spur (Figs. 169-174, 176). Mesonotum polished; sternopleura largely polished, quite devoid of grey dusting. Anal lobe of wing very little developed, the vein closing anal cell very recurrent. Head rather round in profile.

2(3)   No pv bristles on mid femora; the ventral spines in a double row, long and bristle-like (Figs. 169-174). Acr and dc usually differentiated (Figs. 52, 53) .... I. _ciliaris_-group

3(2)   Mid femora with pv bristles behind the two rows of short black spines beneath (Fig. 176). Acr and dc always evenly distributed over mesonotum (Fig. 71) .... II. _pallipes_-group

4(1)   1 or 2 pairs of erect vt bristles which are always distinct from the other pubescence (cf. Fig. 4). Acr, if present, distinctly separated from dc by bare areas (Fig. 70).

5(6)   Humeri weakly developed, humeral bristle absent (Fig. 70). Antennae with elongate segment 3 and long whitish, rather thick arista (Figs. 66-69). No apical spur on mid tibiae or only a very small one (Figs. 177-180). Mesonotum polished or at most thinly dusted (Fig. 70), sternopleura largely polished. Anal lobe of wing weakly developed, the vein closing anal cell very recurrent and S-shaped (Figs. 680-683). Head rather round in profile; frons very narrow. Mid femora with long pv bristles (Figs. 177-180). III. _albiseta_-group

6(5)   Humeri well-developed (Figs. 94-98). Antennal arista dark (so far as the Fennoscandian fauna is concerned).

7(12)   Mid femora without pv bristles or bristly-hairs behind the two rows of dark ventral spine-like bristles. Mid tibia without or with only a very small apical spur (cf. Figs. 181, 187).

8(9)   No humeral bristle, or only a fine hair present (Figs. 94-96). Mesonotum polished and almost bare, the small hairs very indistinct, acr usually absent. Sternopleura largely polished, up to and including its hind-margin. IV. _unguiculatus_-group

9(8)   A distinct humeral bristle (Figs. 97-98). Mesonotum dusted at least anteriorly, acr and dc distinct.

10(11)   Mesonotum partly polished or rather thinly grey dusted, sternopleura with a large polished patch. Anal lobe of wing weakly developed and the vein closing anal cell re-

current (cf. Fig. 693). Humeri moderately large, no posthumeral bristle (Figs. 97-98). Legs with mid femora
thickened, at least as thick as fore femora (cf. Fig. 189).......
.... V. longicornis-group

11(10) Very densely light grey dusted species, sternopleura usually dusted. Anal lobe of wing distinct, the vein closing
anal cell at right-angles to vein Cu (Fig. 747). Humeri
conspicuously large, a posthumeral bristle. Fore femora
very thickened; mid femora (Fig. 244) slender, with only
fine ventral bristles in two rows. Dysaletria-like species
with broad frons and linear face (Fig. 156). ... XI. hackmani-group

12(7)   Mid femora with distinct pv bristles behind the double row
of short black spines beneath (cf. Figs. 198, 199, 206, 230).

13(16) Mesonotum largely polished, sternopleura with a bare polished patch extending to, and including, its hind-margin.
Mid tibia with a very small apical spur (cf. Figs. 198, 199).

14(15) Acr multiserial, narrowly separated from numerous dc.
Proboscis conspicuously large and strong, almost as long
as head is high. Fore tarsi with conspicuously dilated segments 3 and 4. .... VI. luteus-group

15(14) Acr narrowly biserial and broadly separated from dc. Proboscis small, and all tarsi slender. .... VII. nigritarsis-group

16(13) Mesonotum densely grey dusted; if polished, then mid
tibia always with a large, sharply pointed apical spur.

17(18) Mesonotum more or less polished, only seldom very thinly
dusted (Figs. 112-115). Humeral bristle very small; if
longer, then always only fine. Mid tibia with a large, sharply
pointed apical spur (Figs. 202-206)............ VIII. minutus-group

18(17) Mesonotum with a dense covering of greyish dust. A distinct humeral bristle, or several shorter bristles, always
present.

19(20) Humeri with a single long bristle. Thorax rather narrow,
sternopleura with a small bare patch or entirely dusted.
Tibial spur long and sharply pointed, much longer than
tibia is deep; if short or blunt, then rather slender species
with the other characters of this section. ....
.... IX. pallidiventris - cursitans-group

20(19) Humeri with several shorter bristles. Thorax robust, sternopleura with a large bare patch extending to, and including,

50

its hind-margin. Apical spur on mid tibia small, at most as
long as tibia is deep, blunt. 2 pairs of vt bristles and rather
deep jowls below eyes (Fig. 155) .... X. brevicornis-group

Key to species of Platypalpus

The species without numbering have not yet been found in Fennoscandia, or
Denmark, but their occurrence there is expected.

1	1 pair of convergent vt bristles (cf. Fig. 81), or none distinct from the hairs on vertex and upper part of occiput 2
-	2 pairs of distinct vt bristles (cf. Fig. 72), inner convergent, outer divergent 66
2(1)	Yellow species, at least sides of thorax yellow.................... 3
-	Black species, thorax always black in ground-colour even if densely light grey dusted 8
3(2)	Mesonotum unicolourous yellow without any pattern, or whole of thorax indefinitely obscured............................ 4
-	Mesonotum with a narrow dark median stripe, or mesonotum extensively darkened... 6
4(3)	Head including occiput yellow, a very long and strong yellowish proboscis. Mesonotum polished with 8-serial acr separated from dc; large bristles yellowish. Mid femora (Fig. 198) very stout, pv bristles long, brownish; mid tibia with a small blunt spur. Fore tarsi with apical 2 or 3 segments dilated .... 35. luteus (Meig.)
-	At least occiput greyish-black; proboscis smaller, at most half as long as head is high. No pv bristles on mid femora (cf. Fig. 193); apical tarsal segments not dilated........... 5
5(4)	Mesonotum shining; acr and dc numerous, evenly distributed over mesonotum; no humeral bristle; large thoracic bristles darkened. A very small, pointed yellow spur on mid tibia (Fig. 175) .... 10. nonstriatus Str.
-	Mesonotum thinly grey dusted; acr biserial, dc uniserial; humeral bristle present; large thoracic bristles yellowish. A slender yellow spur on mid tibia about as long as tibia is deep (Fig. 193)....................... 32. exilis (Meig.)
6(3)	No distinct vt and humeral bristles. Mid tibia (Figs. 173,

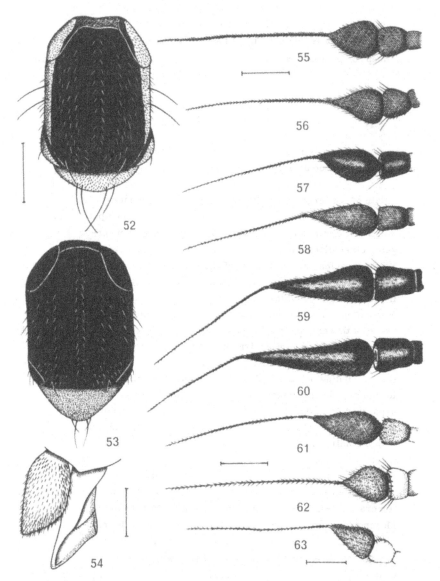

Fig. 52-53. Thorax in dorsal view of Platypalpus. - 52: ciliaris (Fall.), ♂;
53: stigmatellus (Zett.), ♂. Scale: 0.3 mm.

Fig. 54. Palpus and proboscis of Platypalpus mikii (Beck.), ♂. Scale: 0.1 mm.
Figs. 55-63. Antennae of Platypalpus. - 55: ciliaris (Fall.), ♂; 56: confiformis
Chv., ♂; 57: confinis (Zett.), ♀; 58: confinis (Zett.), ♂; 59: stigmatellus (Zett.),
♀; 60: stigmatellus (Zett.), ♂; 61: pectoralis (Fall.), ♂; 62: mikii (Beck.), ♂;
63: nonstriatus Strobl, ♂. Scale: 0.1 mm.

174) often apically stouter with a covering of dense pile, apical spur often practically absent 7

\- A pair of vt bristles and a humeral bristle present. Mid tibia (Fig. 186) not dilated apically, apical spur very small, pointed; mid femora rather slender. Mesonotum with a dark median stripe, and indistinct biserial acr and uniserial dc .... 22. _sahlbergi_ (Frey)

7(6) Acr and dc numerous, evenly and densely distributed over mesonotum which has a rather narrow, dark median stripe.....
.... 8. _pectoralis_ (Fall.)

\- Acr irregularly biserial, rather long and diverging, distinctly separated from multiserial dc at sides. Mesonotum uniformly blackish-brown; or with a dark median stripe in var. _tristriolata_ .... 9. _mikii_ (Beck.)

8(2) No distinct vt bristles, vertex and upper part of occiput with numerous subequal hairs; in doubtful cases, the species with thoracic hairs evenly distributed over mesonotum belong here. Mesonotum (Figs. 52, 53, 71) mostly polished; no humeral bristle. Sternopleura largely polished, the bare patch extending to, and including, its hind-margin; or pleura mostly polished. At most a very small apical spur on mid tibia 9

\- A pair of distinct vt bristles, which, even if small, is distinct from other pubescence (cf. Fig. 73).......................14

9(8) Mid femora (cf. Fig. 170) without pv bristles or bristly-hairs behind the double row of spines beneath, the spines in posterior row usually longer. Acr at most 4-serial, separated from dc by bare stripes. Ocellar bristles generally small (except for _confiformis_). Mid femora rather slender and palpi small...10

\- Mid femora (Fig. 176) with a row of pv bristles behind the double row of small black spines beneath. Thoracic hairs (acr and dc) evenly and densely distributed over mesonotum (Fig. 71). Ocellar bristles usually long. Mid femora very stout and palpi large. Pleura dusted except for sternopleura 13

10(9) Acr and dc numerous, acr bi- to quadriserial, and a considerable pubescence between dc and sides of thorax. Pleura dusted except for bare sternopleura. Frons and occiput always dusted. Antennae (Figs. 55, 56) blackish, with short segment 3 and long arista.................................... 11

- Mesonotum almost bare, the few acr and dc minute, and only a few hairs outside the dc. Pleura mainly polished. Antennae (Figs. 57-60) blackish (if basal segments yellow and fine vt bristles, see paragraph 19) 12

11(10) Mid tibia (Fig. 169) slightly swollen apically and covered on at least apical half by dense covering of whitish pile. Mid femora (Fig. 169) as stout as fore femora, the black spines in posterior row beneath shorter and more numerous on apical half. Ocellar bristles small. Male genitalia (Figs. 245-248) very large, globose; segment 8 of female abdomen

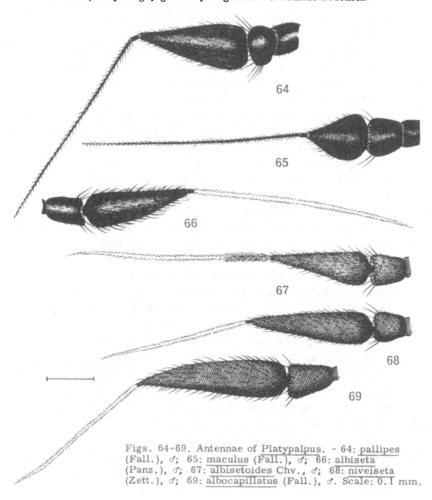

Figs. 64-69. Antennae of Platypalpus. - 64: pallipes (Fall.), ♂; 65: maculus (Fall.), ♂; 66: albiseta (Panz.), ♂; 67: albisetoides Chv., ♂; 68: niveiseta (Zett.), ♂; 69: albocapillatus (Fall.), ♂. Scale: 0.1 mm.

(Fig. 255) long and slender, almost 3 times as long as deep, polished black .... 4. ciliaris (Fall.)

- Mid tibia (Fig. 170) not swollen apically, with only sparse hairs. Mid femora (Fig. 170) stouter, almost twice as deep as fore femora, the black spines in posterior row beneath more uniformly long and thinner towards apex. Ocellar bristles large. Male genitalia (Figs. 249-251) small; segment 8 of female (Fig. 256) short, dull.......... 5. confiformis Chv.

12(10) Antennae (Figs. 57, 58) short, segment 3 about 1.5 - 2 times as long as deep in both sexes. Frons and vertex greyish dusted. Abdomen of male with sparse hairs. Mid femora (Fig. 171) rather slender, scarcely 1.5 times as deep as fore femora; mid tibia apically with a covering of whitish pile ....
.... 6. confinis (Zett.)

- Antennae (Figs. 59, 60) long, segment 3 about 3 - 4 times as long as deep in male, 2 - 3 times in female. Frons and vertex polished black. Abdomen of male with dense hairs. Mid femora (Fig. 172) stouter, about twice as deep as fore femora; mid tibia at tip without a covering of whitish pile .....
.... 7. stigmatellus (Zett.)

13(9) Frons shining. Antennal segment 3 (Fig. 65) short, slightly longer than deep. Femora more or less darkened; mid femora with fine, less distinct yellow pv bristles. Abdominal segment 8 of ♀ large and stout (Fig. 278)......... 11. maculus (Zett.)

- Both frons and occiput dusted. Antennal segment 3 (Fig. 64) longer, twice (or more) as long as deep. Legs yellow; mid femora (Fig. 176) with a row of distinct, long blackish pv bristles. Segment 8 of ♀ smaller (Fig. 279)....12. pallipes (Fall.)

14(8) Mesonotum more or less polished black; or subshining, very slightly and uniformly dusted (if there is doubt as to the presence of greyish dust, the species will be found in this section)..15

- Mesonotum densely grey to light grey dusted (care must be taken with rubbed specimens)................................. 35

15(14) No apical spur on mid tibia (cf. Figs. 180, 181), or only a very small one. Mesonotum (cf. Figs. 94-96) entirely polished black, at most sides narrowly dusted (if entirely but rather thinly dusted, see paragraph 37)......................16

- A large, sharply pointed apical spur on mid tibia (cf. Figs. 202-205); if smaller (fenestella), then mesonotum only

polished on a large patch posteriorly (Fig. 113) and pv bristles on mid femora pale 23

16(15)   Mid femora (cf. Figs. 181, 184) without pv bristles. Arista
always dark 17

-   Mid femora (cf. Figs. 177-180) with long blackish pv bristles. Arista whitish and rather stout. The vein closing anal cell
recurrent and S-shaped (albiseta-group)........................20

17(16)   Antennal segment 3 (Fig. 84) very long, 5 - 6 times as long
as deep; arista much shorter, 1/4 as long as segment 3.
Pleura dusted except for sternopleura; thoracic bristles
black; a distinct humeral bristle. Acr and dc biserial, rather long, pale, diverging; a considerable pubescence between dc and sides of thorax................27. longicornioides Chv.

-   Antennal segment 3 (cf. Figs. 78, 79) much shorter, at most
2.5 times as long as deep; arista longer than segment 3.
Large thoracic bristles fine, pale. Mesonotum (Figs. 94-96)
almost bare; no acr, dc uniserial and only minute 18

18(17)   Pleura dusted except for sternopleura. Mid femora (Fig.
184) thickened, almost twice as stout as fore femora. Antennae (Fig. 79) dark; head grey dusted. All dc minute, no
distinct humeral bristle (Fig. 95). Male genitalia (Fig. 312)
large, apically with very long pale bristly-hairs; segment 8

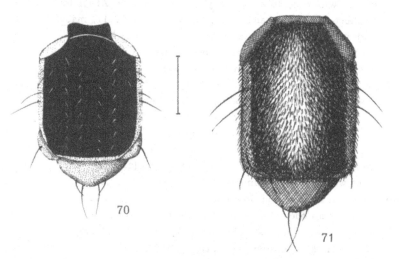

Figs. 70-71. Thorax in dorsal view of Platypalpus. - 70: niveiseta (Zett.), ♂;
71: maculus (Zett.), ♂. Scale: 0.3 mm.

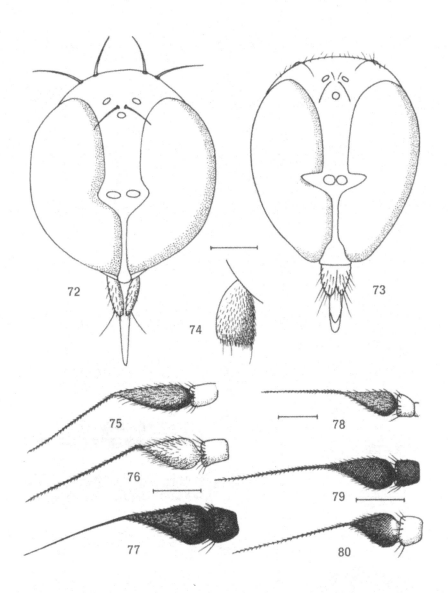

Figs. 72-73. Heads in frontal view of Platypalpus. - 72: alter (Coll.), ♂; 73: unguiculatus (Zett.), ♀. Scale: 0.1 mm.

Fig. 74. Palpus of Platypalpus sahlbergi (Frey), ♂. Scale: 0.1 mm.

Figs. 75-80. Antennae of Platypalpus. - 75: unguiculatus (Zett.), ♀; 76: alter (Coll.), ♂; 77: lapponicus Frey, ♂; 78: zetterstedti Chv., ♂; 79: laestadianorum (Frey), ♂; 80: sahlbergi (Frey), ♂. Scale: 0.1 mm.

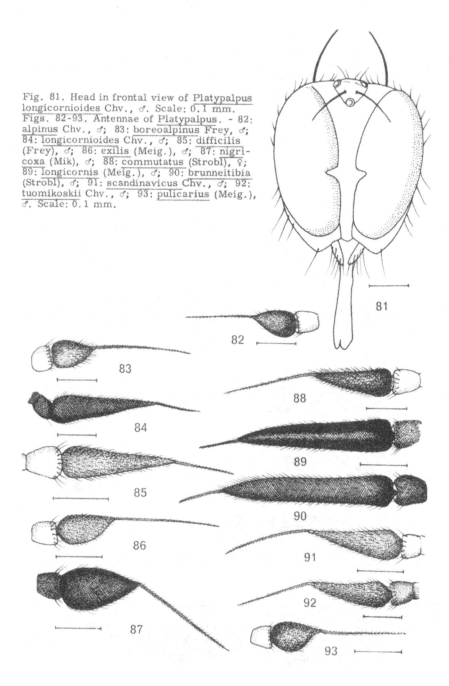

Fig. 81. Head in frontal view of Platypalpus longicornioides Chv., ♂. Scale: 0.1 mm. Figs. 82-93. Antennae of Platypalpus. - 82: alpinus Chv., ♂; 83: boreoalpinus Frey, ♂; 84: longicornioides Chv., ♂; 85: difficilis (Frey), ♂; 86: exilis (Meig.), ♂; 87: nigricoxa (Mik), ♂; 88: commutatus (Strobl), ♀; 89: longicornis (Meig.), ♂; 90: brunneitibia (Strobl), ♂; 91: scandinavicus Chv., ♂; 92: tuomikoskii Chv., ♂; 93: pulicarius (Meig.), ♂. Scale: 0.1 mm.

of ♀ elongated (Fig. 306). Generally larger species, body
about 2.5 mm in length .... 20. laestadianorum (Frey)
- Pleura mainly polished. Mid femora (Figs. 181, 182) slender, only slightly stouter than fore femora. Basal antennal
segments (Figs. 75, 78) usually pale. Male genitalia (cf.
Figs. 296, 301) small and almost bare; segment 8 of ♀
(Figs. 308, 309) short. Generally smaller, polished black
species 19
19(18) Head entirely polished black. Basal antennal segments
yellow (Fig. 75). No humeral bristle, and all dc minute
(Fig. 94). Veins R4+5 and M almost parallel (Fig. 684)....
.... 17. unguiculatus (Zett.)
- Frons and occiput greyish dusted, only vertex polished
black. Basal antennal segments (Fig. 78) yellowish in ♂,
dark in ♀. A humeral bristle and last pair of prescutellar
dc long, pale (Fig. 96). Veins R4+5 and M evenly divergent (Fig. 685)............................... 18. zetterstedti Chv.
20(16) Fore femora with a row of long blackish pv bristles. Humeri very weakly developed; pleura extensively polished 21
- Fore femora without pv bristles, only two rows of longer
dark hairs beneath. Humeri (Fig. 70) more distinct; pleura
densely dusted with a polished patch on sternopleura. Antennal segment 3 (Fig. 68) about 4 times as long as deep,
arista subequal .... 16. niveiseta (Zett.)
21(20) Legs extensively blackish. Frons on the upper half broader
than front ocellus. Antennal segment 3 (Fig. 69) longer, about
4 - 5 times as long as deep, arista subequal .15. albocapillatus (Fall.)
- Legs mostly yellowish. Frons very narrow, not wider above
than front ocellus. Antennal segment 3 (Figs. 66, 67) shorter,
about 3 times as long as deep, arista much longer 22
22(21) Mid femora (Fig. 177) rather slender, about 1.5 times as deep
as fore femora. Pleura mostly polished. Acr biserial, distinct even though minute. Male genitalia (Fig. 284) rather short
and broad; slender cerci overlap the lamellae ... 13. albiseta (Panz.)
- Mid femora (Fig. 178) shorter and thicker, about twice as
deep as fore femora. Pleura very thinly dusted but sternopleura, central area of mesopleura and hind part of hypopleura polished. Acr practically absent. Male genitalia
(Fig. 287) not conspicuously convex; cerci shorter, concealed in lamellae .... 14. albisetoides Chv.

23(15) Mesonotum (Figs. 112-115) at least on posterior half in
front of scutellum polished black 24
- Mesonotum very thinly grey or yellow-grey dusted, sometimes with only a small polished patch at middle 31
24(23) Legs including coxae yellow. 25
- Legs (Fig. 204) extensively darkened, at least four posterior coxae blackish. Antennae (Figs. 106, 107) always dark,
segment 3 short. 30
25(24) Antennae (Fig. 102) blackish-brown, segment 3 slightly longer
than deep. Frons polished black. Pleura dusted except for
sternopleura; mesonotum (Fig. 112) including humeri polished black. Thoracic hairs and bristles pale; acr irregularly
4-serial; dc biserial, rather long with similar hairs at side....
.... 39. fuscicornis (Zett.)
- Antennae (Figs. 104, 105) yellowish at least on basal segments ... 26
26(25) Mid femora very thickened. Thoracic hairs and bristles
dark; acr and dc minute; humeral bristle small. Frons
polished, occiput weakly dusted. Mesonotum largely polished; pleura dusted except for sternopleura. Antennal segment
3 brownish, twice as long as deep, arista 1.5 times as long....
.... pseudociliaris (Str.)

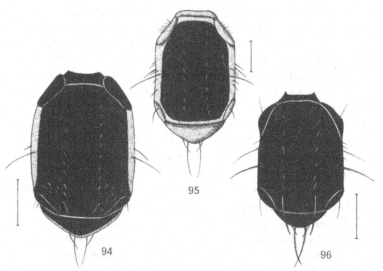

95

94

96

Figs. 94-96. Thorax in dorsal view of Platypalpus. - 94: unguiculatus (Zett.),
♂; 95: laestadianorum (Frey), ♂; 96: zetterstedti Chv., ♀. Scale: 0.2 mm.

\-   Mid femora only slightly deeper than fore femora; if in
        doubt as to its thickness, then pleura extensively polished.
        Thoracic hairs and bristles pale. Frons and occiput at least
        weakly dusted. Antennal segment 3 short, about 1.5 times as
        long as deep. 27

27(26) A distinct humeral bristle (Fig. 113). Mid tibiae rather
        slender with a small apical spur which is not longer than
        tibia is deep. Acr and dc longer, about as long as antennal
        segment 2, rather distinct. 28

\-   Humeral bristle minute (Figs. 114, 115). Mid tibiae (Fig.
        205) rather thickened with a large pointed apical spur. Acr
        and dc minute, rather indistinct 29

28(27) Pleura dusted except for a small polished patch on sterno-
        pleura; mesonotum (Fig. 113) with a large polished patch
        on posterior third. Frons and occiput densely dusted. Anten-
        nae (Fig. 105) yellow, arista brownish . . . . .41. _fenestella_ Kov.

\-   Pleura extensively polished; mesonotum mainly polished
        in both sexes, dusted at sides and on humeri. Frons and oc-
        ciput only weakly dusted. Basal antennal segments somewhat
        reddish, segment 3 darkened . . . . _ingenuus_ (Coll.)

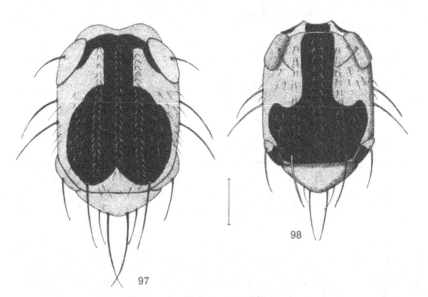

Figs. 97-98. Thorax in dorsal view of _Platypalpus_. - 97: _longicornis_ (Meig.),
♂; 98: _boreoalpinus_ Frey, ♀. Scale: 0.2 mm.

29(27) Antennae (Fig. 104) yellow; segment 3 broad at base;
        arista dark. Prothorax including humeri greyish dusted.
        Mesonotum (Figs. 114, 115) polished black on posterior
        half in ♂, and also on a median stripe right up to front in
        ♀ .... 40. ruficornis (v. Ros.)

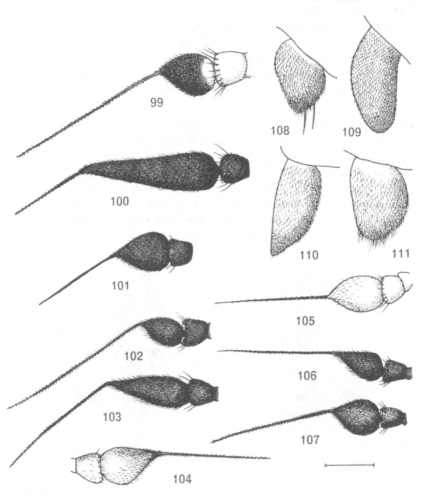

Figs. 99-107. Antennae of Platypalpus. - 99: luteus (Meig.), ♂; 100: nigritarsis (Fall.), ♂; 101: sylvicola (Coll.), ♂; 102: fuscicornis (Zett.), ♂; 103: minutus (Meig.), ♂; 104: ruficornis (von Ros.), ♂; 105: fenestella Kov., ♀; 106:
niger (Meig.), ♂; 107: ater (Wahlbg.), ♂. Scale: 0.1 mm.
Figs. 108-111. Palpi of Platypalpus. - 108: fenestella Kov., ♀; 109: ingenuus
(Coll.), ♀; 110: ruficornis (von Ros.), ♂; 111: same species, ♀. Scale: 0.1 mm.

62

-    Antennal segment 3 darker at tip and narrower at base.
Upper part of prothorax including lower half of humeri
polished black .... politus (Coll.)
30(24) Legs mostly black, only tibiae dark brown; mid femora
(Fig. 203) thickened. Thoracic hairs and bristles pale; acr
and dc rather longer. Pleura dusted except for polished
sternopleura and mesopleura. Male genitalia (Figs. 386-
388) with very long sclerotised curved cerci which are as
long as a very long process on right lamella. Larger species, body about 2.5 mm in length .... 42. ater (Wahlbg.)
-    Fore coxae and all femora largely yellow at base; mid
femora (Fig. 204) rather slender. Thoracic bristles dark;
acr and dc paler and inconspicuous. Pleura extensively polished black. Male genitalia (Figs. 392-394) small, closed.
Smaller species, body less than 2 mm in length (Atlantic
form) .... 44. niger (Meig.)
31(23) Legs yellow, tarsi darkened. Occiput polished black. No
humeral bristle. Basal antennal segments yellow; segment
3 dark, very short; arista long, whitish .... leucothrix (Str.)
-    Legs extensively darkened. Occiput dusted. Humeral bristle present, even if fine. Antennae and arista always dark 32
32(31) A distinct humeral bristle; large thoracic bristles yellow.
Frons very narrow. The vein closing anal cell recurrent
and S-shaped (Fig. 709). Mesonotum metallic-black, very
thinly dusted; pleura dusted except for sternopleura...........
....45. aeneus Macq.
-    Humeral bristle small and fine, large thoracic bristles dark.
Frons broader. The vein closing anal cell slightly recurrent (Figs. 707, 708)... 33
33(32) Pleura extensively polished black. All femora (Fig. 204)
largely yellow at base, black towards tip. Fore tibiae without a rim-like projection at tip beneath. Smaller species,
with mesonotum more or less thinly dusted (Continental form)..
....44. niger (Meig.)
-    Pleura silvery-grey dusted except for polished sternopleura. Femora (Fig. 206) with another pattern 34
34(33) Fore tibiae with a slight rim-like projection at tip beneath.
Femora (Fig. 206) black with base and tip yellowish. Mesonotum thinly dusted with a bare, narrow median stripe.
Larger species, body usually over 2 mm in length .43. minutus (Meig.)

- Without a rim-like projection at tip of fore tibiae. Femora all black. Mesonotum uniformly thinly dusted. Smaller species, body 1.7 - 2 mm in length .... albifacies (Coll.)

35(14) Antennae (cf. Figs. 122, 129, 134-136) black, basal segments sometimes somewhat blackish-brown 36

- At least basal antennal segments (cf. Figs. 145-150) yellow to yellowish-brown, or antennae entirely yellow 48

36(35) Apical spur on mid tibia (Figs. 215, 216) very small; or if almost as long as tibia is deep, then blunt-tipped; or in ♂ of cothurnatus with a tiny curved hair at tip (Fig. 215) 37

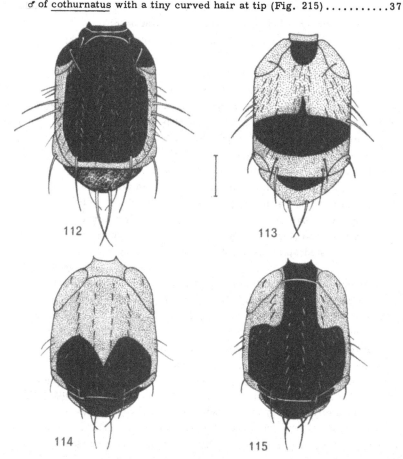

Figs. 112-115. Thorax in dorsal view of Platypalpus. - 112: fuscicornis (Zett.), ♂; 113: fenestella Kov., ♀; 114: ruficornis (von Ros.), ♂; 115: same species, ♀. Scale: 0.2 mm.

–   Mid tibia (Figs. 217-222) with a large, sharply pointed
        apical spur which is distinctly longer than tibia is deep............39

37(36) Dc bristles few in number, distinct and bristle-like, lon-
        ger than acr. A very small apical spur on mid tibia (Fig.
        216), much shorter than tibia is deep. Antennal segment 3
        (Fig. 125) small, slightly longer than deep. Very small spe-
        cies, body 1.0 - 1.7 mm in length 38

–   Dc bristles numerous, short and hair-like, scarcely longer
        than acr. Apical spur on mid tibia (Fig. 215) larger and
        blunt-tipped (in ♂ with a tiny curved hair at tip), almost as
        long as tibia is deep. Antennal segment 3 (Fig. 122) longer,
        about twice as long as deep, narrowly pointed. Not so small
        a species, body 1.7 - 2.5 mm in length........ 52. cothurnatus Macq.

38(37) Tarsi with distinct dark annulations. The two rows of acr
        bristles close together. Left lamella of ♂ genitalia (Fig.
        421) with long pale hairs on its outer margin...53. cryptospina (Frey)

–   Tarsi yellow, or apical two segments slightly darkened,
        without annulations. Acr wider apart. Left lamella of ♂
        genitalia with only short fine hairs on its outer margin........
        .... aliterolamellatus Kov.

39(36) Large thoracic bristles black. Acr irregularly quadriseri-
        al on a broad median stripe. Antennal segment 3 (Figs.
        126, 127) long and slender, almost 3.5 times as long as deep.
        Fore and hind tibiae with small dark bristly hairs above; mid
        femora (Fig. 217) with a row of dark brown bristles posteri-
        orly.. 54. optivus (Coll.)

–   Large thoracic bristles yellow or yellowish-brown............... 40

40(39) Acr bristles quadriserial on a broad median stripe, at
        least on anterior half of mesonotum. Legs (Figs. 218, 219)
        usually extensively darkened, not entirely yellow; anterior
        four femora of almost the same thickness...................... 41

–   Acr bristles biserial throughout................................42

41(40) Acr longer than usual and rather whitish, quadriserial
        throughout. Large thoracic bristles yellowish. Wings (Fig.
        719) almost clear; veins brownish. Legs generally paler;
        usually only posterior four coxae, fore coxae at base, and
        more or less broad rings on femora (Fig. 218) blackish;
        rest of legs yellow. Generally smaller and paler species,
        body 2.3 - 3.3 mm in length .... 55. annulatus (Fall.)

–   Acr shorter, biserial about middle or posteriorly. Large

thoracic bristles brownish. Wings (Fig. 720) brown clouded;
veins very dark. Legs darker; all coxae, trochanters, and
femora (Fig. 219) except for tip blackish; rest of legs rather
yellowish-brown. Generally larger species, body 2.9 - 3.8
mm in length........................... 56. melancholicus (Coll.)

42(40) Antennal segment 3 (Figs. 130, 132, 133) longer, at least
2.5 times as long as deep; arista equal to, or at least not
very much longer than, segment 3.............................43

- Antennal segment 3 (Figs. 134-136) shorter, at most twice
as long as deep; arista much longer than segment 3 46

43(42) Palpi very pale yellow. At least hind tarsi yellow, with only
apical two segments very darkened, without annulations. Vt
bristles widely separated. Wings somewhat yellowish with
pale veins. A smaller, pale species, body 1.5 - 2.1 mm in
length. Abdomen light brownish.....................insperatus Kov.

- Palpi dark in ground-colour, covered with silver pile. All
tarsi with distinct dark annulations. Abdomen black 44

44(43) Vt bristles closer together. Mid femora (Fig. 222) much

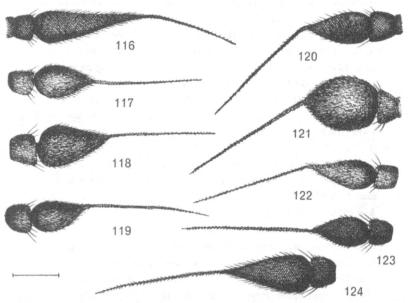

Figs. 116-124. Antennae of Platypalpus. - 116: maculipes (Meig.), ♂; 117:
pallidicoxa (Frey), ♂; 118: agilis (Meig.), ♂; 119: pseudorapidus Kov., ♂;
120: aeneus (Macq.), ♂; 121: rapidus (Meig.), ♂; 122: cothurnatus Macq., ♂;
123: balticus Kov., ♂; 124: nigrosetosus (Strobl), ♂. Scale: 0.1 mm.

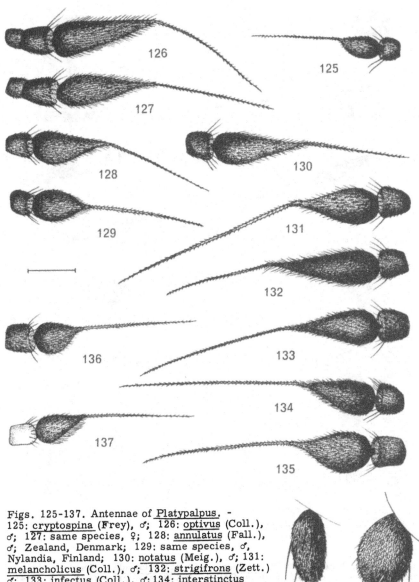

Figs. 125-137. Antennae of Platypalpus. -
125: cryptospina (Frey), ♂; 126: optivus (Coll.),
♂; 127: same species, ♀; 128: annulatus (Fall.),
♂; Zealand, Denmark; 129: same species, ♂,
Nylandia, Finland; 130: notatus (Meig.), ♂; 131:
melancholicus (Coll.), ♂; 132: strigifrons (Zett.)
♂; 133: infectus (Coll.), ♂; 134: interstinctus
(Coll.), ♂; 135: coarctatus (Coll.), ♂; 136:
clarandus (Coll.), ♂; 137: articulatus Macq.,
♂. Scale: 0.1 mm.
Figs. 138-139. Palpi of Platypalpus.- 138: interstinctus (Coll.), ♂; 139: coarctatus (Coll.), ♂.
Scale: 0.1 mm.

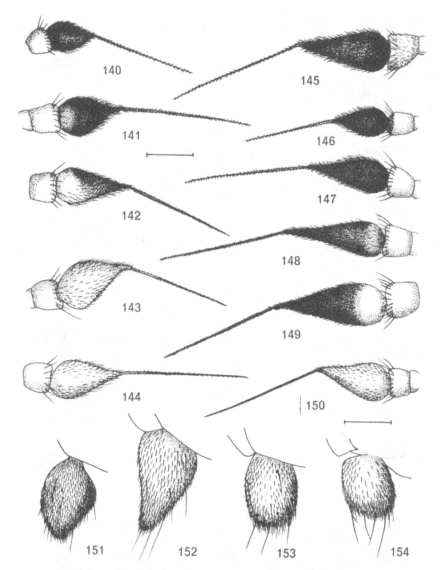

Figs. 140-150. Antennae of Platypalpus. - 140: articulatoides (Frey), ♂; 141: ecalceatus (Zett.), ♂; 142: stabilis (Coll.), ♂; 143: albicornis (Zett.), ♂; 144: pallidicornis (Coll.), ♂; 145: annulipes (Meig.), ♂; 146: calceatus (Meig.), ♂; 147: pallidiventris (Meig.), ♂; 148: longiseta (Zett.), ♂; 149: laticinctus Walk., ♂; 150: flavicornis (Meig.), ♂. Scale: 0.1 mm.
Figs. 151-154: Palpi of Platypalpus. - 151: laticinctus Walk., ♂; 152: albicornis (Zett.), ♂; 153: flavicornis (Meig.), ♂; 154: pallidicornis (Coll.), ♂. Scale: 0.1 mm.

stouter than fore femora; knees black; fore tibiae thickened
and with dark bristles above. Abdomen with small grey lateral
patches on anterior two segments. Usually 3 notopleural
bristles. Arista (Fig. 133) distinctly longer than antennal
segment 3 ···································· 59. infectus (Coll.)
- Vt bristles wider apart. Mid femora (Figs. 220, 221) rather
narrower, not very much stouter than fore femora. Abdomen entirely shining black including basal segments. Only
2 notopleural bristles. Arista (Figs. 130, 132) as long as,
or scarcely longer than, antennal segment 3 45

45(44) Legs extensively darkened, at least fore coxae at base, whole
of posterior coxae and four posterior femora (Fig. 220)
near tip dark brown. Antennae (Fig. 130) with segment 3
usually slightly shorter than arista. Face narrower than frons
in front. Dc bristles weakly developed. Male genitalia (Fig.
432) with long, blunt-tipped cerci. Generally smaller species, body 1.9 - 2.7 mm in length.............. 57. notatus (Meig.)
- Legs extensively yellow, in particular fore coxae and femora only rarely clouded. Antennal segment 3 (Fig. 132) as
long as, or slightly longer than, arista. Face almost as deep
as frons in front (Fig. 4). Dc distinct, bristle-like. Left cercus of male genitalia (Fig. 435) with a sclerotised pointed
hook at tip. Larger species, body 2.5 - 3.5 mm in length......
.... 58. strigifrons (Zett.)

46(42) Generally larger and robust species, body 2.4 - 3.8 mm
in length. Vt bristles closer together. Clypeus grey dusted;
palpi (Figs. 138, 139) moderately large, brownish. Legs
(Fig. 223) often with dark maculations......................... 47
- Smaller species, body 1.5 - 2.6 mm in length. Mesonotum
rather golden-yellow dusted. Vt bristles wider apart. Clypeus black; palpi small, dark in ground-colour. Antennal
segment 3 (Fig. 136) small, only slightly longer than deep.
Legs (Fig. 224) yellow except for annulated tarsi..........
.... 62. clarandus (Coll.)

47(46) Acr bristles less numerous and the two rows conspicuously
wide apart. Palpi (Fig. 138) narrow, much longer than broad.
Fore tibiae in ♂ rather slender, with short hairs beneath .....
.... 60. interstinctus (Coll.)
- Acr bristles more numerous and the two rows close together.
Palpi (Fig. 139) broadly ovate, almost as broad as long.

69

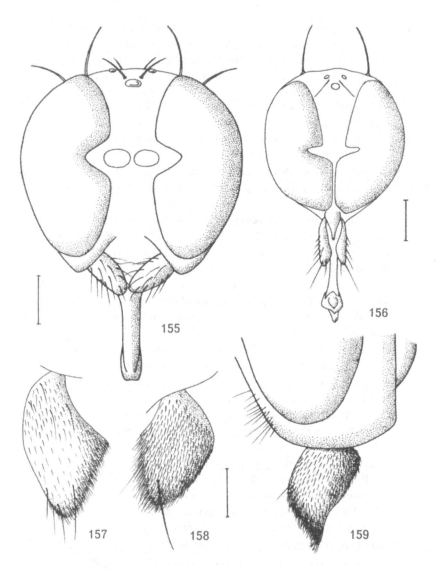

Figs. 155-156. Heads in frontal view of Platypalpus. - 155: brevicornis (Zett.),
♂; 156: hackmani Chv., ♂. Scale: 0.1 mm.
Figs. 157-158. Palpi of Platypalpus. - 157: major (Zett.), ♀; 158: cursitans
(Fabr.), ♂. Scale: 0.1 mm.
Fig. 159: Lower head in lateral view of Platypalpus candicans (Fall.), ♂. Scale:
0.1 mm.

Fore tibiae in ♂ very spindle-shaped dilated, with long hairs
beneath which also continue on to fore metatarsus.......
.... 61. coarctatus (Coll.)

48(35) Thoracic pleura densely grey dusted, including sternopleura.. 49

- Sternopleura with a bare polished patch (in doubtful cases
see paragraph 53) 52

49(48) Veins R4+5 and M almost parallel (Figs. 743, 747). At
least 3 pairs of dc large and bristle-like. Smaller species,
body less than 3 mm in length 50

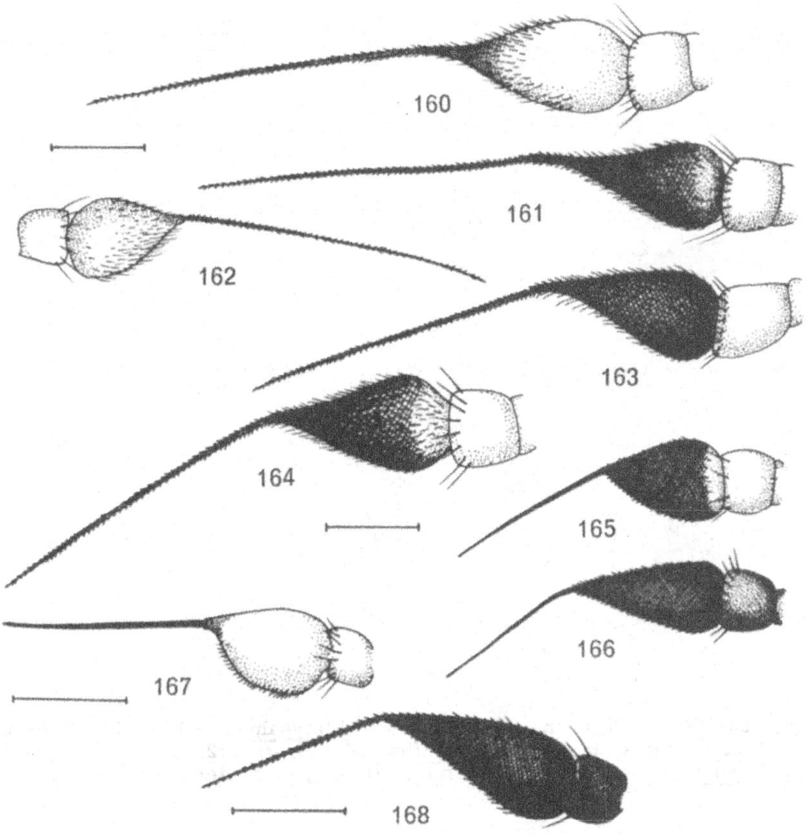

Figs. 160-168: Antennae of Platypalpus. - 160: major (Zett.), ♀; 161: analis
(Meig.), ♂; 162: verralli (Coll.), ♂; 163: candicans (Fall.), ♂; 164: cursitans
(Fabr.), ♂: 165: brevicornis (Zett.), ♂; 166: sordidus (Zett.), ♀; 167: hack-
mani Chv., ♂; 168: subbrevis (Frey), ♀. Scale: 0.1 mm.

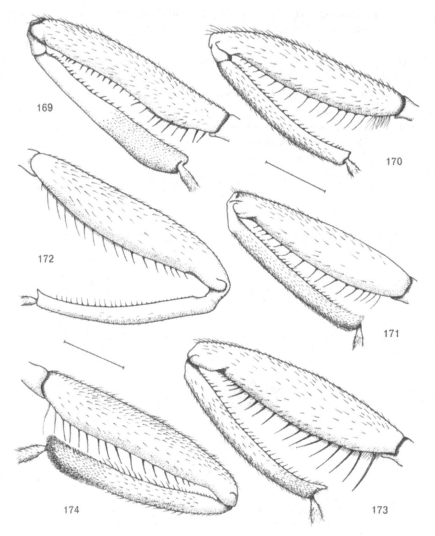

Figs. 169-174. Mid legs in posterior view of Platypalpus. - 169: ciliaris (Fall.),
♂; 170: confiformis Chv., ♂; 171: confinis (Zett.), ♂; 172: stigmatellus (Zett.),
♂; 173: pectoralis (Fall.), ♂; 174: mikii (Beck.), ♂. Scale: 0.3 mm.

- Vein M distinctly bowed (Figs. 741, 742). All dc small and hair-like except for 1 or 2 prescutellar pairs. Larger species, body generally more than 3 mm in length 51

50(49) Apical spur on mid tibia (Fig. 244) and pv bristles on mid femora absent. Frons (Fig. 156) conspicuously broad; face very linear. Anterior notopleural (posthumeral) bristle present; 3 pairs of large dc. Abdomen grey dusted. Tarsi with last segment darkened. Smaller, body 1.4 - 2.2 mm in length ....83. hackmani Chv.

- Mid tibia (Fig. 240) with a large apical spur; distinct pv bristles on mid femora. Frons narrower, only slightly broader than face. Anterior notopleural bristle absent, but 5 - 6 pairs of large dc. Abdomen shining black. Tarsi annulated. Larger, body 2.2 - 2.8 mm in length ....79. verralli (Coll.)

51(49) Vein M very strongly and evenly bowed even towards apex (Fig. 741). Frons very narrow, not broader than front ocellus. Antennal segment 3 (Fig. 163) blackish-brown. Tarsi pale or indistinctly annulated. Jowls (Fig. 159) rather deep, dull grey. Abdomen extensively shining......... 77. candicans (Fall.)

- Vein M not so strongly and evenly bowed, curved up and ending at tip of wing close to end of vein R4+5 (Fig. 742). Frons broader than front ocellus. Antennal segment 3 (Fig. 164) brownish, often yellowish at base, broader. Tarsi with brownish annulations. Jowls very narrow, black. Abdomen greyish dusted at sides ....78. cursitans (Fabr.)

52(48) Anterior notopleural (posthumeral) bristle developed. Fore and hind tibiae with distinct dark bristly-hairs above 53

- Anterior notopleural bristle not developed. Fore and hind tibiae with short hairs (except for major and stabilis); no dark bristly-hairs above 55

53(52) Acr irregularly tri- to quadriserial. Abdomen extensively yellowish. Sternopleura indistinctly polished at middle, rather subshining. Large species, body about 3.5 - 4 mm in length .... 76. analis (Meig.)

- Acr biserial. Abdomen shining black. Sternopleura largely polished black anteriorly. Smaller species, body about 2.5 - 3 mm in length 54

54(53) Antennal segment 3 (Fig. 147) blackish. ♂: tarsi with distinct dark annulations; fore tibiae rather slender. ♀: darker grey dusted with abdomen extensively shining; somewhat longer

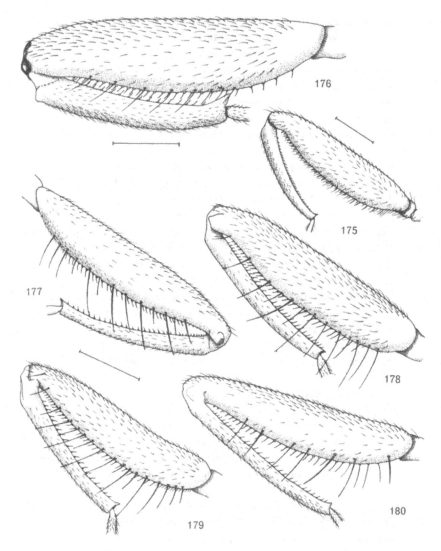

Fig. 175. Mid leg in anterior view of Platypalpus nonstriatus Strobl, ♂.
Figs. 176-180. Mid legs in posterior view of Platypalpus. - 176: pallipes (Fall.),
♂; 177: albiseta (Panz.), ♂; 178: albisetoides Chv., ♂; 179: albocapillatus
(Fall.), ♂; 180: niveiseta (Zett.), ♂. Scale: 0.3 mm.

acr. Generally smaller .... 69. <u>pallidiventris</u> (Meig.)

- Antennal segment 3 (Fig. 148) usually narrowly yellowish
  at base. ♂: fore tarsi yellow except for blackish last segment; fore tibiae spindle-shaped dilated. ♀: lighter grey
  with more greyish banded abdomen; smaller acr through-
  out. Generally larger, more robust species..... 70. <u>longiseta</u> (Zett.)

55(52) Antennae (Figs. 143, 144, 150, 160) entirely yellow, at most
  segment 3 slightly darkened at extreme tip. Palpi (Figs.
  152-154) large, pale (in N.European species). Usually very
  light grey dusted species with pale yellow legs................... 56

- Antennal segment 3 (cf. Figs. 145, 146) dark, at most narrowly yellowish at extreme base; in doubtful cases (<u>stabilis</u>),
  palpi always small. Generally darker grey dusted species 59

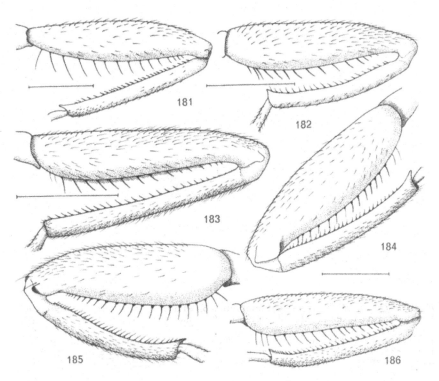

Figs. 181-186. Mid legs in posterior view of <u>Platypalpus</u>. - 181: unguiculatus
(Zett.), ♂; 182: zetterstedti Chv., ♂; 183: alter (Coll.), ♂; 184: laestadianorum (Frey), ♂; 185: lapponicus Frey, ♂; 186: sahlbergi (Frey), ♂. Scale: 0.3 mm.

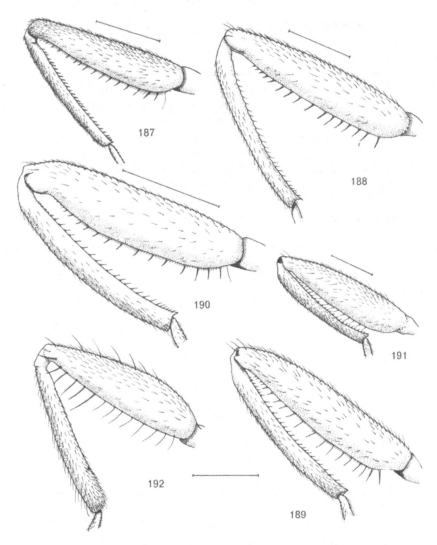

Figs. 187-190. Mid legs in posterior view of Platypalpus. - 187: boreoalpinus Frey, ♂; 188: brunneitibia (Strobl), ♂; 189: longicornis (Meig.), ♂; 190: difficilis (Frey), ♂. Scale: 0.3 mm.
Fig. 191. Mid leg in anterior view of Platypalpus commutatus (Strobl), ♂.
Fig. 192. Fore leg in anterior view of Platypalpus brunneitibia (Strobl), ♂.
Scale: 0.3 mm.

56(55) Vein M very conspicuously evenly bowed (Fig. 739). Mid
femora (Fig. 236) very thickened, about twice as deep as
fore femora; mid tibia with a large, sharply pointed apical
spur. Large species, body usually much over 3 mm in length....
.... 75. major (Zett.) ♀
- Veins R4+5 and M almost parallel (Figs. 736-738). Mid
femora (Figs. 233-235) rather slender, scarcely deeper
than fore femora. Smaller species, body less than 3 mm
in length 57

57(56) Tarsi yellow with only last segment darkened at tip. Palpi
(Fig. 152) not very broad, somewhat pointed at tip. Acr irregularly tri- to quadriserial. Tibial spur (Fig. 233) blunt in
♂, more pointed in ♀. Vt bristles wider apart... 72. albicornis (Zett.)
- Fore tarsi with distinct annulations. Palpi (Figs. 153, 154)
very broadly ovate and blunt-tipped............................. 58

58(57) Frons broader, deeper than antennal segment 2; face silvery-
grey dusted; vt bristles widely separated. Abdomen broadly grey dusted at sides, leaving polished median triangles
on each tergite. Tibial spur (Fig. 234) sharply pointed in
both sexes. Acr biserial .... 73. flavicornis (Meig.)
- Frons narrower; face more or less yellowish-grey dusted.
Vt bristles closer together. Abdomen extensively shining
black, grey lateral margins inconspicuous. Apical spur on
mid tibia (Fig. 235) small and blunt in ♂, sharply pointed in
♀. Acr bi- to triserial .... 74. pallidicornis (Coll.)

59(55) Mid tibia (cf. Figs. 227, 232) with a large apical spur which
is at least as long as tibia is deep 60
- Mid tibia (Figs. 225, 226) with a very small blunt spur which
is much shorter than tibia is deep. Very small species 65

60(59) Wings tinged yellow, veins pale. Only 1 notopleural bristle.
Antennal segment 3 about 2 - 2.5 times as long as deep.
Small species (body 1.5 - 2 mm in length) with yellow legs
and dark annulated tarsi. Tibial spur blunt in ♂, sharply
pointed in ♀ .... subtilis (Coll.)
- Not as above 61

61(60) Antennal segment 3 (Figs. 145, 149) longer, at least 2.5
times as long as deep. Legs often considerably darkened.
Larger species, or acr irregularly triserial and tibial spur
blunt 62
- Antennal segment 3 (Figs. 141, 142, 146) small, slightly

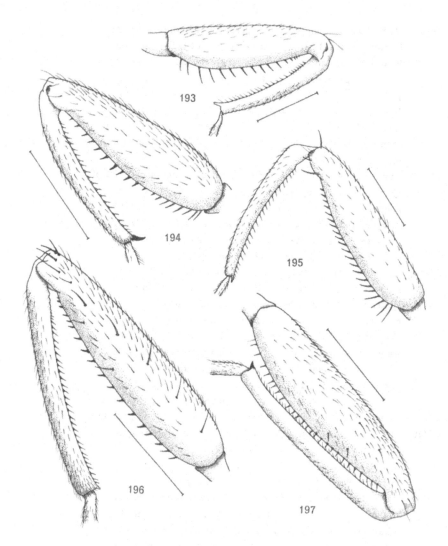

Figs. 193-195. Mid legs in posterior view of Platypalpus. - 193: exilis (Meig.),
♂; 194: pulicarius (Meig.), ♂; 195: nigricoxa (Mik), ♂.
Figs. 196-197: Mid legs in anterior view of Platypalpus. - 196: scandinavicus
Chv., ♂; 197: tuomikoskii Chv., ♂. Scale: 0.3 mm.

longer than deep, or at most twice as long as deep (stabilis).
Medium-sized or small species with biserial acr and sharply
pointed spur 63

62(61) Tarsi yellow, posterior tarsi often weakly annulated; mid
tibia (Fig. 232) with a large, sharply pointed spur. Acr narrowly biserial. Larger, body 3.0 - 4.5 mm in length ....
.... 71. laticinctus Walk.

- Tarsi with distinct dark annulations, and in particular fore
tarsi very sharply so; mid tibia (Fig. 227) with a shorter
and blunt-tipped spur. Acr irregularly triserial. Smaller,
body 2.2 - 3.0 mm in length .... 65. annulipes (Meig.)

63(61) Fore tarsi entirely yellow, without annulations. Antennal
segment 3 (Fig. 141) brownish, short and very broad. Frons
broad, widening above. Basal cells on wing subequal, crossveins contiguous (Fig. 730). Medium-sized species, body
about 2.5 - 3.0 mm in length................. 66. ecalceatus (Zett.)

- Apical tarsal segments darkened, or tarsi annulated. Frons
rather narrow. Crossveins widely separated on wing (Figs.
731, 732)... 64

64(63) Tarsi yellow with apical two segments blackish; anterior
four femora and tibiae slender. Antennal segment 3 (Fig.
146) uniformly blackish-brown, small and slender, about
1.5 times as long as deep. Small, body about 1.5 - 2.0 mm
in length .... 67. calceatus (Meig.)

- Tarsi faintly brown annulated, more darkly so on apical
two segments; anterior four femora and fore tibiae thickened
in ♂. Antennal segment 3 (Fig. 142) yellowish on basal half,
darker towards tip, almost twice as long as deep, not slender. Larger, body generally about 3 mm in length ...68. stabilis (Coll.)

65(59) All tarsi with distinct brownish annulations, or extensively
darkened. ♂ genitalia: left lamella (Fig. 452) almost bare,
without long hairs on its outer (left) margin......63. articulatus Macq.

- Fore tarsi with very sharp black annulation; posterior four
tarsi pale except for dark last segment. ♂ genitalia: left
lamella (Fig. 455) with outstanding long pale hairs on its
outer (left) margin .... 64. articulatoides (Frey)

66(1) Mesonotum polished black at least on posterior third. No
apical spur on mid tibia (cf. Figs. 185, 189, 200) or only
a very small one, not longer than tibia is deep.................. 67

- Mesonotum entirely greyish or yellowish-grey dusted, even

if only thinly so. Apical spur on mid tibia absent or present 75

67(66) Antennal segment 3 (cf. Figs. 77, 82, 101) short, or at
       most 3 times as long as deep; arista longer, or at least
       as long as segment 3... 68

-      Antennal segment 3 (Figs. 89, 100) very long, more than
       4 times as long as deep; arista very short, at most slightly
       longer than half length of segment 3......................... 73

68(67) Antennae (Figs. 76, 82, 83, 88) yellow at least on basal
       segments; in doubtful cases (commutatus) mid femora
       (Fig. 191) slender. No pv bristles on mid femora (Figs. 183,
       187) behind the two rows of black spine-like bristles 69

-      Antennae (Fig. 77) entirely black. Pv bristles on mid femora
       present (Fig. 200) or absent (Fig. 185)....................... 72

69(68) Antennae (Fig. 76) entirely yellow. Pleura polished black;
       no humeral or acr bristles. Legs yellow except for blackish
       last tarsal segment; anterior four femora (Fig. 183) equally

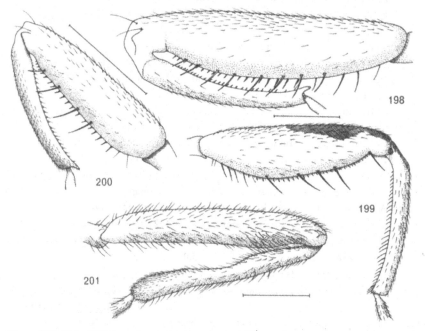

Figs. 198-200. Mid legs in posterior view of Platypalpus. - 198: luteus (Meig.),
♂; 199: nigritarsis (Fall.), ♂; 200: sylvicola (Coll.), ♂. Scale: 0.3 mm.
Fig. 201. Hind leg in anterior view of Platypalpus excisus (Beck.), ♂. Scale:
0.3 mm.

slender ....19. alter (Coll.)

- Antennal segment 3 (Figs. 82, 83, 88) dark. Pleura dusted,
  sternopleura with a bare polished patch; a distinct humeral
  bristle; acr biserial and rather long (Fig. 98) 70

70(69) Antennal segment 3 (Figs. 82, 83) short, slightly longer
  than deep; arista much longer, about twice as long as segment 3 71

- Antennal segment 3 (Fig. 88) longer, 2.5 - 3 times as long
  as deep; arista as long as segment 3 .... 25. commutatus (Str.)

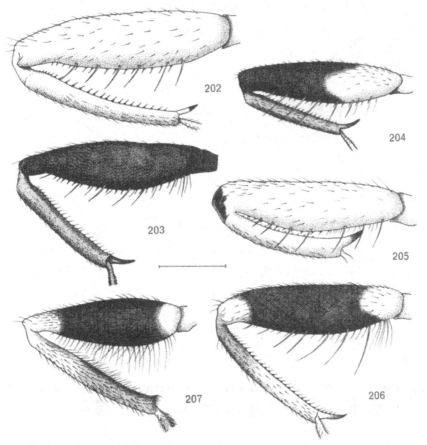

Figs. 202-206. Mid legs in posterior view of Platypalpus. - 202: fuscicornis
(Zett.), ♂; 203: ater (Wahlbg.), ♂; 204: niger (Meig.), ♂; 205: ruficornis (von
Ros.), ♂; 206: minutus (Meig.), ♂. Scale: 0.3 mm.
Fig. 207. Fore leg in anterior view of Platypalpus minutus (Meig.), ♂. Scale:
0.3 mm.

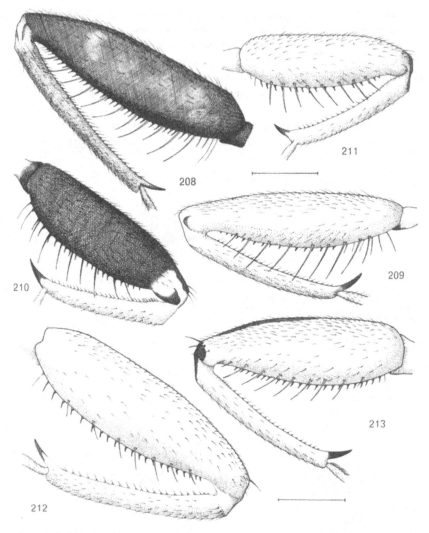

Figs. 208-213. Mid legs in posterior view of Platypalpus. - 208: aeneus (Macq.),
♂; 209: maculipes (Meig.), ♂; 210: rapidus (Meig.), ♀; 211: pallidicoxa (Frey),
♂; 212: agilis (Meig.), ♂; 213: pseudorapidus Kov., ♂. Scale: 0.3 mm.

71(70) At least last tarsal segment extensively darkened; mid
femora (Fig. 187) rather long and slender, as deep as fore
femora. Mesonotum (Fig. 98) polished on posterior half in
♂, also on a broad median stripe extending to front in ♀.......
....23. boreoalpinus Frey
- Legs including tarsi yellow; mid femora considerably shorter and stouter, deeper than fore femora. Mesonotum polished
on a broad median stripe extending anteriorly in both sexes.....
.... 24. alpinus Chv.
72(68) Frons and thoracic pleura dusted. Mid femora (Fig. 185)
thickened, without pv bristles. Antennal segment 3 (Fig. 77)
about twice as long as deep; arista slightly longer. Large
species, body 2.5 - 3 mm in length............. 21. lapponicus Frey
- The very broad frons and whole of thoracic pleura polished
black. Mid femora (Fig. 200) slender, with dark pv bristles.
Antennal segment 3 (Fig. 101) slightly longer than deep; arista about twice as long. Very small species, body 1.0 - 1.5 mm
in length.................................... 38. sylvicola (Coll.)
73(67) Vertex and upper frons polished black. Legs darker yellow
with posterior four coxae, hind femora at tip and all tarsi
blackish; mid femora (Fig. 199) with dark pv bristles. Arista longer, half as long as antennal segment 3 (Fig. 100).
Hypopleura polished black; acr and dc sparse, inconspicuous...... 74
- Frons and vertex greyish dusted. Legs including coxae
yellow, only tarsi darkened; no pv bristles on mid femora
(Fig. 189). Arista very short, scarcely one third as long as
antennal segment 3 (Fig. 89). Hypopleura dusted; acr and dc
rather longer and numerous .... 26. longicornis (Meig.)
74(73) Hind tibiae en ♂ simple, not curved. Fore coxae in ♀ yellow
with only base slightly darkened; mid femora (Fig. 199)
dark at most on apical half above .... 36. nigritarsis (Fall.)
- Hind tibiae (Fig. 201) in ♂ very curved at middle, slightly
thickened near base, and with a shallow excision at about
middle beneath. ♀ with fore coxae except for tip and mid
femora on the whole length above darkened....... 37. excisus (Beck.)
75(66) No apical spur on mid tibia (cf. Figs. 188, 190) or only a
small one, at most as long as tibia is deep. Basal antennal
segments often yellowish....................................... 76
- A long, sharply pointed apical spur on mid tibia (cf. Figs. 209,
210, 212, 214), the spur much longer than tibia is deep. An-

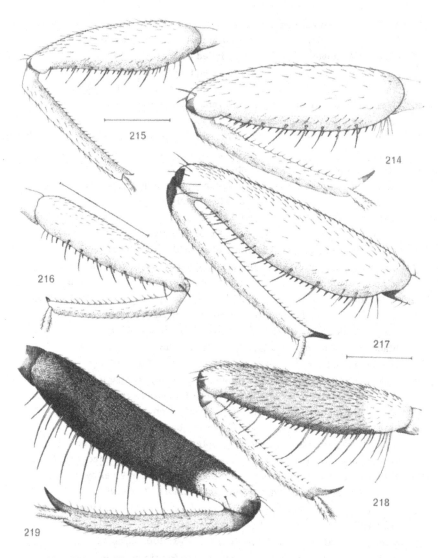

Figs. 214-219. Mid legs in posterior view of Platypalpus. - 214: nigrosetosus Strobl, ♂; 215: cothurnatus Macq., ♂; 216: cryptospina (Frey), ♂; 217: optivus (Coll.), ♂; 218: annulatus (Fall.), ♂; 219: melancholicus (Coll.), ♂. Scale: 0.3 mm.

tennae (cf. Figs. 116, 121) always black 86

76(75) Acr biserial. Mid femora (cf. Figs. 188, 190) rather slender,
at most only slightly deeper than fore femora. Antennal segment 3 (cf. Figs. 85, 90) longer and narrower, at least almost
3 times as long as deep. No pv bristles on mid femora (cf.
Figs. 188, 190) 77

- Acr at least quadriserial. Mid femora (cf. Figs. 194, 241)
rather thickened, much deeper than fore femora; in doubtful
cases pv bristles present on mid femora. Antennal segment
3 (cf. Figs. 87, 93, 165, 166, 168) shorter and broader, at
most twice as long as deep 80

77(76) Antennae (Fig. 90) very long, unicolourous blackish; segment 3 at least 5 times as long as deep; arista very short,
about 1/4 as long as segment 3. Legs extensively brownish,
particularly tibiae, tarsi very dark........... 28. brunneitibia (Str.)

- Antennae (Figs. 85, 91, 92) shorter, segment 3 at most 4
times as long as deep, narrower and pointed towards tip;
arista longer, at least 1/2 as long as segment 3 or equal in
length. Legs yellowish with darkened tarsi...................... 78

78(77) Antennal segment 3 (Fig. 85) almost 4 times as long as deep,
and arista shorter, half as long as segment 3; basal segments
yellowish-brown. Large bristles on head and thorax brownish
or black. Sternite 8 of ♀ (Fig. 349) greyish dusted.....
.... 29. difficilis (Frey)

- Antennae (Figs. 91, 92) shorter, segment 3 at most 3 times
as long as deep, and arista longer, as long as segment 3......... 79

79(78) Basal antennal segments (Fig. 91) yellow. Large bristles
on head and thorax black. Sternite 8 of ♀ (Fig. 350) polished
black, very produced and rounded apically .... 30. scandinavicus Chv.

- Antennae (Fig. 92) uniformly blackish-brown, including
basal segments. Large bristles on head and thorax yellowish-
brown to light brown. Sternite 8 of ♀ (Fig. 351) densely greyish
dusted and not conspicuously produced ....31. tuomikoskii Chv.

80(76) No pv bristles on mid femora (Figs. 194, 195). Not very robust species. Jowls narrow below eyes, and only a single
humeral bristle (nigricoxa posseses fine hairs in front)........... 81

- Distinct pv bristles on mid femora (Figs. 241-243). Species
with robust thorax. Deep jowls below eyes (Fig. 155), and
several bristles on humeri 84

81(80) Basal antennal segments (Fig. 93) yellowish. Acr 4-serial.

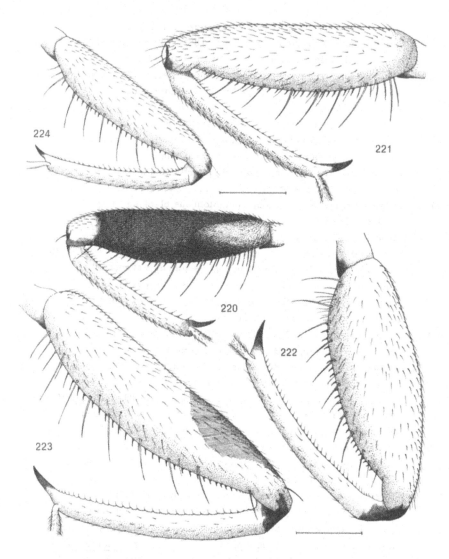

Figs. 220-224. Mid legs in posterior view of Platypalpus. - 220: notatus (Meig.), ♂; 221: strigifrons (Zett.), ♂; 222: infectus (Coll.), ♂; 223: interstinctus (Coll.), ♂; 224: clarandus (Coll.), ♂. Scale: 0.3 mm.

Smaller species, body about 1.5 - 2 mm in length 82
-     Antennae (Fig. 87) black. Acr 6-serial. Legs (Fig, 195) dark
      yellow with posterior four coxae and all tarsi extensively
      darkened, posterior femora with a narrow black stripe beneath.
      Large bristles black. Large species, body about 3 mm in length.....
      .... 34. nigricoxa (Mik)
82(81) Legs yellow except for more or less darkened tarsi. Anten-
      nal segment 3 (Fig. 93) slightly longer than deep. Large
      bristles on head and thorax pale 83
-     Legs yellow with posterior four coxae blackish and posterior
      femora extensively darkened. Antennal segment 3 about twice
      as long as deep. Large bristles blackish. Apical spur on mid
      tibia practically absent. .... incertus (Coll.)
83(82) Apical spur on mid tibia (Fig. 194) black and pointed, only
      slightly shorter than tibia is deep. Tarsi yellow except for
      faintly brownish apical two segments .... 33. pulicarius (Meig.)
-     Apical spur on mid tibia very small, blunt. Tarsi with
      brownish annulations or front pair extensively darkened........
      .... stackelbergi Kov.
84(80) Basal antennal segments (Fig. 165) yellow to brownish,
      segment 3 dark, slightly longer than deep; arista longer
      than segment 3. Palpi pale. Mid tibia (Fig. 241) with a
      very small pointed apical spur; legs brownish in ♂, yel-
      low in ♀ ....80. brevicornis (Zett.)
-     Antennae (Figs. 166, 168) uniformly dark, basal segments
      hardly paler; segment 3 longer, twice as long as deep at
      base, arista subequal. Palpi dark 85
85(84) Legs darkened on coxae and femora except for tip; mid
      femora (Fig. 242) conspicuously thickened; apical tibial
      spur as long as tibia is deep, blunt (particularly in ♀).
      Halteres dark. Generally larger species, ♀ body over 3 mm
      in length .... 81. sordidus (Zett.)
-     Legs also extensively darkened on tibiae and tarsi; mid
      femora (Fig. 243) only slightly deeper than fore femora;
      apical tibial spur small, much shorter than tibia is deep.
      Halteres pale. Generally smaller species, body 2 - 2.5 mm
      in length .... 82. subbrevis (Frey)♀
86(75) Acr bristles quadriserial on a broad median stripe, at
      least in front...87
-     Acr bristles biserial throughout, the two rows close to

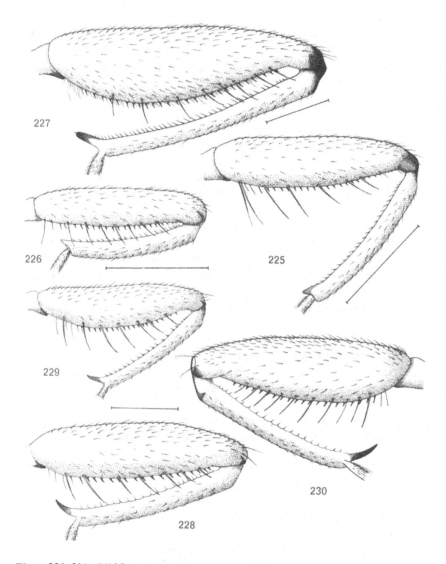

Figs. 225-230. Mid legs in posterior view of Platypalpus. - 225: articulatus
Macq., ♂; 226: articulatoides (Frey), ♂; 227: annulipes (Meig.), ♂; 228: ecal-
ceatus (Zett.), ♂; 229: calceatus (Meig.), ♂; 230: pallidiventris (Meig.), ♂.
Scale: 0.3 mm.

each other (except for nigrosetosus) 88

87(86) Large thoracic bristles and pv bristles on mid femora yellow. Legs including coxae yellow; mid femora (Fig. 209)
slightly deeper than fore femora. Antennal segment 3 (Fig.
116) very slender, thrice as long as deep; arista subequal......
.... 46. maculipes (Meig.)

\- Large thoracic bristles and pv bristles on mid femora black.
Legs with posterior four coxae, almost whole of mid femora,
and hind femora on apical half blackish; mid femora (Fig. 210)
strongly thickened. Antennal segment 3 (Fig. 121) very broad,
slightly longer than deep; arista twice as long as segment 3.....
.... 47. rapidus (Meig.)

88(86) Large thoracic bristles yellowish-brown (see paragraph 92).....
.... balticus Kov.

\- Large thoracic bristles black................................... 89

89(88) Thoracic pleura somewhat subshining, sterno- and hypopleura polished black. Legs yellow but posterior coxae slightly
darkened and all tarsi weakly annulated. Antennal segment
3 (Fig. 117) long-ovate, twice as long as deep. Smaller species, body about 2 mm in length .... 48. pallidicoxa (Frey)

\- Thoracic pleura grey dusted, except for polished sternopleura 90

90(89) Four posterior coxae blackish; legs usually more or less
blackish in colour 91

\- Four posterior coxae yellow; legs uniformly yellow to
yellowish-brown 92

91(90) Legs yellow in ♂ (Fig. 212), extensively blackish in ♀, particularly on femora. Arista shorter, about 1.5 times as long as
antennal segment 3 (Fig. 118). Humeri polished black on the
outside. Robust, generally larger species, body 2.4 - 3.8 mm
in length .... 49. agilis (Meig.)

\- Legs (Fig. 213) yellow in both sexes with a black stripe on
dorsum of mid femora, and apical half of hind femora blackish. Arista longer, at least twice as long as antennal segment 3 (Fig. 119). Humeri wholly grey dusted. Generally
smaller species, body 1.7 - 2.6 mm in length. 50. pseudorapidus Kov.

92(90) Palpi very small, dark. Frons narrower. Acr in two widely separated rows. Mid femora (Fig. 214) considerably
thickened. Antennal segment 3 (Fig. 124) pointed, at least
twice as long as deep at base; arista scarcely twice as long
as segment 3............................. 51. nigrosetosus (Str.)

- Palpi larger, pale yellow. Frons broad. Acr closely biserial. Mid femora rather slender. Antennal segment 3 (Fig.
123) small, ovate, slightly longer than deep; arista longer,
almost thrice as long as segment 3 .... balticus Kov.

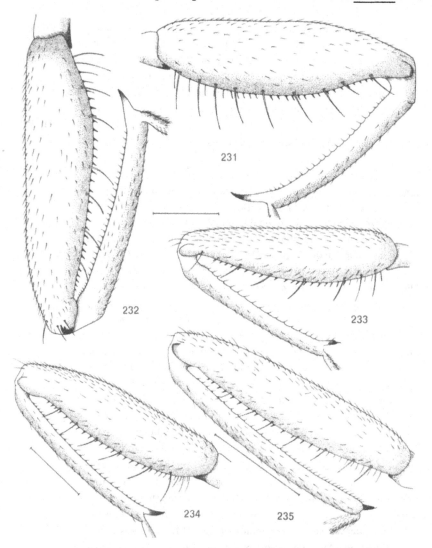

Figs. 231-235. Mid legs in posterior view of Platypalpus. - 231: longiseta
(Zett.), ♂; 232: laticinctus Walk., ♂; 233: albicornis (Zett.), ♂; 234: flavicornis (Meig.), ♂; 235: pallidicornis (Coll.), ♂. Scale: 0.3 mm.

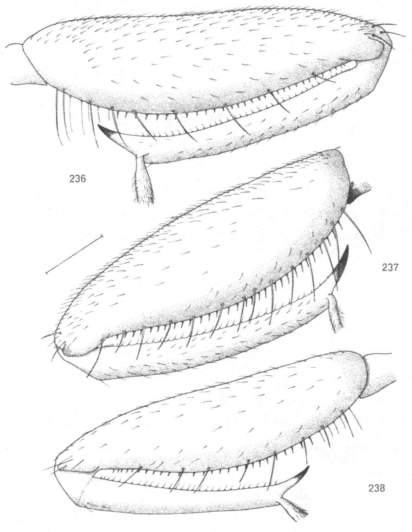

Figs. 236-238. Mid legs in posterior view of Platypalpus. - 236: major (Zett.), ♀; 237: analis (Meig.), ♂; 238: candicans (Fall.), ♀. Scale: 0.3 mm.

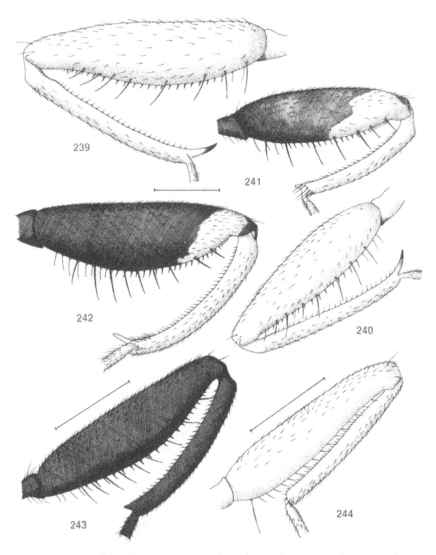

Figs. 239-244. Mid legs in posterior view of Platypalpus. - 239: cursitans (Fabr.), ♂; 240: verralli (Coll.), ♂; 241: brevicornis (Zett.), ♂; 242: sordidus (Zett.), ♀; 243: subbrevis (Frey), ♀; 244: hackmani Chv., ♂. Scale: 0.3 mm.

I. ciliaris - group

4. PLATYPALPUS CILIARIS (Fallén, 1816)

Figs. 52, 55, 169, 245-248, 255, 671.

Tachydromia ciliaris Fallén, 1816: 33.
Platypalpus compungens Walker, 1851: 128.

Mesonotum polished black with narrowly grey dusted margins, thoracic hairs pale, acr biserial, dc uniserial with numerous hairs at sides. Mid tibia thickened apically and covered with whitish pile, terminalia in both sexes very large.

♂. Head black, greyish dusted, vt bristles not differentiated from other pale pubescence, anterior ocellar bristles small. Frons as deep as antennal segment 2, widening above; face linear, silvery-grey. Antennae (Fig. 55) black, segment 3 short, arista about 3 times longer. Palpi pale yellow, with only minute pale hairs apically.

Thorax (Fig. 52) polished black on mesonotum, with narrow lateral margins, humeri and scutellum greyish dusted. Pleura densely grey dusted, sternopleura largely polished, pleura often translucent brownish. Small hairs pale and distinct even though fine; acr biserial, diverging; dc uniserial with numerous hairs at sides. Large bristles pale.

Legs yellowish, last tarsal segment blackish. Mid femora (Fig. 169) rather slender, scarcely deeper than fore femora; the ventral black spines small on apical two-thirds and the rows very close together, very long and bristlelike on basal third and with the two rows wider apart; no pv bristles. Mid tibia (Fig. 169) swollen on apical half, darkened towards the rounded tip and covered with silvery pile.

Wings (Fig. 671) clear with veins R4+5 and M almost parallel, crossveins contiguous; the vein closing anal cell recurrent. Squamae and halteres pale. Abdomen shining black to blackish-brown, covered with sparse fine pale hairs. Genitalia (Figs. 245-248) very large, globose, the very convex right lamella with scattered minute pale hairs and whitish pile.

Length: body 2.4 - 3.4 mm, wing 3.0 - 3.8 mm.

♀. Legs generally paler yellow, including apical half of mid tibia; the black bristles in posterior row on mid femora beneath uniformly longer except on apical third. Venter of abdomen often yellowish, segment 7 polished black, segment 8 very long and slender, almost polished on basal half, its apical half and cerci dull grey.

Length: body 2.4 - 3.6 mm, wing 2.9 - 4.0 mm.

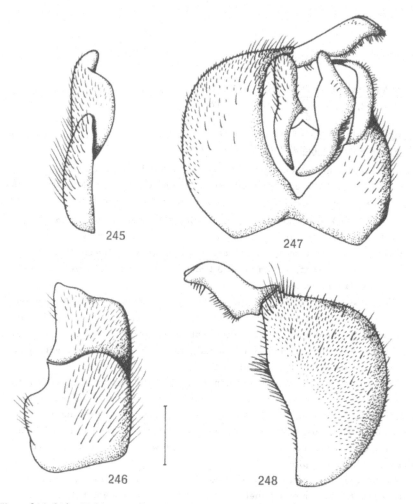

Figs. 245-248. Male genitalia of <u>Platypalpus ciliaris</u> (Fall.). - 245: left periandrial lamella in lateral view; 246: same in frontal view; 247: periandrium
with cerci; 248: right periandrial lamella. Scale: 0.2 mm.

Common in Denmark and southern parts of Fennoscandia up to 65° north,
in Sweden north to Ly. Lpm., in Finland to Om. - Great Britain, European
USSR incl. south, C. Europe. - June-early October. Common in late summer
on ground-vegetation and bushes, generally in shaded and humid biotopes but
also in mountains.

Note. The closely related P. parvicaudus (Collin, 1926), known from Great
Britain and Spain, has somewhat brownish basal antennal segments, longer a-

rista, acr bi- to triserial, uniformly short black spines in anterior row on mid femora beneath, much smaller male genitalia, and apical two abdominal segments of ♀ short and dull grey.

5. PLATYPALPUS CONFIFORMIS Chvála, 1971
Figs. 56, 170, 249-251, 256, 672.

Coryneta macula - obscuripes Engel, 1939: 82 (nec obscuripes Strobl, 1899).
Platypalpus confiformis Chvála, 1971: 11.

Resembling ciliaris but acr and dc less distinct, and both frons and face broader. Mid femora thickened, mid tibia slender apically and not silvery pilose. Terminalia small in both sexes.

♂. Head greyish dusted with frons almost as deep below as antennal segment 3, widening above; face narrower than frons. Anterior pair of ocellar bristles very long, pale; vt bristles not differentiated. Antennae (Fig. 56) black to blackish-brown, segment 3 small; palpi whitish-yellow with only minute pale hairs at tip.

Thorax as in ciliaris, with the same pale hairing and bristling, but acr and dc less distinct, the former very narrowly biserial and 1 or 2 pairs of prescutellar dc longer.

Legs yellow, tarsi uniformly darkened towards tip. Mid femora (Fig. 170) thickened, almost twice as stout as rather slender fore femora; ventral spines rather longer, particularly in posterior row, and more uniformly long on

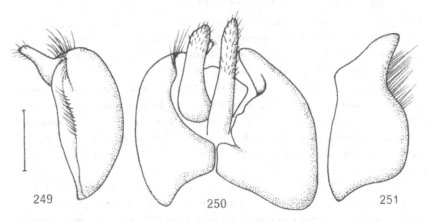

249   250   251

Figs. 249-251. Male genitalia of Platypalpus confiformis Chv. - 249: right periandrial lamella; 250: periandrium with cerci; 251: left periandrial lamella. Scale: 0.1 mm.

basal two-thirds, becoming thinner and quite pale at extreme base. Mid tibia (Fig. 170) slender, not swollen or silvery pilose on apical half to third as in ciliaris.

Wings (Fig. 672) large, clear, veins blackish; venation as in ciliaris. Squamae dirty yellow, halteres whitish-yellow with stalk darker at base. Abdomen shining blackish-brown, dorsum with rather long, densely-set pale hairs. Genitalia (Figs. 249-251) not broader than end of abdomen, lamellae polished black and almost bare.

Length: body 2.1 - 2.4 mm, wing 3.0 - 3.3 mm.

♀. Generally larger, with palpi larger and the ventral spines in posterior row on mid femora somewhat shorter. Abdomen not so densely haired on dorsum, apical two segments (Fig. 256) small and densely greyish dusted.

Length: body 2.4 - 3.0 mm, wing 3.4 - 4.0 mm.

Rather uncommon in the north of Fennoscandia, in Sweden from Jmt. north to T.Lpm., in Finland from Ks to Li; also from the Kola Peninsula; the southernmost record is in Norway in STi. - North of European USSR, Alps. - July-August. A boreoalpine species in distribution.

6. PLATYPALPUS CONFINIS (Zetterstedt, 1842)
   Figs. 57, 58, 171, 252-254, 257, 673.

Tachydromia pygmaea Zetterstedt, 1838: 551 (nec pygmaea Macquart, 1823 and Meigen, 1838).
Tachydromia confinis Zetterstedt, 1842: 307.

Shining black species with scattered minute pale hairs on mesonotum, head grey dusted. Antennae black, segment 3 about 2.5 times as long as deep, slightly shorter in ♀. Legs yellow with slender mid femora, and mid tibia slightly swollen apically and covered with silver pile.

♂. Head black, densely grey dusted; frons somewhat subshining, almost as broad as antennal segment 3, widening above; face very linear, silvery-grey, clypeus polished black. Vt bristles not differentiated from the other pubescence. Antennae (Fig. 58) blackish, segment 3 about 2.5 times as long as deep, arista almost 1.5 times as long as segment 3. Palpi rather small, pale yellow, with several fine pale hairs at tip.

Thorax polished black on mesonotum, but with narrow lateral margins, humeri and scutellum thinly grey dusted. Pleura rather densely grey dusted but sternopleura and most of mesopleura polished. Mesonotum almost bare, biserial acr and uniserial dc with some additional hairs at sides which are pale and very minute. Large bristles pale and rather fine, including a single pair of prescutellar dc placed just in front of scutellum.

Legs yellow, posterior four femora towards tip and apical tarsal segments darkened. Mid femora (Fig. 171) slender, scarcely stouter than fore femora, without pv bristles but the ventral spine-like bristles in posterior row much longer than those in anterior row. Mid tibia (Fig. 171) slightly stouter on apical quarter as in ciliaris and with a similar covering of silver pile.

Wings (Fig. 673) very faintly brownish, veins R4+5 and M almost parallel but slightly curved downwards before tip. Crossveins contiguous, the vein closing anal cell recurrent. Squamae and halteres pale. Abdomen shining black to blackish-brown, clothed with scattered fine pale hairs. Genitalia (Figs. 252-254) somewhat larger with lamellae shining and almost bare.

Length: body 2.0 - 2.5 mm, wing 3.2 - 3.8 mm.

♀. Antennal segment 3 (Fig. 57) slightly shorter, about twice as long as deep, arista twice as long. Abdomen almost bare, shining, apical two segments (Fig. 257) and cerci greyish dusted.

Length: body 2.2 - 3.0 mm, wing 3.0 - 3.6 mm.

Common in northern Fennoscandia including the extreme north, less common towards the south; in Norway south to HOi and Ak, in Sweden to Jmt. and Ång., in Finland to the Baltic coast; common also in the Kola Peninsula, but rare in Russian Carelia. - Scotland, north of European USSR. Very probably absent in C.Europe, since the records of "confinis" from C.Europe refer to a new, still undescribed, species. - End June-August.

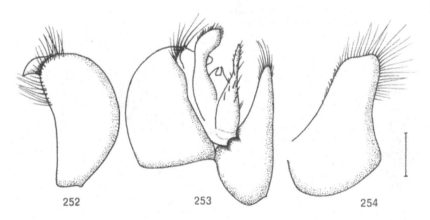

Figs. 252-254. Male genitalia of Platypalpus confinis (Zett.). - 252: right periandrial lamella; 253: periandrium with cerci; 254: left periandrial lamella. Scale: 0.1 mm.

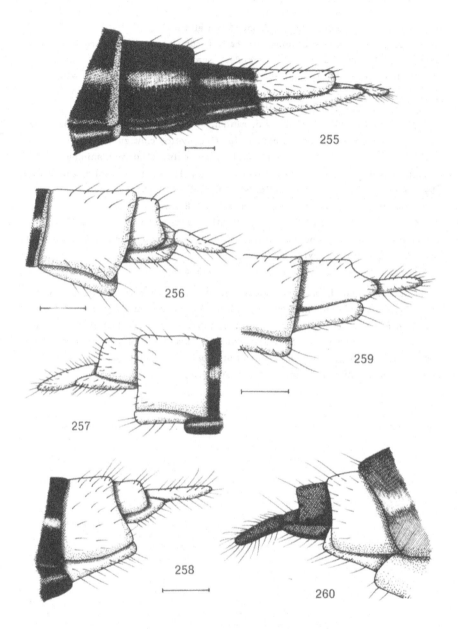

Figs. 255-260. Terminal segments of female abdomens in <u>Platypalpus</u>. - 255: <u>ciliaris</u> (Fall.); 256: <u>confiformis</u> Chv., 257: <u>confinis</u> (Zett.); 258: <u>stigmatellus</u> (Zett.); 259: <u>pectoralis</u> (Fall.); 260: <u>nonstriatus</u> Strobl. Scale: 0.1 mm.

7. PLATYPALPUS STIGMATELLUS (Zetterstedt, 1842)
Figs. 53, 59, 60, 172, 258, 261-263, 674.

Tachydromia stigmatella Zetterstedt, 1842: 306.
Tachydromia lateralis Becker, 1887: 137 (nec lateralis Loew, 1864).
Tachydromia Beckeri Mik, 1894: 166 (nom. n. for lateralis Becker, nec Loew).

Resembling confinis but frons and vertex polished black, antennae with much longer segment 3, and mid tibia slender towards tip, without a covering of silver pile.

♂. Head as in confinis but frons and vertex polished black, and occiput above neck only thinly dusted. Antennae (Fig. 60) black, segment 3 very long, 3 to almost 4 times as long as deep, arista shorter than segment 3. Palpi pale yellow, ovate, apically with several fine pale hairs.

Thorax largely polished black, whole of mesonotum including humeri and almost whole of mesopleura and sternopleura shining; scutellum thinly dusted. Thoracic hairs and bristles as in confinis but acr irregularly bi- to triserial.

Legs yellow, mid femora above and hind femora towards tip slightly darkened, and tarsi brownish on apical segments. Mid femora (Fig. 172) distinctly stouter and larger than fore femora, mid tibia (Fig. 172) not in the least swollen and without any whitish pile towards tip, but with a small rim-like projection at tip.

Wings (Fig. 674) very large and clear, costal section between veins R1 and R2+3 deeper black, and crossveins widely separated, 2nd basal cell thus longer. Abdomen on dorsum with densely-set long pale hairs, venter with scattered fine hairs. Genitalia (Figs. 261-263) small, polished black and almost bare.

Length: body 2.2 - 2.6 mm, wing 3.3 - 3.8 mm.

♀. Antennae (Fig. 59) shorter, segment 3 about 2.5 times as long as deep and arista slightly longer. Abdomen with only scattered fine pale hairs even on dorsum, shining black, but apical two segments (Fig. 258) and cerci densely grey dusted.

Length: body 2.4 - 3.2 mm, wing 3.4 - 4.3 mm.

Common in northern Fennoscandia but absent in the southern parts and in Denmark; in Norway south to Ak but no records available from the extreme north, in Sweden from T.Lpm. and Nb., south to Hls., and throughout Finland; also in the Kola Peninsula and Russian Carelia. - Scotland, N part of European USSR E to Ural, C.Europe. - End June-early September. Flying slowly above, and resting on, vegetation, preferring shaded humid places; in mountainous regions in C.Europe.

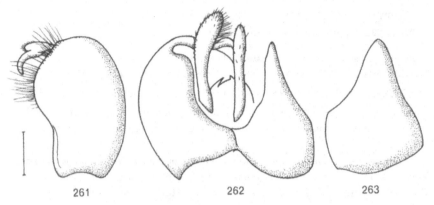

Figs. 261-263. Male genitalia of Platypalpus stigmatellus (Zett.). - 261: right periandrial lamella; 262: periandrium with cerci; 263: left periandrial lamella. Scale: 0.1 mm.

8. PLATYPALPUS PECTORALIS (Fallén, 1815)
Figs. 61, 173, 259, 264-266, 675.

Tachydromia pectoralis Fallén, 1815: 9.
Tachydromia gilvipes Meigen, 1822: 87.
Tachydromia varipes Meigen, 1822: 88.
Tachydromia taeniata Meigen, 1822: 88.
Tachydromia straminipes Zetterstedt, 1842: 296.
Platypalpus pulchellus Walker, 1851; 130.

Yellow species, mesonotum with a dark median stripe disappearing before scutellum, and small hairs evenly distributed on mesonotum. Legs usually yellowish, mid tibia stouter apically, with a dense covering of silver pile.

♂. Head black with greyish dusting, frons narrow in front but widening greatly above, face linear. Ocellar bristles minute, vt bristles not differentiated. Antennae (Fig. 61) blackish, often brownish on basal segments, segment 3 about twice as long as deep, with longer arista. Palpi small-ovate, whitish-yellow, with a few longer pale hairs at tip.

Thorax yellow to yellowish-brown, mesonotum polished except for narrow dusted margins and scutellum, pleura grey dusted but sternopleura largely polished dark brown. Mesonotum with a fairly broad dark median stripe, disappearing in posterior third; acr and dc not differentiated, evenly distributed over disc but less numerous than in maculus or pallipes. All hairs and bristles rather pale.

100

Legs variable in colour, from dirty yellow to extensively darkened, last tarsal segment almost blackish. Mid femora (Fig. 173) slightly thickened, the ventral spines in posterior row longer and thinner, becoming paler and bristle-like towards base, without pv bristles. Mid tibia (Fig. 173) silvery pilose apically, but not as swollen as in ciliaris.

Wings (Fig. 675) clear, venation as in the ciliaris-group. Abdomen polished black on dorsum, venter more or less yellowish, apical two or three segments greyish dusted. Genitalia (Figs. 264-266) rather small and closed, polished black.

Length: body 1.8 - 2.3 mm, wing 2.6 - 3.3 mm.

♀. Resembling male, apical two abdominal segments (Fig. 259) and rather small but slender cerci greyish dusted.

Length: body 1.9 - 2.6 mm, wing 2.8 - 3.3 mm.

Very common in Denmark and southern Fennoscandia; in Norway north to HOi, in Sweden to Upl., in Finland to Oa and Sb; also in Russian Carelia - Widespread in Europe except for southern parts. - End May-early October. On ground-vegetation and bushes, often in deciduous forests; it prefers shady and humid biotopes.

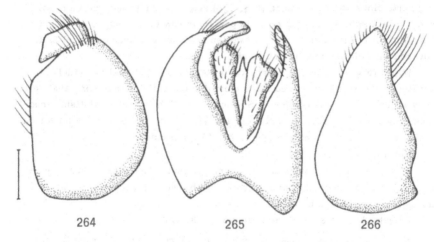

Figs. 264-266. Male genitalia of Platypalpus pectoralis (Fall.). - 264: right periandrial lamella; 265: periandrium with cerci; 266: left periandrial lamella. Scale 0.1 mm.

9. PLATYPALPUS MIKII (Becker, 1890)
   Figs. 54, 62, 174, 267-270, 676.

101

Tachydromia Mikii Becker, 1890: 67.

Coryneta Miki - tristriolata Engel, 1939: 84 - syn.n.

Reddish-yellow species with mesonotum polished and extensively darkened, thoracic hairs fine and pale, acr 4-serial and diverging, dc multiserial. Antennae yellow at base; mid tibia dilated apically and with a covering of whitish pile.

♂. Head blackish, rather densely grey dusted, frons narrow in front, widening above, face very linear. A pair of fine dark ocellar bristles, vt bristles not differentiated. Antennae (Fig. 62) small, yellow on basal segments, segment 3 brownish and very short, arista much longer. Palpi (Fig. 54) whitish, large and pointed apically.

Thorax yellowish-brown on pleura, humeri and postalar calli, but mesonotum and scutellum polished dark brown to almost black; pleura mostly shining. Acr irregularly 4-serial, diverging, separated by a bare stripe from multiserial dc, all fine and whitish. Humeri with a bristle-like hair; large bristles pale including a pair of prescutellar dc, but apical scutellars very long and dark.

Legs yellowish-brown, often darkened on posterior two pairs, last tarsal segments blackish. Mid femora (Fig. 174) scarcely thickened, with a double row of black spines beneath, those in posterior row much longer, no pv bristles. Mid tibia (Fig. 174) distinctly swollen and velvety brown on apical half, covered with whitish pile as in ciliaris.

Wings (Fig. 676) large, clear, veins R4+5 and M parallel but widely separated, crossveins separated for a short distance; the vein closing anal cell very recurrent, vein A complete. Squamae brownish, halteres whitish. Abdomen brown to dark brown, basal sternites yellowish, fine pale hairs longer on posterior segments. Genitalia (Figs. 267-270) darkened.

Length: body 2.3 - 2.7 mm, wing 3.0 - 3.4 mm.

♀. Thorax at sides and abdomen more yellowish, and legs rather pale yellow, leaving apical third of mid tibia velvety brown. Apical two abdominal segments and cerci greyish dusted.

Length: body 2.3 - 2.8 mm, wing 3.3 - 3.4 mm.

Rare. Sweden: Sk., Esperöd, 2 ♀ (Zetterstedt); Skäralid, ♀ (Ringdahl). - NW of European USSR, Caucasus, and in the mountains and hilly areas of C. and S. Europe (Poland, W. Germany, Austria, Switzerland, Romania, Bulgaria). - June-August.

Note. The species has been correctly identified by study of the holotype ♂ (coll. Berlin), and the syntypes of var. tristriolata (coll. Eberswalde) only represent paler specimens with the original dark colour of the mesonotum restricted to a broad median stripe, leaving the sides more reddish-brown.

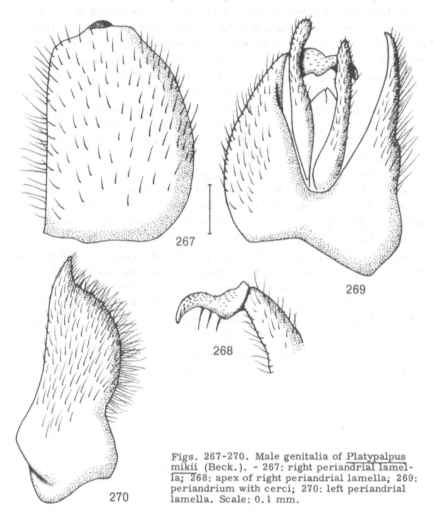

Figs. 267-270. Male genitalia of Platypalpus mikii (Beck.). - 267: right periandrial lamella; 268: apex of right periandrial lamella; 269: periandrium with cerci; 270: left periandrial lamella. Scale: 0.1 mm.

10. PLATYPALPUS NONSTRIATUS Strobl, 1901
Figs. 63, 175, 260, 271-273, 677.

Platypalpus pectoralis var. nonstriatus Strobl, 1901: 203.

Yellow species, small pale thoracic hairs evenly distributed on unicolourous yellow mesonotum. Mid femora thickened, no pv bristles; mid tibia slender apically, without a covering of silver pile.

♂. Head black and rather densely grey dusted, frons as deep below as antennal segment 2, widening above, face slightly narrower. A pair of long black ocellar bristles, vt bristles weakly differentiated even if fine, widely separated. Antennae (Fig. 63) pale yellow on basal segments, segment 3 dark brown, almost twice as long as deep, arista dark and much longer. Palpi ovate, pale, with 1 - 3 longer black hairs at tip.

Thorax uniformly yellow, mesonotum shining except for narrow lateral margins, pleura densely silvery-grey dusted leaving sternopleura largely polished. Acr and dc small, pale, uniformly distributed over mesonotum, large bristles blackish; no humeral bristle.

Legs unicolourous yellow but last segment on all tarsi darkened. Mid femora (Fig. 175) very thickened, more than twice as deep as slender fore femora, ventrally with a double row of short black spines becoming longer towards base, no pv bristles. Mid tibia (Fig. 175) slender towards tip, with a small pointed yellow spur apically.

Wings (Fig. 677) large, clear, veins dark brown. Veins R4+5 and M parallel but slightly curved downwards, crossveins contiguous. Squamae and halteres whitish-yellow. Abdomen more or less shining black on dorsum, venter yellowish, finely pale pubescent. Genitalia (Figs. 271-273) not broader than end of abdomen, lamellae yellowish with concolourous pale hairs, cerci brownish.

Length: body 2.1 - 2.5 mm, wing 3.2 - 3.5 mm.

♀. Antennal segment 3 generally smaller, sometimes almost as long as deep. Abdomen rather brown on dorsum, segment 7 (Fig. 260) yellowish and

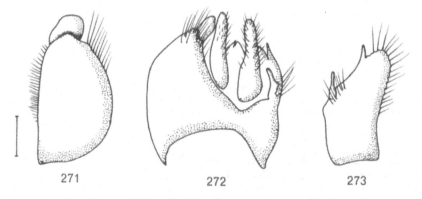

Figs. 271-273. Male genitalia of Platypalpus nonstriatus Strobl. - 271: right periandrial lamella; 272: periandrium with cerci; 273: left periandrial lamella. Scale: 0.1 mm.

thinly dusted, segment 8 and rather short cerci extensively darkened, thinly grey dusted.

Length: body 2.3 - 2.6 mm, wing 3.2 - 3.5 mm.

Rare. Finland: Ab, Vichtis, 3 ♂ (Frey); Sb, Tuovilanlaks, ♂ 4 ♀ (Lundström, Palmén), Kiuruvesi, ♀ (Lundström); Kb, Suojärvi, ♀ (Tuomikoski); also in Russian Carelia: Ib, Walkjärvi, ♀ (Sahlberg); Kr, Suistamo, ♂ ♀ (Tuomikoski). - NW of European USSR (Estonia, Kovalev in litt.) and Austria.

II. pallipes - group

11. PLATYPALPUS MACULUS (Zetterstedt, 1842)
Figs. 65, 71, 274-276, 278, 678.

Tachydromia macula Zetterstedt, 1842: 289.

Large blackish species, mesonotum polished black and entirely covered by short pale hairs, large thoracic bristles pale. Antennae black with very short segment 3, frons and face shining black. Posterior femora with black markings; pv bristles on mid femora pale, indistinct.

♂. Head with shining black rather broad frons, widening above, face slightly narrower and almost as shining as frons. Occiput thinly grey dusted, a pair of fine vt bristles, hardly distinguishable from other pubescence. Antennae (Fig. 65) black, segment 3 short, very slightly longer than deep, arista about 3 times longer. Palpi large, pale.

Thorax (Fig. 71) polished black on mesonotum except for narrowly dusted margins and scutellum, pleura densely silvery-grey dusted leaving sternopleura largely polished. Small thoracic hairs pale and uniformly densely distributed over mesonotum, large bristles pale.

Legs yellow, posterior four femora with a distinct blackish patch at tip above, extreme tips of tibiae and whole of tarsi very darkened. Mid femora very large and thickened, a few pv bristles pale and rather inconspicuous. Apical spur on mid tibia somewhat pointed but very small.

Wings (Fig. 678) large, clear, with blackish veins. Veins R4+5 and M slightly convergent before tip, crossveins almost contiguous. The vein closing anal cell very recurrent, vein A distinct throughout. Squamae and halteres pale. Abdomen shining blackish-brown, covered with fine pale hairs. Genitalia (Figs. 274-276) large, left bifid cercus and the polished black process of right lamella conspicuously overlapping the thinly grey dusted lamellae.

Length: body 2.8 - 3.4 mm, wing 3.5 - 4.3 mm.

♀. Larger, with longer palpi and usually much less distinct dark patches

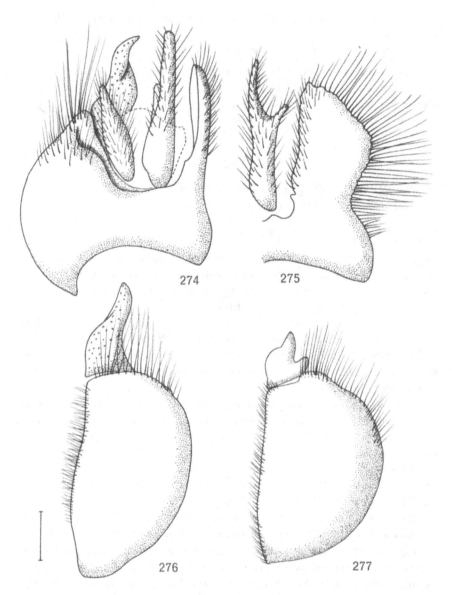

Figs. 274-276. Male genitalia of Platypalpus maculus (Zett.). - 274: periandrium with cerci ; 275: left periandrial lamella with left cercus; 276: right periandrial lamella.
Fig. 277. Right periandrial lamella of Platypalpus wuorentausi Frey. Scale: 0.1 mm.

on posterior four femora. Abdominal segment 7 (Fig. 278) thinly grey dusted, segment 8 large and densely silvery-grey dusted like the very small cerci.

Length: body 3.3 - 4.0 mm, wing 4.2 - 4.8 mm.

Common and widespread in Fennoscandia but more abundant in the north; no material from Denmark available, although recorded from Zealand by Lundbeck (1910). - Great Britain, N and NW parts of European USSR, C.Europe (Alps, Carpathians). - End May-August.

Note. The Asiatic P.wuorentausi Frey, 1943 is closely related to maculus: I was unable to find any distinguishing characters except for the strikingly different male genitalia (Fig. 277).

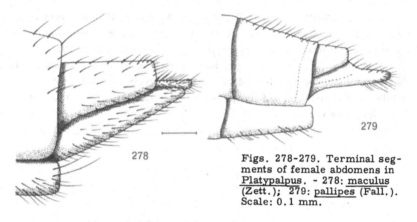

Figs. 278-279. Terminal segments of female abdomens in Platypalpus. - 278: maculus (Zett.); 279: pallipes (Fall.). Scale: 0.1 mm.

12. PLATYPALPUS PALLIPES (Fallén, 1815)
Figs. 64, 176, 279, 280-282, 679.

Tachydromia pallipes Fallén, 1815: 8.
Tachydromia flavipalpis Meigen, 1822: 74.

Resembling maculus but frons and face silvery-grey dusted, antennae with longer segment 3, legs yellow except for darkened tarsi, and mid femora with distinct blackish pv bristles.

♂. Head rather silvery-grey but thinly dusted on frons and face, a long pair of anterior ocellar bristles and indistinctly separated vt bristles blackish. Antennae (Fig. 64) black with segment 3 longer than in maculus, about 2 to 2.3 times as long as deep, and arista about 1.5 times as long as segment 3.

Thorax as in maculus but large thoracic bristles almost dark brown. Legs extensively yellow except for darkened tarsi, pv bristles on mid femora black and very distinct but only few in number. Wings (Fig. 679) and abdomen as in

107

<u>maculus</u> but genitalia (Figs. 280-282) with cerci and dorsal process on right la-
mella much shorter.

Length: body 2.5 - 3.0 mm, wing 3.0 - 3.6 mm.

♀. Antennal segment 3 usually shorter, about twice as long as deep, and
arista longer. Abdominal segment 7 (Fig. 279) thinly greyish dusted at sides,
segment 8 rather small and slender, densely dusted.

Length: body 2.8 - 3.2 mm, wing 3.3 - 4.0 mm.

Rather common, but unlike <u>maculus</u> with a southern distribution, not reach-
ing far beyond 65° north; Denmark, in Norway north to HEn, in Sweden to Ås.
Lpm. and in Finland to Ok; common in Russian Carelia. - W. and C. Europe,
NW of European USSR. - Mid May-mid October.

Note. The North American P. pectinator Melander, 1924 is a very closely
related but distinct species, having dark pv bristles on mid femora as in <u>palli-
pes</u> but shorter antennae as in <u>maculus</u>. P. longimanus (Corti, 1907), known
from England and C. and S. Europe, has a very long antennal segment 3 with a
shorter thick arista, smaller palpi, more extensively polished pleura, no pv
bristles on mid femora, and conspicuously elongated last tarsal segment on anterior four legs in male.

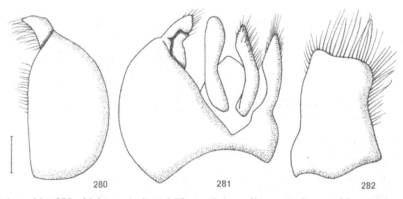

Figs. 280-282. Male genitalia of <u>Platypalpus pallipes</u> (Fall.). - 280: right periandrial lamella; 281: periandrium with cerci; 282: left periandrial lamella.
Scale: 0.1 mm.

III. albiseta - group

13. <u>PLATYPALPUS ALBISETA</u> (Panzer, 1806)

Figs. 66, 177, 283-285, 680.

Tachydromia albiseta Panzer, 1806: 17.

Tachydromia castanipes Meigen, 1822: 79.

Tachydromia vivida Meigen, 1838: 97.

Tachydromia fuscimana Zetterstedt, 1842: 292.

Extensively polished black species, antennae black with segment 3 almost 4 times as long as deep, white arista longer than rest of antennae. Legs yellow with brown pattern, mid femora rather slender; frons very linear.

♂. Head almost circular in profile, frons dull grey and very linear, narrower on lower half than front ocellus, eyes almost touching below antennae. A pair of long black vt bristles. Antennae (Fig. 66) black, segment 3 about 3 to almost 4 times as long as deep, white arista distinctly longer than rest of antenna. Palpi small, dark,

Thorax mainly polished black including pleura, humeri very weakly developed. Mesonotum almost bare, widely biserial acr and uniserial dc scarcely visible, large bristles black but fine.

Legs mostly yellow on coxae, femora and hind tibiae, fore femora above and hind femora at tip often darkened, anterior four tibiae and all tarsi extensively darkened. Both fore and mid femora (Fig. 177) with very long black bristles posteroventrally, mid femora slightly stouter. Mid tibia (Fig. 177) practically without apical spur.

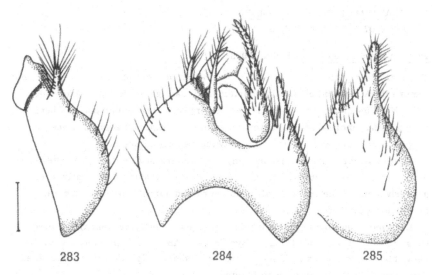

283   284   285

Figs. 283-285. Male genitalia of Platypalpus albiseta (Panz.). - 283: right periandrial lamella; 284: periandrium with cerci; 285: left periandrial lamella. Scale: 0.1 mm.

Wings (Fig. 680) slightly brownish clouded, veins blackish; veins R4+5 and M conspicuously bowed, almost twice as wide apart at middle as at tip, the vein closing anal cell very recurrent and S-shaped. Squamae brownish, halteres pale yellow with stalk brown at base. Abdomen polished black and almost bare, genitalia (Figs. 283-285) rather small with grey dusted cerci, and lamellar appendages small and slender.

Length: body 2.2 - 2.5 mm, wing 2.8 - 3.0 mm.

♀. Resembling male, apical two abdominal segments thinly grey dusted posteriorly, long slender cerci densely dusted.

Length: body 2.0 - 3.0 mm, wing 2.4 - 2.8 mm.

Rather common in Denmark and southern Fennoscandia up to 62° north; in Norway to SFi, in Sweden to Sdm., in Finland to Ta and Sa. - Widespread in Europe, even in the south; ? N. Africa. - June-early September, in the south from April. On ground-vegetation and bushes in shaded and rather humid places, often in forest clearings.

Note. P. pygialis Chvála, 1973, known from England and C. Europe, is closely allied to albiseta but the antennae are smaller, with longer white arista, and the male genitalia are much larger with conspicuously elongated left cercus and lamellar appendages overlapping lamellae.

14. PLATYPALPUS ALBISETOIDES Chvála, 1973
Figs. 67, 178, 286-288, 681.

Platypalpus albisetoides Chvála, 1973: 121.

The main diagnostic features as in albiseta but thoracic pleura subshining, mid femora thickened and obviously shorter, and male genitalia quite different.

♂. Head as in albiseta, with the two silvery-grey triangular patches behind vertex rather more distinct, antennal segment 3 (Fig. 67) about 3 times as long as deep, and white arista almost twice as long as segment 3.

Thorax almost bare on mesonotum, only 2 pairs of fine dark prescutellar dc distinct. Pleura extensively thinly grey dusted, leaving practically whole of sternopleura, central area of mesopleura and narrow posterior margin of hypopleura polished.

Legs more uniformly yellowish-brown or extensively darkened, but fore tibiae towards tip and all tarsi always darker. Anterior four femora with the same black bristling as in albiseta but mid femora (Fig. 178) more thickened, more than twice as deep as fore femora.

Wings (Fig. 681) similarly brownish tinged but veins R4+5 and M less converging towards wing-tip, ending rather wide apart. Abdomen polished blackish-brown, genitalia (Figs. 286-288) large with cerci rather short and enclosed within lamellae.

Length: body 2.5 mm, wing 2.8 - 2.9 mm.

♀. Resembling male; abdominal segment 7 polished black, darker in contrast to rather brownish abdomen, but most of segment 8 and long slender cerci densely grey dusted.

Length: body 2.6 - 3.0 mm, wing 2.7 - 3.0 mm.

Rare. Sweden: Bohuslän, ♂ (Boheman); Finland: Ab, Runsala, ♂ (holotype) 2 ♀ (Frey). - No further data available.

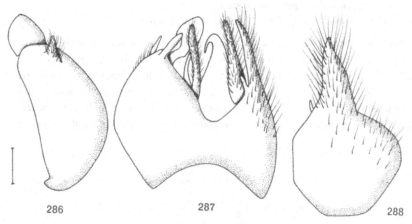

Figs. 286-288. Male genitalia of Platypalpus albisetoides Chv. - 286: right periandrial lamella; 287: periandrium with cerci; 288: left periandrial lamella. Scale: 0.1 mm.

15. PLATYPALPUS ALBOCAPILLATUS (Fallén, 1815)
Fig. 69, 179, 289-291, 682.

Tachydromia albo-capillata Fallén, 1815: 9.

Resembling albiseta but frons not so narrow, antennal segment 3 longer with white arista equal in length, and legs very extensively blackish.

♂. Head greyish dusted, frons narrow though not as much as in albiseta, slightly broader above than front ocellus. Upper part of occiput with rather bristle-like black hairs, and a pair of ocellar bristles longer, almost as long as vt bristles. Antennae (Fig. 69) with segment 3 longer, about 4 - 5 times as long as deep; arista brownish at extreme base, as long as segment 3. Palpi blackish.

Thorax as in <u>albiseta</u> but legs mostly black, usually only knees and fore tibiae towards base translucent brownish, in paler coloured specimens legs somewhat dark brown. Mid femora (Fig. 179) rather thicker and larger than in <u>albiseta</u> but with the same bristling.

Wings (Fig. 682) very faintly tinged brown especially on costal half, squamae blackish, halteres greyish-yellow. Abdomen shining black with scattered dark hairs; genitalia (Figs. 289-291) large with lamellae thinly dusted and with fine dark hairs, left cercus long and abruptly bent apically.

Length: body 2.0 - 2.4 mm, wing 2.4 - 2.6 mm.

♀. Resembling male, apical two abdominal segments subshining black; · cerci very long, slender, greyish dusted.

Length: body 2.3 - 2.8 mm, wing 2.4 - 2.6 mm.

Not uncommon in Denmark and Fennoscandia except for the extreme north; in Norway to NTi, in Sweden along the Gulf of Bothnia to Nb., in Finland to Ob and Ks; also from the Kola Peninsula. - Great Britain, and along the Baltic coast to NW and C parts of European USSR; the records from C.Europe are very probably in error. - June-September. In Scandinavia mainly a coastal species, and along rivers and in lake districts far inland, in northern regions.

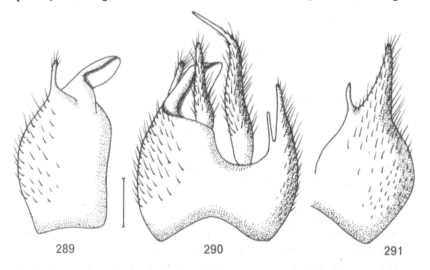

Figs. 289-291. Male genitalia of Platypalpus albocapillatus (Fall.). - 289: right periandrial lamella; 290: periandrium with cerci; 291: left periandrial lamella. Scale: 0.1 mm.

16. PLATYPALPUS NIVEISETA (Zetterstedt, 1842)

Figs. 68, 70, 180, 292-294, 683.

Tachydromia niveiseta Zetterstedt, 1842: 311.

Tachydromia albiseta var.brunnipes Strobl, 1906: 312.

Differing from other species of this group by the broader frons, fairly well developed humeri, greyish dusted pleura, and fore femora without black pv bristles. White arista about as long as segment 3.

♂. Head greyish dusted, frons rather narrow but about as deep in front as front ocellus, widening above; face as deep as frons in front. A pair of ocellar and a pair of vt bristles blackish, equal in length. Antennae (Fig. 68) with segment 3 about 4 times as long as deep, white arista as long as, or slightly longer than segment 3. Palpi dark, very small.

Thorax (Fig. 70) polished black on mesonotum but with narrow margins, scutellum and distinct humeri greyish dusted, pleura rather densely grey dusted, sternopleura largely polished. Mesonotum with very fine and minute pale biserial acr and uniserial dc, large bristles blackish.

Legs yellow but tibiae and tarsi often darkened and femora sometimes brownish above or with a darkening before tip. Fore femora rather slender, ventrally with two rows of longer dark bristly-hairs, no black pv bristles. Mid femora (Fig. 180) thickened, almost twice as stout as fore femora, with long black pv bristles.

Wings (Fig. 683) faintly brownish clouded with venation as in albiseta but veins R4+5 and M more parallel. Abdomen shining black, covered with short

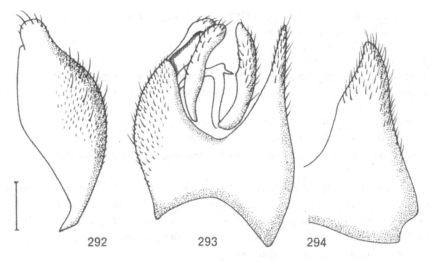

Figs. 292-294. Male genitalia of Platypalpus niveiseta (Zett.). - 292: right periandrial lamella; 293: periandrium with cerci; 294: left periandrial lamella. Scale: 0.1 mm.

pale hairs; genitalia (Figs. 292-294) rather large, cerci enclosed within lamellae; right lamella very thinly dusted but cerci and left lamella densely dusted.

Length: body 2.0 - 2.5 mm, wing 2.6 - 2.8 mm.

♀. Resembling male, apical two abdominal segments and long slender cerci greyish dusted, otherwise abdomen shining.

Length: body 2.2 - 2.5 mm, wing 2.7 - 2.8 mm.

Rare. Sweden: Sk., ♂ (holotype) (Zetterstedt), ♀ (Boheman). - Widespread in Europe but rather rare everywhere, known with certainty from England, Czechoslovakia, Austria and Spain. - May-August.

Note. The most related species to niveiseta is the C.European P.obscurus (von Roser, 1840) which might be found in south Fennoscandia; it has thinly dusted mesonotum, only narrowly polished sternopleura, and is generally darker. P.leucothrix (Strobl, 1910) (syn. P.acroleucus Frey, 1943), known so far from England and C.Europe, is another species with white arista, but belongs to a different group of species; it has a large apical spur on mid tibia, short antennae with yellow basal segments, polished black occiput, thinly dusted mesonotum but pleura extensively polished, and the vein closing anal cell at right-angles to vein Cu.

IV. unguiculatus - group

17. PLATYPALPUS UNGUICULATUS (Zetterstedt, 1838)
    Figs. 73, 75, 94, 181, 295-297, 309, 684.

Tachydromia unguiculata Zetterstedt, 1838: 551 (p.p.).
Tachydromia gilvipes Coquillett, 1900: 422 (nec gilvipes Meigen, 1822).
Platypalpus xanthopodus Melander, 1928: 367 (nom.n. for gilvipes Coq.).

A small, polished black and almost bare species; head polished with a pair of fine pale vt bristles, no humeral bristle. Antennae yellow on basal segments; legs yellow, no pv bristles on mid femora, and very small apical spur on mid tibia.

♂. Head (Fig. 73) entirely polished black except for silvery pilose clypeus; frons rather broad and widening above, face very linear. Anterior pair of ocellar and a pair of vt bristles fine, pale. Antennae (Fig. 75) yellow on basal segments, segment 3 dark brown, more than twice as long as deep, arista longer. Palpi small, whitish-yellow.

Thorax (Fig. 94) polished black to blackish-brown on mesonotum except for narrow dusted margins, pleura mostly polished. Mesonotum almost bare, acr and humeral bristle absent, dc uniserial but very minute, a single pre-scutellar pair longer and pale as are the few longer bristles.

Legs slender, pale yellow, but last tarsal segment extensively darkened. Mid femora (Fig. 181) slightly thickened, the black ventral spine-like bristles in posterior row longer, no pv bristles; practically no spur on mid tibia (Fig. 181).

Wings (Fig. 684) clear with rather pale veins, veins R4+5 and M parallel on apical half, crossveins contiguous; the vein closing anal cell slightly re-current and like vein A only fine. Squamae and halteres pale. Abdomen enti-rely polished, with scattered minute pale hairs; genitalia (Figs. 295-297) rather small and almost bare, polished black.

Length: body 1.7 - 2.0 mm, wing 2.3 - 2.4 mm.

♀. Apical tarsal segments somewhat darkened, particularly on front legs; abdomen more brownish, apical two segments (Fig. 309) and small cerci grey-ish dusted, but last sternite subshining black above.

Length: body 1.8 - 2.2 mm, wing 2.3 - 2.4 mm.

Very common in northern parts of Fennoscandia; in Norway from HOi to extreme north, throughout Sweden including Sk., not yet found in Denmark, and throughout Finland; also in Russian Carelia and on Kola Peninsula. - Holarctic species; Fennoscandia, N and C parts (Moscow region) of European USSR; in N. America from Alaska to Ont., Que. and Labr. - June-August.

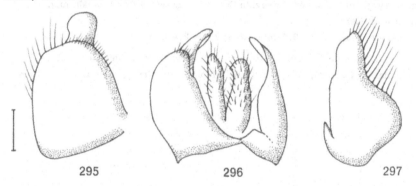

Figs. 295-297. Male genitalia of Platypalpus unguiculatus (Zett.). - 295: right periandrial lamella; 296: periandrium with cerci; 297: left periandrial lamel-la. Scale: 0.1 mm.

18. PLATYPALPUS ZETTERSTEDTI Chvála, 1971

Figs. 78, 96, 182, 298-302, 308, 685.

Platypalpus zetterstedti Chvála, 1971: 21.

A small polished black species resembling unguiculatus but frons and occiput thinly grey dusted, antennae with shorter segment 3 and females with basal segments dark; humeral bristle present.

♂. Head as in unguiculatus but frons on lower two-thirds and occiput thinly grey dusted, vertex and upper part of frons polished black. A pair of ocellar and a pair of vt bristles somewhat darker. Antennal segment 3 (Fig. 78) twice as long as deep, or slightly longer than deep, basal segments yellowish. Palpi small, pale, with a dark hair at tip.

Dc uniserial, pale, but rather longer than in unguiculatus, last prescutellar pair very long, as long as upper notopleural, and a pair of scutellars; a humeral and a postalar bristle, smaller.

Legs yellow, but apical two tarsal segments and hind femora towards tip darkened, last tarsal segment very dark. Anterior four femora (Fig. 182) of the same thickness. Wings (Fig. 685) clear with brownish veins; veins R4+5 and M straight throughout, distinctly and evenly diverging. Otherwise as in unguiculatus, including abdomen, but genitalia (Figs. 298-302) with much shorter hairs on both lamellae, and cerci different.

Length: body 1.9 mm, wing 2.3 mm.

♀. Basal antennal segments extensively darkened but still somewhat paler than segment 3. Legs not so pale yellow, rather yellowish-brown on tibiae and tarsi, posterior four femora sometimes darker towards tip. Abdominal segment 7 (Fig. 308) and cerci greyish dusted, segment 8 (last) subshining, sternite often polished.

Length: body 1.8 - 2.5 mm, wing 2.3 - 2.4 mm.

Rare. Sweden: T.Lpm., Vittangi, ♀ (holotype) (H. Andersson); Abisko, ♂ (Ringdahl), ♀ (Pont); Finland: LkW, Muonio, 3♀ (Frey); also from the Kola

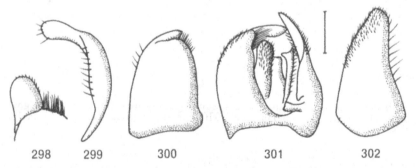

Figs. 298-302. Male genitalia of Platypalpus zetterstedti Chv. - 298: Apex of right periandrial lamella; 299: right periandrial lamella in lateral view; 300: the same in frontal view; 301: periandrium with cerci; 302: left periandrial lamella. Scale: 0.1 mm.

Peninsula: Lr, Kantalaks, ♀ (Frey); Bjäloguba, ♀ (Hellén). - No further records available. - June-July.

19. PLATYPALPUS ALTER (Collin, 1961)
Figs. 72, 76, 183, 303-305, 307, 686.

Tachydromia altera Collin, 1961: 212.

A small polished black species resembling unguiculatus but vertex with 2 pairs of fine pale vt bristles. Mesonotum polished and almost bare, acr practically absent, no humeral and postalar bristles. Legs slender, yellow, tibial spur absent.

♂. Head (Fig. 72) with broad, thinly grey dusted frons which is as deep in front as antennal segment 3, widening above; face almost linear. Vertex almost polished, occiput subshining. Anterior pair of ocellar and 2 pairs of vt bristles long but fine, pale. Antennae (Fig. 76) yellow, segment 3 often brownish, almost twice as long as deep; dark arista more than twice as long as segment 3. Palpi pale, ovate, terminal bristly-hair long, pale.

Thorax including mesonotum and pleura mostly polished black; acr practically absent, dc uniserial, fine and pale, becoming longer posteriorly, and last prescutellar pair (placed in posterior quarter of mesonotum) very long, pale, similar to 2 notopleural bristles and to apical scutellars. Humeral and postalar bristles absent.

Legs quite yellow except for almost blackish apical two tarsal segments. Anterior four femora rather slender, the ventral black spines on mid femora (Fig. 183) slightly longer in posterior row, considerably so towards base, no

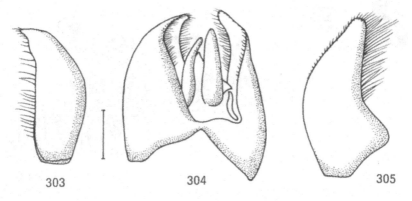

303   304   305

Figs. 303-305. Male genitalia of Platypalpus alter (Coll.). - 303: right periandrial lamella; 304: periandrium with cerci; 305: left periandrial lamella. Scale: 0.1 mm.

pv bristles. Mid tibia (Fig. 183) without apical spur, hind femora with longer pale pv hairs.

Wings (Fig. 686) clear with rather pale veins, veins R4+5 and M almost parallel; crossveins contiguous or first basal cell slightly longer; the vein

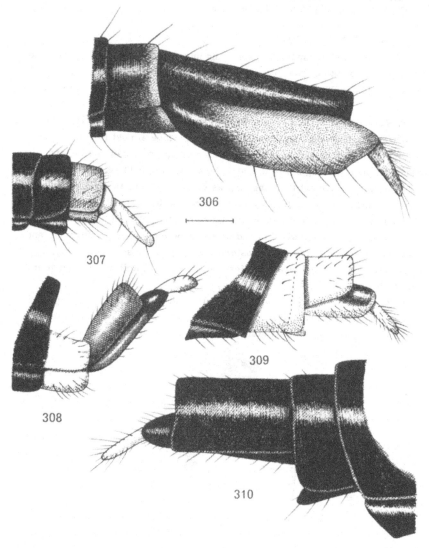

Figs. 306-310. Terminal segments of female abdomens in Platypalpus. - 306: laestadianorum (Frey); 307: alter (Coll.); 308: zetterstedti Chv.; 309: unguiculatus (Zett.); 310: lapponicus Frey. Scale: 0.1 mm.

closing anal cell slightly recurrent, squamae and halteres whitish-yellow. Abdomen shining black, covered with fine pale hairs becoming longer posteriorly; genitalia (Figs. 303-305) small, polished black and almost bare, cerci greyish, left lamella with long fine hairs.

Length: body 1.8 - 2.0 mm, wing 2.1 - 2.2 mm.

♀. Resembling male, abdomen polished with segment 8 very thinly dusted; cerci long and slender, more densely grey dusted.

Length: body 2.1 - 2.4 mm, wing 2.2 - 2.3 mm.

Rare. Norway: TRy, Giebostad, ♀ (Zetterstedt); Finland: Ks, Kuusamo, 2♀ (Frey, Sahlberg), Ponoj, ♂♀ (Frey); "Lapponia", ♀ (Zetterstedt); also from the Kola Peninsula: Lapponia rossica, ♂ (Sahlberg). - Scotland (Inverness). - June.

20. PLATYPALPUS LAESTADIANORUM (Frey, 1913)

Figs. 79, 95, 184, 306, 311-313, 687.

Tachydromia laestadianorum Frey, 1913: 83 (p.p.).

Larger species with polished black and almost bare mesonotum, pleura dusted except for sternopleura, and terminalia very large in both sexes. Legs yellow with thickened mid femora, no pv bristles.

♂. Head black in ground-colour and greyish dusted, leaving only clypeus polished. Frons broad and distinctly widening above, face as deep as frons in front. Anterior pair of ocellar bristles very long, a pair of vt bristles smaller, all pale. Antennae (Fig. 79) uniformly dark-brown, segment 3 more than twice as long as deep, arista longer. Palpi pale, very small, with a long pale bristle-like hair at tip.

Thorax (Fig. 95) polished black to blackish-brown on mesonotum except for rather densely grey dusted sides, pleura dusted leaving sternopleura largely polished. No humeral and acr bristles; dc uniserial, pale and very fine, even posteriorly; large bristles pale but practically only upper notopleural and apical pair of scutellars large.

Legs yellow, apical tarsal segments somewhat brownish. Fore femora slightly thickened but mid femora (Fig. 184) almost twice as deep as fore femora, ventrally with a double row of short black spines (those in posterior row hardly longer), no pv bristles. Mid tibia (Fig. 184) with a very small apical spur.

Wings (Fig. 687) faintly yellowish-brown tinged or almost clear, veins rather pale; veins R4+5 and M slightly convergent apically, crossveins contiguous; the vein closing anal cell almost at right-angles to vein Cu, but di-

119

stinct at base only. Squamae and halteres pale. Abdomen polished black to blackish-brown, covered with sparse fine pale hairs; genitalia (Figs. 311-313) conspicuously large and long, polished black; right lamella with a conspicuous tuft of long yellowish bristly-hairs apically, and similar hairs also on left lamella towards tip.

Length: body 2.1 - 2.5 mm, wing 2.6 - 3.2 mm.

♀. Larger, and abdomen almost bare; last segment (Fig. 306) very long and stout, and polished like the preceding segment 7, only sternite 8 on apical half and cerci thinly greyish dusted.

Length: body 2.4 - 3.0 mm, wing 2.8 - 3.4 mm.

Common in Lapland and Kola Peninsula; in Norway in Fø, in Sweden from T.Lpm. and Nb. south to Ås.Lpm., and in Finland from Li to Ks. - Extreme north of European USSR; the record from the Carpathians (Tatra Mts.) given by Engel (1939) and accepted by Frey (1943) and Ringdahl (1951) is almost certainly an error. - July-early August.

Note. P.laestadianorum is generally confused with confiformis and lapponicus and all the three species are included in the original syntypic series of laestadianorum; one male of the species described above (Muonio, Frey - Spec.typ.No.4447) has been selected as lectotype.

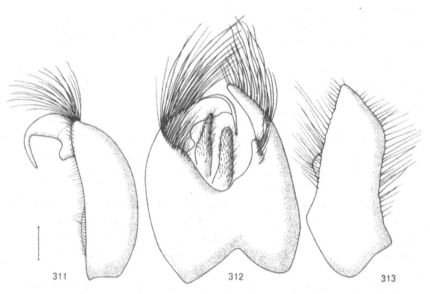

Figs. 311-313. Male genitalia of Platypalpus laestadianorum (Frey). - 311: right periandrial lamella; 312: periandrium with cerci; 313: left periandrial lamella. Scale: 0.1 mm.

21. PLATYPALPUS LAPPONICUS Frey, 1943

Figs. 77, 185, 310, 314-316, 688.

Platypalpus lapponicus Frey, 1943: 7, 15.

Larger species with 2 pairs of vt bristles, mesonotum mostly polished and almost bare, large thoracic bristles pale; antennae black. Legs dark yellow with tarsi and posterior coxae darkened; tibial spur very small, no pv bristles on mid femora.

♂. Head black, rather densely grey dusted but clypeus and narrow jowls below eyes polished. Frons broad, widening above, face still broader than frons in front. 2 pairs of vt bristles and a pair of ocellar bristles blackish, rather short. Antennae (Fig. 77) black, segment 3 twice as long as deep at base and very pointed, arista slightly longer. Palpi whitish-yellow, small-ovate.

Thorax polished black on mesonotum, but rather broad margins including humeri and scutellum densely light grey dusted. Pleura with the same dusting, sternopleura largely polished. Mesonotum almost bare, closely biserial acr and uniserial dc pale and very inconspicuous; large bristles pale including a small curved humeral bristle, large apical scutellars darkened.

Legs somewhat dark yellow, posterior four coxae and all tarsi darkened, apical 2 or 3 tarsal segments almost black. Fore femora slightly thickened, mid femora (Fig. 185) distinctly stouter but scarcely longer, the black ventral spines in posterior row longer, no pv bristles. Fore tibiae spindle-sha-

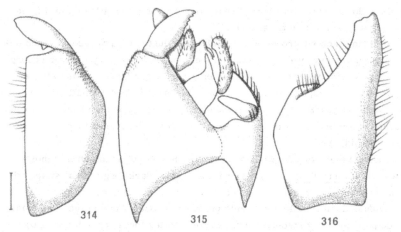

Figs. 314-316. Male genitalia of Platypalpus lapponicus Frey. - 314: right periandrial lamella; 315: periandrium with cerci; 316: left periandrial lamella. Scale: 0.1 mm.

ped dilated, mid tibia (Fig. 185) with a small pointed projection at tip; fore and hind femora with pale hairs beneath.

Wings (Fig. 688) clear with brownish veins, veins R4+5 and M almost parallel but wider apart, crossveins usually contiguous and the vein closing anal cell recurrent. Squamae and halteres pale. Abdomen polished black and almost bare; genitalia (Figs. 314-316) large and similarly polished.

Length: body 2.3 - 2.7 mm, wing 2.8 - 3.0 mm.

♀. Resembling male; abdomen polished black including apical two segments (Fig. 310), segment 8 long and narrow, cerci small, dull.

Length: body 2.6 - 3.3 mm, wing 2.8 - 3.2 mm.

A north Fennoscandian species with the same pattern of distribution as laestadianorum, but much less common; Norway in F, Sweden in T.Lpm., and in Finland from Li south to Ks; also from the Kola Peninsula. - N of European USSR. - July-August. Along brooks and on the coast (Frey, 1943).

22. PLATYPALPUS SAHLBERGI (Frey, 1909)
    Figs. 74, 80, 186, 317-319, 689.

Tachydromia sahlbergi Frey, 1909: 7.
Tachydromia Sahlbergi var.nigricollis Frey, 1913: 82 (p.p.).

Yellowish-brown species with a pair of pale vt bristles, mesonotum polished with a dark median stripe and inconspicuous pale acr and dc. Antennae mostly pale yellow, legs yellow with slender femora, no pv bristles on mid femora and a very small mid tibial spur.

♂. Head blackish-brown, very thinly grey dusted on a broad frons and occiput, but much narrower face including clypeus silvery-grey. A pair of long but fine pale widely separated vt bristles directed forwards, as are similar long pale hairs on upper part of occiput. Antennae (Fig. 80) small, pale yellow, segment 3 darker brownish towards tip, short, about 1.5 times as long as deep, brownish arista much longer. Palpi (Fig. 74) rather large and blunt-tipped, whitish-yellow.

Thorax yellow to yellowish-brown, mesonotum polished, often almost brown (var.nigricollis) but always with a distinct dark longitudinal stripe widening in front and on scutellum; narrow mesonotal margins and pleura thinly grey dusted, sternopleura largely polished. Closely biserial acr and uniserial dc pale and very inconspicuous, large bristles including a humeral fine and pale.

Legs uniformly pale yellow but last tarsal segment dark. Anterior four

122

femora equally slender; no pv bristles on mid femora (Fig. 186), but the ventral spines in two rows long and bristle-like, brownish. Fore tibiae slightly spindle-shaped, mid tibia (Fig. 186) with a very small but pointed spur-like projection at tip.

Wings (Fig. 689) clear with pale veins, veins R4+5 and M indistinctly diverging or almost equally bowed, crossveins practically contiguous; the vein closing anal cell very recurrent, hardly visible. Squamae and halteres whitish-yellow. Abdomen subshining blackish-brown, rather densely covered by short pale hairs. Genitalia (Figs.317-319) rather small and compact, lamellae polished with long pale hairs.

Length: body 1.7 - 2.2 mm, wing 2.3 - 2.5 mm.

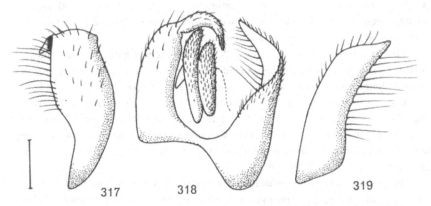

Figs. 317-319. Male genitalia of <u>Platypalpus sahlbergi</u> (Frey). - 317: right periandrial lamella; 318: periandrium with cerci; 319: left periandrial lamella. Scale: 0.1 mm.

♀. Larger; fore tibiae rather slender and hind tibiae less dilated towards tip. Abdomen very pointed with less densely-set hairs; apical two segments slender and grey dusted like the cerci.

Length: body 2.0 - 2.8 mm, wing 2.5 - 3.0 mm (holotype body 2.0 mm, wing 2.8 mm).

Uncommon in the north; in Norway in Fn, in Finland from Ks north to Le; also from the Kola Peninsula. - N of European USSR; the records from the Alps (Frey, 1943; ? after Strobl, 1893) and Germany (Engel, 1939) are probably errors. - End May-July.

V. longicornis-group

23. PLATYPALPUS BOREOALPINUS Frey, 1943
Figs. 83, 98, 187, 320-322, 690.

Tachydromia Sahlbergi var.nigricollis Frey, 1913: 82 (p.p.).
Platypalpus boreoalpinus Frey, 1943: 5, 15 (p.p.).

A species with 2 pairs of vt bristles, mesonotum polished black on posterior half in ♂, and on a broad median stripe in ♀ that also extends in front; acr and dc biserial, pale. Antennae yellow on basal segments, segment 3 dark. Legs yellow with dark tarsi.

♂. Head black with greyish dusting, frons broad and slightly widening above, face very narrow, silvery-grey. A pair of ocellar and 2 pairs of vt bristles yellowish-brown, fine and long. Antennae (Fig. 83) yellow on basal segments; segment 3 brownish, short, slightly longer than deep, dark arista about twice as long. Palpi pale, small-ovate.

Thorax (cf.Fig. 98 of ♀) polished black on posterior half of mesonotum, rest of thorax densely light grey dusted, sternopleura largely polished. All thoracic hairs and bristles pale, acr and dc rather longer.

Legs rather long and slender, yellow but tarsi somewhat brownish, last segment very darkened. Anterior four femora equally slender, mid femora (Fig. 187) with a double row of rather thin black ventral spines and with 2 long brownish anterodorsal bristly-hairs in anterior third; no pv bristles. No apical spur on mid tibia (Fig. 187).

Wings (Fig. 690) clear with indistinct pale veins, veins R4+5 and M parallel or slightly diverging, crossveins contiguous, the vein closing anal cell fine, recurrent. Squamae and halteres whitish-yellow. Abdomen and rather small genitalia (Figs. 320-322) shining black.

Length: body 1.8 - 2.2 mm, wing 2.4 - 2.7 mm.

♀. The black polished patch on posterior half of mesonotum (Fig. 98) reaches front of mesonotum as a broad median stripe, humeri and rather broad lateral margins densely dusted right up to notopleural depression. Abdomen shining except for apical two segments, segment 7 and sometimes tip of sternite 8 subshining; cerci dull.

Length: body 2.0 - 2.5 mm, wing 2.4 - 2.8 mm.

Rather common in the northern parts of Fennoscandia; in Norway from HOi north to Tr, in Sweden from Jmt. to T.Lpm. and Nb., and throughout Finland including the Baltic coast; also in Russian Carelia, north to Kola Peninsula. - N and NW of European USSR, mountains of C.Europe (Austria, Czechoslovakia). - End June-August. On ground-vegetation, in mountain meadows.

Note. The 2 ♀ taken by Tuomikoski at Suistamo and mentioned by Collin (1961: 209) under <u>unica</u> are undoubtedly <u>boreoalpinus</u>, but the females in the BM from Styria (Collin 1.c.) should be <u>alpinus</u>.

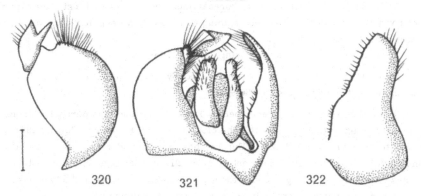

Figs. 320-322. Male genitalia of <u>Platypalpus boreoalpinus</u> Frey. - 320: right periandrial lamella; 321: periandrium with cerci; 322: left periandrial lamella. Scale: 0.1 mm.

24. PLATYPALPUS ALPINUS Chvála, 1971

Figs. 82, 323-325, 691.

<u>Coryneta</u> <u>unguiculata</u> Zetterstedt; Engel, 1939: 103.
<u>Platypalpus</u> boreoalpinus Frey, 1943: 5, 15 (p.p.).
<u>Platypalpus</u> alpinus Chvála, 1971: 24.

Closely resembling <u>boreoalpinus</u> but legs entirely yellow including tarsi, antennal segment 3 extensively darkened and mesonotum polished black on a broad median stripe right up to the front in both sexes.

♂. Head as in <u>boreoalpinus</u> but vertex subshining near ocellar tubercle, a pair of ocellar and 2 pairs of vt bristles very pale. Antennae (Fig. 82) yellow on basal segments but segment 3 entirely dark brown, not brownish or sometimes yellowish at extreme base as in <u>boreoalpinus</u>.

Thorax extensively polished black on mesonotum, leaving humeri, a small patch at sides of humeri and narrow mesonotal margins including scutellum densely grey dusted. Acr and dc pale and somewhat longer, the acr irregularly triserial in front but biserial posteriorly, all hairs considerably diverging.

Legs yellow including all tarsi, anterior four femora somewhat shorter and slightly thickened. Wings (Fig. 691) rather broad apically and blunt-tipped, venation as in <u>boreoalpinus</u> but the vein closing anal cell less recurrent. Geni-

125

talia (Figs. 323-325) polished black with left lamella strikingly large, overlapping the whole hypopygium, and only a single process on right lamella.

Length: body 1.8 - 2.1 mm, wing 2.1 - 2.4 mm.

♀. No sexual differences as in boreoalpinus, but posterior tarsi often slightly brownish towards tip. Abdomen shining with apical two segments and slender cerci densely greyish dusted.

Length: body 2.0 - 2.5 mm, wing 2.1 - 2.4 mm.

Rare. Finland: N, Hoplax, ♂ (Frey); Ta, Sysmä, ♂ (Hellén). - Austrian, Italian and Jugoslavian Alps. - May-July.

Note. This species was described by Engel (1939) and also partly by Strobl (1893, 1910) as unguiculatus Zett. The closely related P. unicus (Collin, 1961), known to me from England and C. Europe, has much broader face, deeper black and slightly longer antennal segment 3, generally stouter legs with the ventral spines on mid femora longer and without the whitish pile between the two rows, and mesonotum more extensively polished even between humeri.

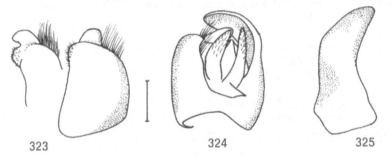

323   324   325

Figs. 323-325. Male genitalia of Platypalpus alpinus Chv. - 323: right periandrial lamella; 324: periandrium with cerci; 325: left periandrial lamella. Scale: 0.1 mm.

25. PLATYPALPUS COMMUTATUS (Strobl, 1893)
    Figs. 88, 191, 326-328, 345, 692.

Tachydromia commutata Strobl, 1893: 111.
? Tachydromia interpola Collin, 1961: 212.

Resembling boreoalpinus but antennal segment 3 long and dark, 2.5 - 3 times as long as deep, and arista as long as segment 3. Mesonotum polished black on posterior two-thirds in both sexes, thorax otherwise rather whitish-grey dusted.

♂. Head including occiput rather densely whitish-grey dusted, frons broad

and slightly widening above, face much narrower than frons in front. A pair
of ocellar and 2 pairs of vt bristles long, light brownish. Antennae (Fig. 88)
yellow to darker yellowish-brown on basal segments, segment 3 blackish-brown,
almost 3 times as long as deep with subequal arista. Palpi whitish-yellow, very small.

Thorax polished black on a large patch on posterior two-thirds of mesonotum and on sternopleura, otherwise whitish-grey dusted. Small hairs rather
long, pale: acr biserial and slightly diverging, dc irregularly biserial with
numerous hairs at sides. Large bristles pale, a humeral, 2 notopleural, a
postalar, 1 or 2 pairs of large prescutellar dc, and a pair of scutellars.

Legs yellow but apical tarsal segments somewhat brownish, last segment
almost blackish. Mid femora (Fig. 191) only slightly thickened, no pv bristles,
the ventral spines in posterior row longer, and about 4 - 5 brownish bristly-
hairs anterodorsally on apical half; similar hairs also on fore femora dorsally and anteroventrally; no tibial spur. Hind tibiae slightly dilated distally and
with a round rim-like apical projection armed with a circlet of pale marginal
fringes.

Wings (Fig. 692) clear, venation as in boreoalpinus. Squamae and halteres
whitish-yellow. Abdomen shining blackish, with minute pale hairs; genitalia
(Figs. 326-328) rather small with enlarged left lamella.

Length: body 2.1 - 2.5 mm, wing 2.6 - 2.8 mm.

♀. Mesonotum sometimes weakly dusted on a narrow stripe in front of scutellum and on a median stripe between the two rows of acr. Abdominal segments
7 and 8 (Fig. 345) including slender cerci grey dusted.

Length: body 2.4 - 2.9 mm, wing 2.6 - 3.2 mm.

Rare; only from the Kola Peninsula: Lr, Ponoj, ♂ 2 ♀ (Hellén). - Scot-

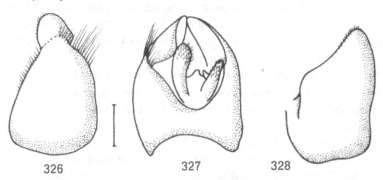

Figs. 326-328. Male genitalia of Platypalpus commutatus (Strobl). - 326: right
periandrial lamella; 327: periandrium with cerci; 328: left periandrial lamella. Scale: 0.1 mm.

land, N of European USSR, and mountains of C.Europe; Alps (Austria, Italy) and Krkonoše Mts. (Czechoslovakia). - End June-early September. On ground-vegetation, also in open places and in meadows.

Note. P.interpolus (Coll.) is very probably identical with commutatus, an error due to Strobl's inadequate description of commutatus; I have seen the type specimens of commutatus but not of interpolus, but I possess a female of commutatus from Inverness, Scotland.

26. PLATYPALPUS LONGICORNIS (Meigen, 1822)
Figs. 89, 97, 189, 329-331, 346, 693.

Tachydromia longicornis Meigen, 1822: 73.
Tachydromia pubicornis Zetterstedt, 1838: 553.

Larger species with 2 pairs of dark vt bristles, mesonotum polished black. Antennae pale on basal segments, segment 3 conspicuously long, arista very short. Legs extensively yellow, no tibial spur.

♂. Head black, entirely thinly grey dusted; with a broad frons that widens above, and much narrower parallel face. Anterior pair of ocellar and 2 pairs of vt bristles long, blackish. Antennae (Fig. 89) yellowish-brown to dark brown on basal segments, segment 3 black and very long, .at least 5 times as long as deep, arista scarcely 1/4 as long as segment 3. Palpi very small, pale.

Thorax (Fig. 97) polished black on mesonotum except for humeri, rather broad side margins, scutellum and often a patch in front of it which are grey dusted. Pleura dusted, sternopleura with a polished bare patch at middle. Thoracic hairs and bristles as in commutatus but large bristles brownish to black-ish-brown.

Legs yellow except for brownish 2 or 3 apical tarsal segments. Fore femora with longer brownish hairs beneath; mid femora (Fig. 189) scarcely deeper, no pv bristles, the black spines in posterior row beneath longer, and with a row of 6 - 8 brown bristly-hairs anteriorly.

Wings (Fig. 693) clear with brownish veins, venation as usual in this group of species but the vein closing anal cell less recurrent. Abdomen shining black with longer pale hairs on venter; genitalia (Figs. 329-331) polished black, not broader than end of abdomen.

Length: body 1.8 - 2.3 mm, wing 2.2 - 3.0 mm.

♀. Generally larger, and hind femora with finer brownish hairs beneath. Apical two abdominal segments (Fig. 346) and slender cerci rather densely grey dusted but sternite 8 mostly polished.

Length: body 2.2 - 3.3 mm, wing 2.5 - 3.0 mm.

Very common in Denmark and throughout Fennoscandia, including the extreme north. - Common in Europe, from Faroe Isl. in the north, south to the Mediterranean. - (April) May-October. On bushes and ground-vegetation, usually in shaded places but also in meadows.

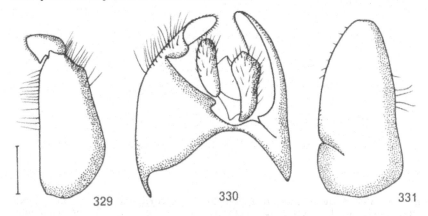

Figs. 329-331. Male genitalia of _Platypalpus longicornis_ (Meig.). - 329: right periandrial lamella; 330: periandrium with cerci; 331: left periandrial lamella. Scale: 0.1 mm.

27. PLATYPALPUS LONGICORNIOIDES Chvála, 1972
Figs. 81, 84, 332-335.

Platypalpus longicornioides Chvála, 1972: 4.

Very closely resembling _longicornis_ but only 1 pair of black widely separated vt bristles and antennae uniformly dark with shorter segment 3.

♂. Head (Fig. 81) black, rather densely grey dusted; frons narrower in front, about as deep as the small antennal segment 2, distinctly widening above; face not narrower than frons in front. 1 pair of widely separated long black vertical bristles, as long as anterior pair of ocellar bristles. Antennae (Fig. 84) unicolourous dark blackish-brown even on basal segments, segment 3 about 4 times as long as deep, arista 1/3 as long as segment 3.

Thorax as in _longicornis_ with extensively polished mesonotum, but large bristles black, and biserial acr and dc distinctly diverging. Legs unicolourous yellowish but hind tibiae somewhat brownish towards tips, and hind tarsi with indefinite and faint brown annulations. Wings clear, veins R4+5 and M straight throughout and indistinctly diverging. Abdomen shining black with scattered minute pale hairs; genitalia (Figs. 332-335) polished black, small, lamellae almost bare.

Length: body 1.7 mm, wing 2.5 mm.

♀. Unknown.

Very rare. Sweden: Sk., Arkelstorp, ♂ (H.Andersson). - No further data available. - June.

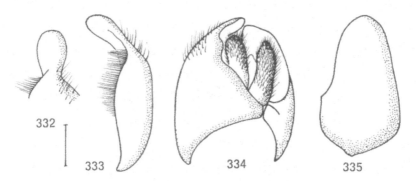

Figs. 332-335. Male genitalia of Platypalpus longicornioides Chv. - 332: apex of right periandrial lamella; 333: right periandrial lamella; 334: periandrium with cerci; 335: left periandrial lamella. Scale: 0.1 mm.

28. PLATYPALPUS BRUNNEITIBIA (Strobl, 1899)

Figs. 90, 188, 192, 336-338, 347, 694.

Tachydromia pubicornis var.brunneitibia Strobl, 1899: 78.

Very much like longicornis but mesonotum entirely dark grey dusted, basal antennal segments blackish, large bristles on head and thorax black, and legs extensively brownish.

♂. Head greyish dusted, very thinly on vertex, frons broad in front and almost twice as deep opposite hind ocelli, face distinctly narrower than frons in front. Anterior pair of ocellar and 2 pairs of vt bristles long, black. Antennae (Fig. 90) as in longicornis but basal segments blackish, concolourous with segment 3.

Thorax entirely thinly blackish-grey dusted on mesonotum, often somewhat subshining anteriorly between humeri. Pleura lighter grey and more densely dusted, sternopleura largely polished. Thoracic hairs and bristles as in longicornis but acr and dc dark brown and large bristles black.

Legs yellowish-brown, tibiae brown and tarsi extensively darkened with apical 2 or 3 segments almost black. Fore femora (Fig. 192) strikingly thickened on basal half, ventrally with a double row of long darker brown hairs and with similar bristly-hairs in a posteroventral row. Fore tibiae (Fig. 192) weakly thickened, dorsally with densely-set brownish hairs. Mid femora (Fig. 188) as deep as fore femora on basal half, the ventral black spines becoming shorter towards tip and those in posterior row longer, and anteriorly with a row of longer dark brown bristles; no pv bristles; mid tibia (Fig. 188) with a very small, pointed apical projection.

Wings (Fig. 694) clear or very weakly brownish clouded, veins blackish-brown. Squamae dark brown, halteres yellowish with brown stalk. Abdomen shining black with fine pale hairs, becoming longer on venter and on posterior segments; genitalia (Figs. 336-338) small, dull grey cerci enclosed within polished lamellae.

Length: body 1.8 - 2.8 mm, wing 2.5 - 2.9 mm.

♀. Larger, with finer hairs on hind tibia beneath, abdomen shining black with segment 7 (Fig. 347), tergite 8 and slender cerci rather densely grey dusted, but sternite 8 polished black and produced apically, rounded at tip.

Length: body 2.5 - 3.1 mm, wing 2.7 - 3.4 mm (holotype body 2.5 mm, wing 2.7 mm).

Rather rare in S.Sweden, north to Boh., and in Finland north to ObS. - Widespread in Europe but nowhere common; European USSR (Estonia, Crimea), W.Germany, Czechoslovakia, Austria, Jugoslavia and Spain. - (May) June-August.

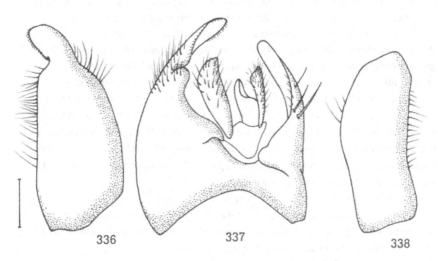

Figs. 336-338. Male genitalia of Platypalpus brunneitibia (Strobl). - 336: right periandrial lamella; 337: periandrium with cerci; 338: left periandrial lamella. Scale: 0.1 mm.

29. PLATYPALPUS DIFFICILIS (Frey, 1907)
Figs. 85, 190, 339-341, 349, 695.

Tachydromia difficilis Frey, 1907: 410.
Tachydromia interjecta Lundbeck, 1910: 295.
Coryneta commutata Strobl; Engel, 1939: 65.

A smaller black species with 2 pairs of vt bristles and thinly dusted mesonotum; acr and dc pale, biserial and diverging. Antennae yellowish on basal segments, segment 3 dark, almost 4 times as long as deep, arista half as long. Legs yellow, no tibial spur.

♂. Head greyish dusted, frons rather narrower in front, widening above; face very narrow, as deep as front ocellus. 2 pairs of long vt bristles, brown to blackish-brown. Antennae (Fig. 85) yellowish on basal segments, segment 3 very dark, almost 4 times as long as deep, arista half as long. Palpi small, pale, with a long dark bristly-hair at tip.

Thorax evenly but rather thinly dark grey dusted on mesonotum, pleura similarly dusted, sternopleura with a large polished patch. Acr and dc pale or very light brownish, biserial, not very long but distinctly diverging; large bristles brownish to blackish-brown, in full number with 1 pair of large prescutellar dc.

Legs yellow or dark yellow with last tarsal segment almost blackish, preceding 2 or 3 segments sometimes indefinitely brownish. Fore femora rather slender, dorsally towards tip with several brown bristly-hairs and with a circlet of anterior preapical bristles. Mid femora (Fig. 190) slightly thickened, the ventral black spines in posterior row longer, anteriorly with several longer brownish hairs; no pv bristles. Apical spur on mid tibia (Fig. 190) very short; hind femora with longer brownish bristly-hairs anteroventrally.

Wings (Fig. 695) clear with brownish veins, veins R4+5 and M almost parallel, crossveins practically contiguous, and the vein closing anal cell recurrent. Squamae and halteres light brown, knob pale yellow. Abdomen shining black but segment 1 dull at sides; genitalia (Figs. 339-341) polished, very small and pointing somewhat towards right, with lamellae almost bare and small cerci concealed within lamellae.

Length: body 1.8 - 2.3 mm, wing 2.2 - 2.6 mm.

♀. Tarsi usually extensively darkened and apical spur on mid tibia practically absent. Apical two abdominal segments (Fig. 349) small and densely grey dusted like the long slender cerci.

Length: body 2.2 - 2.8 mm, wing 2.3 - 2.8 mm.

Not uncommon in Denmark and southern parts of Fennoscandia; in Sweden north to Sdm., in Finland along the Baltic coast, north to Om; not recorded from Norway. - Great Britain, NW of European USSR and through Estonia to its central areas, in C.Europe mostly in mountainous districts. - June-early September. On ground-vegetation; Collin (1961) took it by sweeping conifers.

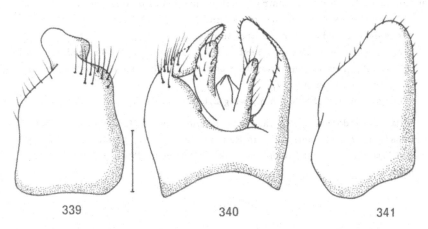

Figs. 339-341. Male genitalia of <u>Platypalpus difficilis</u> (Frey). - 339: right periandrial lamella; 340: periandrium with cerci; 341: left periandrial lamella. Scale: 0.1 mm.

30. PLATYPALPUS SCANDINAVICUS Chvála, 1972
Figs. 91, 196, 342-344, 350, 696.

<u>Platypalpus scandinavicus</u> Chvála, 1972: 5.

Resembling <u>difficilis</u> but antennal segment 3 shorter, about 3 times as long as deep, and arista equal in length; basal segments yellowish. Male genitalia much larger and sternite 8 in ♀ polished black, produced and rounded apically.

♂. Head as in <u>difficilis</u> but frons broader in front, as deep as antennal segment 3; 2 pairs of vt bristles black; antennae (Fig. 91) yellowish on basal segments but segment 3 distinctly shorter, about 3 times as long as deep at base, blackish-brown; arista rather densely pubescent, as long as segment 3. Palpi with pale terminal bristly-hair.

Thorax thinly dark grey dusted on mesonotum, more densely and slightly paler on pleura. Acr and dc rather smaller, pale, biserial with the two rows distinctly divergent and numerous similar hairs at sides; large bristles blackish.

Legs (Fig. 196) yellow with apical 2 or 3 tarsal segments faintly brownish, last segment not darker; otherwise as in difficilis, but ventral bristly-hairs on fore femora and hairs beneath hind femora paler. Wings (Fig. 696) and abdomen as in difficilis but genitalia (Figs. 342-344) strikingly large, polished, lamellae rather bare except for several long brownish hairs on right lamella at tip, and a fringe of pale hairs on outer margins of both lamellae.

Length: body 2.2 mm, wing 2.6 mm.

♂ is described here for the first time.

♀. Larger; abdominal segment 7 (Fig. 350) and tergite 8 including small cerci densely grey dusted; sternite 8 polished black, produced apically and broadly rounded in side view.

Length: body 2.3 - 3.0 mm, wing 2.6 - 3.1 mm.

Rare. Sweden: Nb., Råneå, Hogsön, ♀ (holotype) (H.Andersson); Jmt., Undersåker, ♀ (Ringdahl); also in Russian Carelia: Kr, Suistamo, 2♀ and Salmi, Hiisjärvi, ♂ 4♀ (Tuomikoski). - NW of European USSR. - July-early September.

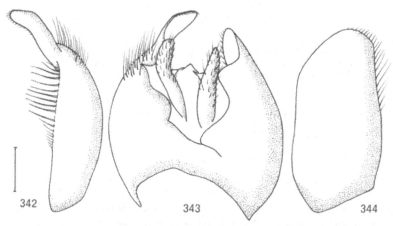

Figs. 342-344. Male genitalia of Platypalpus scandinavicus Chv. - 342: right periandrial lamella; 343: periandrium with cerci; 344: left periandrial lamella. Scale: 0.1 mm.

31. PLATYPALPUS TUOMIKOSKII Chvála, 1972
   Figs. 92, 197, 351, 353-355, 697.

Platypalpus tuomikoskii Chvála, 1972: 6.

Closely resembling scandinavicus but basal antennal segments blackish, large

134

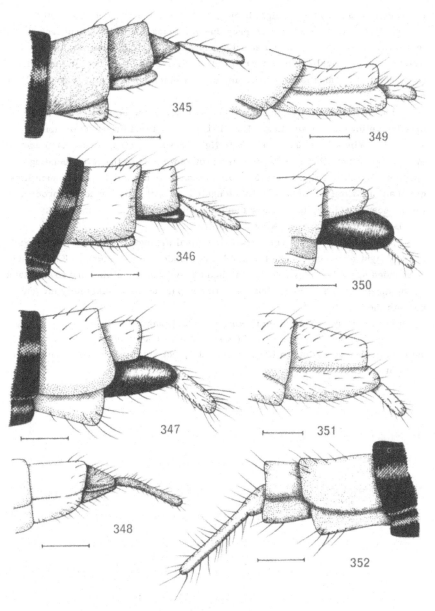

Figs. 345-352. Terminal segments of female abdomens in Platypalpus. - 345:
commutatus (Strobl); 346: longicornis (Meig.); 347: brunneitibia (Strobl);
348: exilis (Meig.); 349: difficilis (Frey); 350: scandinavicus Chv.; 351: tuomikoskii Chv.; 352: nigricoxa (Mik). Scale: 0.1 mm.

135

bristles on head and thorax light brown, and sternite 8 in ♀ densely dusted.

♂. Head rather densely dark grey dusted, frons not so broad, narrower than antennal segment 2 and distinctly widening above, face almost linear. Anterior pair of ocellar and 2 pairs of vt bristles long, light brown. Antennae (Fig. 92) as in scandinavicus but unicolourous blackish-brown even on basal segments.

Thorax as in difficilis or scandinavicus but acr and dc very pale and large bristles yellowish-brown. Legs (Fig. 197) yellow, tarsi brownish on apical segments. Wings (Fig. 697) clear with light brownish veins; veins R4+5 and M very indistinctly but evenly divergent; otherwise as in the two preceding species. Genitalia (Figs. 353-355) polished black and almost bare; much larger than in difficilis, right lamella elongated but with smaller dorsal process. Unlike scandinavicus, lamellae without any long hairs on outer margins.

Length: body 2.0 mm, wing 2.6 mm.

♀. Resembling male; apical two abdominal segments (Fig. 351) small and densely light grey dusted like the small but slender cerci; sternite 8 not broadly rounded apically and polished as in scandinavicus, sometimes narrowly black on the upper margin at base, but the polished part usually completely hidden beneath the tergite.

Length: body 2.2 - 2.8 mm, wing 2.5 - 2.8 mm.

Rare. Finland: N, Helsinki, ♂ (holotype) 3 ♀ (Tuomikoski). - No further data available, according to Kovalev (in litt.), not yet found in the adjacent parts of the USSR.

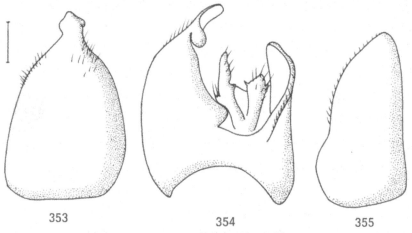

Figs. 353-355. Male genitalia of Platypalpus tuomikoskii Chv. - 353: right periandrial lamella; 354: periandrium with cerci; 355: left periandrial lamella. Scale: 0.1 mm.

32. PLATYPALPUS EXILIS (Meigen, 1822)
Figs. 86, 193, 348, 356-358, 698.

Tachydromia exilis Meigen, 1822: 90.
Platypalpus flavipennis Walker, 1851: 130.
Tachydromia exilis var. nigroterminata Strobl, 1910: 77.

Yellow species with 1 pair of vt bristles and small pale antennae. Head blackish, mesonotum thinly dusted with acr biserial; legs yellow, a small tibial spur, and mid femora without pv bristles.

♂. Head black, densely grey dusted, frons broad and distinctly widened above, face narrow. A pair of fine pale widely separated vt bristles, and a similar pair also on postocular margin. Antennae (Fig. 86) pale yellow on basal segments, segment 3 yellow to brown (in var. nigroterminata), about twice as long as deep, narrow; dark arista about twice as long as segment 3. Palpi small, pale; proboscis yellowish, dark at tip.

Thorax uniformly yellow with weakly grey dusted mesonotum, pleura more densely silvery-grey dusted, sternopleura largely polished. All thoracic hairs and bristles pale, acr broadly biserial or irregularly triserial posteriorly, dc uniserial, last pair bristle-like.

Legs yellow with brownish tarsi, last segment almost blackish. Mid femora (Fig. 193) slightly stouter and larger than fore femora, the black ventral spines in posterior row longer, no pv bristles. Mid tibia (Fig. 193) with a blunt short yellow apical spur.

Wings (Fig. 698) faintly yellowish, veins pale, the vein closing anal cell

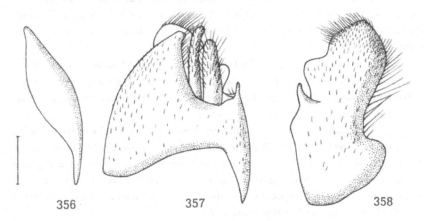

356   357   358

Figs. 356-358. Male genitalia of Platypalpus exilis (Meig.). - 356: right periandrial lamella; 357: periandrium with cerci; 358: left periandrial lamella. Scale: 0.1 mm.

137

slightly recurrent. Squamae and halteres whitish. Abdomen shining yellow with fine pale hairs but middle segments often blackish; genitalia (Figs. 356-358) rather small, lamellae often darkened, cerci blackish.

Length: body 2.0 - 2.1 mm, wing 2.5 - 2.7 mm.

♀. Closely resembling male; abdomen mostly yellow and slightly dusted, segment 8 (Fig. 348) and very slender cerci usually darker.

Length: body 2.1 - 2.5 mm, wing 2.5 - 2.9 mm.

Rather uncommon in Denmark and southern parts of Fennoscandia; in Norway north to Nnø, in Sweden to Äng., in Finland to Sb; also in Russian Carelia. - Great Britain, European USSR from 63°N. south to Transcarpathia and the Caucasus, and through C. Europe to Jugoslavia. - End May-early August. Mostly on bushes and ground-vegetation in shaded places, in spring and early summer.

33. PLATYPALPUS PULICARIUS (Meigen, 1830)
Figs. 93, 194, 359-361, 699.

Tachydromia pulicaria Meigen, 1830: 343.

A small dark species with 2 pairs of vt bristles; mesonotum greyish dusted with 4-serial pale acr, large bristles pale; antennae small with basal segments yellowish. Legs yellow with a small tibial spur.

♂. Head rather densely lighter grey dusted, frons narrow in front, narrower than antennal segment 2 is deep, and only slightly widening above; face narrower than frons in front. 2 pairs of vt bristles long, pale. Antennae (Fig. 93) small, basal segments yellow, segment 3 scarcely 1.5 times as long as deep, blackish-brown; dark arista at least twice as long as segment 3. Palpi small-ovate, very pale.

Thorax rather densely grey dusted on mesonotum and pleura, sternopleura largely polished. Small hairs fine, pale; acr 4-serial, separated by a narrow stripe from multiserial dc, large bristles yellow to yellowish-brown including a long pair of prescutellar dc.

Legs short and rather thickened, yellow with only apical two tarsal segments slightly brownish. Mid femora (Fig. 194) considerably thickened, deeper than fore femora, ventrally with double black spines on apical two-thirds, those in posterior row longer and thinner, basal third with long pale bristly-hairs beneath; no pv bristles. Apical tibial spur (Fig. 194) black, small but distinctly pointed, slightly shorter than tibia is deep.

Wings (Fig. 699) clear with darker brownish veins, the vein closing anal cell very slightly recurrent. Squamae and halteres pale yellow. Abdomen shining black except for posterior sternites, covered with scattered pale hairs; genitalia (Figs. 359-361) small and very thinly dull greyish.

Length: body 1.3 - 2.0 mm, wing 1.8 - 2.2 mm.

♀. Resembling male; abdomen with apical two segments and small cerci densely grey dusted.

Length: body 1.7 - 2.3 mm, wing 2.0 - 2.3 mm.

Rare in Denmark and the southern parts of Sweden north to Sdm. and Finland north to Ta.-England, NW of European USSR (Leningrad region) and C. Europe. - June-early September. Rather rare everywhere, on leaves of bushes and on ground-vegetation.

Note. Recorded here for the first time from Fennoscandia, but there are several very closely related species which may occur in the southern part. P.stackelbergi Kovalev, 1971, described from the adjacent Leningrad region of the European USSR, has only a very small dark blunt spur on mid tibia and faintly brownish annulated or extensively darkened tarsi, and there are also slight differences in the male genitalia (Kovalev 1971: 213). P.incertus (Collin, 1926), known from England and recently recorded from Spain, has posterior four coxae and their corresponding femora except for base extensively darkened, large bristles almost black, and slightly longer antennal segment 3 which is twice as long as deep. Other related species are P.cilitarsis Frey, 1943; P. vegrandis Frey, 1943 and P.novaki (Strobl, 1893) which all have much broader frons.

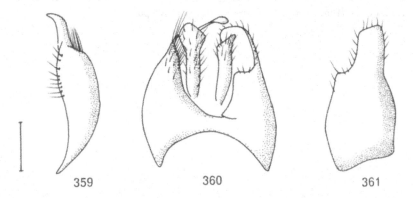

359   360   361

Figs. 359-361. Male genitalia of Platypalpus pulicarius (Meig.). - 359: right periandrial lamella; 360: periandrium with cerci; 361: left periandrial lamella. Scale: 0.1 mm.

34. PLATYPALPUS NIGRICOXA (Mik, 1884)
Figs. 87, 195, 352, 362-364, 700.

Tachydromia nigricoxa Mik, 1884: 82.
Tachydromia Poppiusi Frey, 1913a: 11.

Larger dark grey species with 2 pairs of vt bristles and large thoracic bristles black; acr small, pale, irregularly 6-serial. Antennae black, legs dark yellow with black pattern, mid femora thickened, without pv bristles; tibial spur very small.

♂. Head rather densely grey dusted, more thinly on frons which is broad below and widens above, face much narrower than frons in front. 2 pairs of vt bristles long, black, occiput with dark hairs above. Antennae (Fig. 87) black with segment 3 broadly ovate and slightly longer arista. Palpi blackish-brown with apical hair pale, proboscis strong, black.

Thorax rather thinly dark grey dusted on mesonotum, more densely grey on pleura, sternopleura largely polished. Acr and dc very small, light brown; former irregularly 6-serial on a broad median stripe, narrowly separated from uniserial dc and numerous similar hairs at side. Large bristles long, black, including 2 pairs of prescutellar dc.

Legs dark yellow to yellowish-brown but all coxae and trochanters black, apical tarsal segments and often hind legs brown, and anterior four femora with a narrow black stripe on at least basal half beneath. Mid femora (Fig. 195) distinctly thickened, the black ventral spines in posterior row longer, no pv bristles. Mid tibia (Fig. 195) with a small, flat yellow spur.

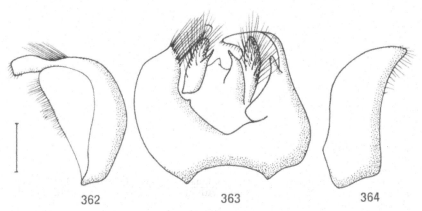

362   363   364

Figs. 362-364. Male genitalia of Platypalpus nigricoxa (Mik). - 362: right peri-andrial lamella; 363: periandrium with cerci; 364: left periandrial lamella. Scale: 0.1 mm.

Wings (Fig. 700) clear with dark veins, the vein closing anal cell slightly recurrent; veins R4+5 and M almost parallel, crossveins contiguous. Squamae blackish, halteres pale yellow. Abdomen shining black with extreme sides of anterior tergites slightly dusted; genitalia (Figs. 362-364) polished black, small and strikingly globose with very convex and almost bare lamellae; small cerci with very long hairs.

Length: body 2.4 - 3.0 mm, wing 2.9 - 3.6 mm.

♀. Closely resembling male, abdomen shining but apical two segments and long slender cerci densely light grey dusted.

Length: body 2.8 - 3.3 mm, wing 2.8 - 3.4 mm.

Not uncommon in extreme north of Fennoscandia; Sweden in T. Lpm., in Finland south to Ks; also from the Kola Peninsula. - Extreme north of European USSR, Alps; a boreoalpine species in distribution. - (May) June-July. On small willow bushes (Frey, 1913a).

Note. The synonymy poppiusi = nigricoxa was verified by study of the types. P. alpigenus (Strobl, 1893) and P. lesinensis (Strobl, 1893) are related species.

VI. luteus - group

35. PLATYPALPUS LUTEUS (Meigen, 1804)
Figs. 99, 198, 365-367, 382, 701.

Tachydromia lutea Meigen, 1804: 238.
Tachydromia glabra Meigen, 1822: 89.
Tachydromia pallida Meigen, 1822: 90.
Platypalpus formalis Walker, 1851: 130.

A larger yellow species with entirely yellow head and strong proboscis. Mesonotum shining with numerous small acr; legs yellow, apical three segments on fore tarsi dilated.

♂. Head entirely yellow and somewhat subshining; frons very broad, slightly darkened above antennae, and ocellar tubercle usually black; face as deep as frons in front. A pair of vt bristles brownish and only fine, ocellar bristles much longer and darkened. Antennae (Fig. 99) yellow with segment 3 almost blackish except for base and very short, scarcely longer than deep, arista almost 4 times as long. Palpi very small, pale yellow; proboscis very strong and not very much shorter than head is high, yellow, darkened at tip.

Thorax unicolourous yellow, mesonotum almost shining but its margins and pleura thinly silvery dusted, sternopleura largely polished. Acr and dc pale and minute, former 8-serial, latter uniserial with numerous hairs at side; large bristles brownish.

Legs pale yellow but claws black, apical three tarsal segments on fore legs conspicuously dilated. Mid femora (Fig. 198) thickened, ventrally with two rows of short black spines on apical two-thirds, and not very distinct yellowish-brown pv bristles. Mid tibia (Fig. 198) with short, blunt pale apical spur, shorter than tibia is deep.

Wings (Fig. 701) clear, veins dark; veins R4+5 and M almost parallel, the vein closing anal cell distinctly recurrent. Squamae and halteres whitish-yellow. Abdomen subshining pale yellow with posterior tergites often darkened; genitalia (Figs. 365-367) yellow with cerci blackish.

Length: body 2.4 - 3.2 mm, wing 3.5 - 3.7 mm.

♀. Larger; abdomen often somewhat brownish, apical two segments (Fig. 382) except for tip of sternite 8, and long slender cerci, silvery dusted.

Length: body 3.0 - 3.5 mm, wing 3.6 - 3.8 mm.

Very common in Denmark and southern parts of Fennoscandia, reaching up to 64° north; in Norway to NTi, in Sweden to Jmt., in Finland to Tb, Sb and Kb; also in Russian Carelia. - Widespread in Europe but less common in C. and S.Europe. - May-mid September. Mainly on bushes in shaded places.

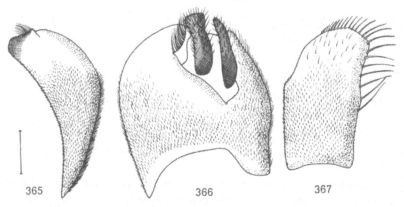

Figs. 365-367. Male genitalia of Platypalpus luteus (Meig.). - 365: right peri-andrial lamella; 366: periandrium with cerci; 367: left periandrial lamella. Scale: 0.1 mm.

VII. nigritarsis - group

36. PLATYPALPUS NIGRITARSIS (Fallén, 1816)

Figs. 100, 199, 368-370, 381, 702.

142

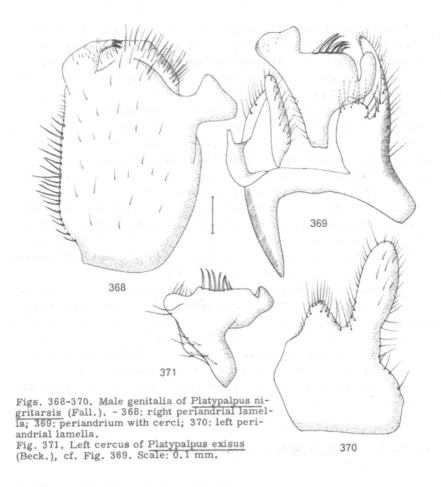

Figs. 368-370. Male genitalia of Platypalpus nigritarsis (Fall.). - 368: right periandrial lamella; 369: periandrium with cerci; 370: left periandrial lamella.
Fig. 371. Left cercus of Platypalpus exisus (Beck.), cf. Fig. 369. Scale: 0.1 mm.

Tachydromia nigritarsis Fallén, 1816: 34.

Platypalpus comptus Walker, 1837: 228.

A species with 2 pairs of black vt bristles; mesonotum, most of pleura and vertex polished black. Antennae very long, black. Legs darker yellow with posterior coxae, all tarsi and hind femora at tip blackish; mid femora thickened with dark pv bristles, no tibial spur.

♂. Head greyish dusted but polished black on upper part of frons and on vertex; frons very broad in front and almost twice as broad opposite hind ocelli; face very narrow, parallel. 2 pairs of long vt bristles black. Antennae (Fig. 100) black, very long; segment 3 about 4 times as long as deep, arista half as long. Palpi small, silvery pilose with a long dark apical bristly-hair.

143

Thorax polished black on mesonotum, its lateral margins including hume-ri, postalar calli, scutellum and upper part of pleura light grey dusted, ster-nopleura and hypopleura polished. All thoracic hairs and bristles blackish; acr and dc fine and inconspicuous, former closely biserial, latter uniserial; a long curved humeral bristle.

Legs rather darker yellow to light reddish-yellow, but posterior four coxae, most of fore tibiae, a patch on mid femora (Fig. 199) above, apical third of hind femora, and all tarsi except fore metatarsi largely blackish. Fore femora rather slender, mid femora (Fig. 199) thickened with small black spines be-neath becoming hair-like towards base, and long black pv bristles. No apical spur on mid tibia (Fig. 199), fore tibia spindle-shaped.

Wings (Fig. 702) almost clear with dark veins, the vein closing anal cell recurrent. Squamae dark, halteres yellowish. Abdomen shining black except for dull sides of segment 1; genitalia (Figs. 368-370) polished black, large, distinctly pointing towards right and with large, abnormally shaped left cercus.

Length: body 1.7 - 2.4 mm, wing 2.2 - 2.7 mm.

♀. Resembling male, but the black pattern on legs usually lighter brownish. Abdominal segment 7 (Fig. 381) short, polished black, segment 8 elongate and thinly grey dusted like the short cerci.

Length: body 2.2 - 2.6 mm, wing 2.3 - 2.6 mm.

A very common species in Denmark and the whole of Fennoscandia inclu-ding the extreme north. - Common and widely distributed throughout Europe, also on Faroe Isl., but perhaps absent in the southernmost areas. - May-early November. Common on ground-vegetation, mainly in late summer.

37. PLATYPALPUS EXCISUS (Becker, 1907)
Figs. 201, 371.

Tachydromia excisa Becker, 1907: 114 (nec excisa Becker, 1908: 39).

Very much like nigritarsis but legs extensively darkened and hind tibia in ♂ with a slight excision at middle and distinctly curved; there are also slight differences in the male genitalia.

♂. Head and thorax as in nigritarsis. Legs extensively dark brown, coxae and apical half of hind femora almost black, leaving tip of fore coxae and base of hind femora paler. Hind tibiae (Fig. 201) slightly dilated in anterior third and at tip, with a large shallow excision at middle beneath, and distinctly cur-ved in posterior view.

Wings and abdomen as in nigritarsis but dorsum of abdomen rather sub-shining, and there are slight differences in the genitalia: the left enlarged cercus (Fig. 371) with less developed lateral lobes and the strong spine-like bristles at middle more numerous (at least 6) and rather slender.

Length: body 1.9 - 2.5 mm, wing 2.2 - 2.8 mm.

♀. Difficult to distinguish from that of nigritarsis so far as the simple hind tibia is concerned; however, the legs are extensively darkened in excisus, coxae all black except for pale tip of fore coxae, and mid femora dark above on almost the whole length.

Length: body 2.4 - 2.8 mm, wing 2.4 - 3.0 mm.

Rare. Sweden: Sk., Illstorp, ♂ (Roth); Lapland, ♂ (Becker). - European USSR, ? Great Britain; C. and S. Europe. - (March), May-September.

Note. Becker described this species twice, in 1907 from N. Africa and Europe, and in 1908 from the Canary Isl. However, the type specimens from the Canary Isl. represent a distinct species, with different shorter antennae (segment 3 about 3 times as long as deep, with arista subequal), paler palpi and other differences including genitalia. Collin (1961) did not separate excisus and nigritarsis as distinct species, but perhaps excisus also occurs in the British Isles.

38. PLATYPALPUS SYLVICOLA (Collin, 1926)
    Figs. 101, 200, 372-374, 703.

Tachydromia sylvicola Collin, 1926: 153.

A tiny polished black species including whole of thorax and a very conspicuously broad frons, with 2 pairs of blackish vt bristles and small black antennae. Legs short, yellow with brown pattern, with distinct dark pv bristles on mid femora but tibial spur very small.

♂. Head rather thinly grey dusted on occiput and on very narrow face, but frons very conspicuously broad, almost triangular in shape, polished black. 2 pairs of vt bristles long and fine, blackish. Antennae (Fig. 101) black, small; segment 3 very broad, almost 1.5 times as long as deep, arista twice as long. Palpi very small, dark, apical bristly-hair dark.

Thorax entirely polished black leaving only scutellum and a short stripe between humeri and roots of wings greyish dusted. Acr and dc brownish and rather longer (as long as antennal segment 2), former biserial, latter uniserial. Large bristles fine and almost black.

Legs rather short, yellow, but posterior four coxae, a faint patch on mid femora above, hind femora apically, and last segment on all tarsi brownish. Mid femora (Fig. 200) slightly thickened, the ventral black spines small, and about 4 - 5 long dark pv bristles. Mid tibia (Fig. 200) with a very small, pointed apical spur, fore tibia weakly spindle-shaped.

Wings (Fig. 703) clear with pale veins, veins R4+5 and M straight but slightly diverging, crossveins contiguous; the vein closing anal cell strikingly recurrent. Squamae and halteres brownish, knob of latter yellowish. Abdomen entirely polished black; genitalia (Figs. 372-374) very small and similarly polished, pointing towards right, lamellae with long fine pale hairs.

Length: body 1.1 - 1.3 mm, wing 1.6 - 1.8 mm.

♀. Resembling male; apical two abdominal segments and very long slender cerci greyish dusted.

Length: body 1.3 - 1.5 mm, wing 1.6 - 1.8 mm.

Rare. Sweden: Sk., ♂ (Boheman); Hall., 2♀ (Holmgren): Sm., Sandreda, ♂♀ (Ringdahl); Finland: Ab, Vichtis, ♂ (Frey); Ta, Sysmä, ♀ (Hellén). - Great Britain, NW of European USSR (Leningrad region). - Mid May-September. Collin (1961) took adults on and about pony-dung.

Note. Not previously recorded from North Europe. Superficially it resembles a small Drapetis species, but the main differential features including the structure of male genitalia show its close relationship to nigritarsis.

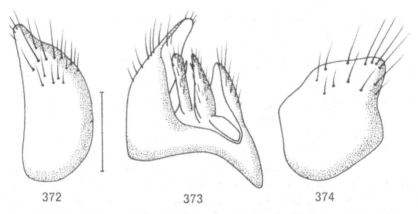

Figs. 372-374. Male genitalia of Platypalpus sylvicola (Coll.). - 372: right periandrial lamella; 373: Periandrium with cerci; 374: left periandrial lamella. Scale: 0.1 mm.

VIII. minutus - group

39. PLATYPALPUS FUSCICORNIS (Zetterstedt, 1842)
Figs. 102, 112, 202, 375-377, 384, 704.

146

Tachydromia fuscicornis Zetterstedt, 1842: 291.

Mesonotum polished black with rather long, numerous pale acr and dc, a pair of long pale vt bristles. Frons polished black, antennae small, blackish. Legs entirely yellow with a large, sharply pointed apical spur on mid tibia.

♂. Head thinly greyish dusted on vertex and occiput, but frons polished black, rather broad in front and distinctly widening above; face very linear, silvery-grey dusted. A pair of long, widely separated pale vt bristles. Antennae (Fig. 102) unicolourous blackish-brown, segment 3 scarcely longer than deep, arista about 3 times as long. Palpi small, pale, somewhat pointed at tip, with long pale terminal bristle.

Thorax (Fig. 112) polished black on mesonotum, lateral margins including scutellum and pleura densely silvery-grey dusted, sternopleura largely polished. Small thoracic hairs pale and moderately long, acr irregularly quadriserial, dc biserial with additional hairs at sides. Large bristles very long, pale, including a humeral and a pair of prescutellar dc.

Legs uniformly rather pale yellow even on tarsi, fore femora slightly thickened, fore tibia distinctly spindle-shaped dilated. Mid femora (Fig. 202) slightly deeper than fore femora with short ventral black spines and long pale pv bristles. Apical tibial spur (Fig. 202) large and sharply pointed.

Wings (Fig. 704) yellowish with brown veins, veins R4+5 and M almost parallel, crossveins contiguous but 2nd basal cell distinctly longer; the vein closing anal cell at right-angles to vein Cu. Squamae and halteres pale yellow. Abdomen subshining blackish-brown, covered with rather long and densely-set pale hairs. Genitalia (Figs. 375-377) rather small, blackish.

Length: body 1.9 - 2.3 mm, wing 2.3 - 2.7 mm.

♀. Generally larger, and abdomen with less densely-set hairs, apical two segments (Fig. 384) and long slender cerci densely greyish dusted.

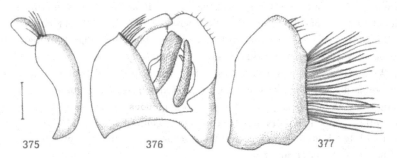

Figs. 375-377. Male genitalia of Platypalpus fuscicornis (Zett.). - 375: right periandrial lamella; 376: periandrium with cerci; 377: left periandrial lamella. Scale: 0.1 mm.

147

Length: body 2.4 - 2.6 mm, wing 3.0 - 3.1 mm.

Uncommon in Denmark and S. Sweden, north to Ög. - NW of European USSR (Estonia), S to its central parts, and C. Europe. The record from Amur, East Asia (Frey, 1943), seems unlikely. - June-July. On ground-vegetation in deciduous forests.

40. PLATYPALPUS RUFICORNIS (von Roser, 1840)
Figs. 104, 110, 111, 114, 115, 205, 378-380, 705.

Tachydromia ruficornis von Roser, 1840: 54.
Tachydromia thoracica Lundbeck, 1910: 317.

Thorax densely light silvery-grey dusted, mesonotum polished black on posterior half in ♂, and also on a median stripe anteriorly in ♀. Antennae small, yellow, with tip of segment 3 and arista darkened. Legs yellow with black knees, conspicuously thickened mid femora and tibiae, tibial spur large and strong, pointed apically.

♂. Head densely grey dusted, frons broad, slightly widening above, face as broad as frons in front. A pair of pale vt bristles very fine, close together. Antennae (Fig. 104) yellow to light orange-yellow, extreme tip of segment 3 and arista darkened; segment 3 about 1.5 times as long as deep, arista almost twice as long. Palpi (Fig. 110) conspicuously large, pale yellow.

Thorax (Fig. 114) densely silvery-grey dusted but posterior half of mesonotum including anterior part of scutellum, and most of mesopleura, pteropleura and sternopleura polished black. Biserial acr and uniserial dc pale and minute, only few in number; other thoracic bristles pale but rather small and fine, including last pair of prescutellar dc and a very small humeral bristle.

Legs yellow but mid knees very black and apical 2 or 3 tarsal segments darkened. Fore femora slightly thickened, fore tibiae slender. Mid femora (Fig. 205) very conspicuously thickened, with a row of about 7 long pale pv bristles in addition to the small ventral black spines. Mid tibia (Fig. 205) distinctly thickened, with a strong sharply pointed spur apically.

Wings (Fig. 705) clear with dark veins, veins R4+5 and M almost parallel; crossveins distant, 2nd basal cell distinctly longer; the vein closing anal cell at right-angles to vein Cu, vein A very inconspicuous. Squamae and halteres pale yellow. Abdomen polished black including genitalia (Figs. 378-380), latter small with left lamella enlarged and very broad, and with a tuft of longer reddish hairs on its outer margin.

Length: body 1.6 - 2.3 mm, wing 2.1 - 2.3 mm.

♀. Larger; with broader and more blunt-tipped palpi (Fig. 111); mesonotum (Fig. 115) also polished on a broad median stripe right up to front; apical two abdominal segments and rather short but slender cerci very grey dusted.

Length: body 2.4 - 2.7 mm, wing 2.3 - 2.4 mm.

Rare. Denmark: SJ, Sönderborg, ♂ (Wüstnei); F, Lohals, ♀ (holotype of thoracica) (Lundbeck); Sweden: Sk., Lund, ♂ (Zetterstedt). - England, C.Europe. - June-August. Uncommon everywhere, on leaves of bushes and on ground-vegetation.

Note. At least two very closely related species may well be found in the south of Fennoscandia. P.politus (Collin, 1926), known from England and also from Spain, has a narrower and apically more darkened antennal segment 3, more extensively polished pleura, particularly on upper part of prothorax and most of humeri, and mesonotum in ♀ almost entirely polished black. P.pseudociliaris (Strobl, 1910) (syn. calcaratus Collin, 1926), known from England, Estonia and C.Europe, has a longer and brownish antennal segment 3, small palpi, frons and whole of mesonotum polished, all thoracic hairs and bristles dark, and pleura dusted except for polished sternopleura.

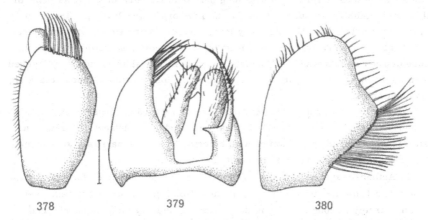

378   379   380

Figs. 378-380. Male genitalia of Platypalpus ruficornis (von Roser). - 378: right periandrial lamella; 379: periandrium with cerci; 380: left periandrial lamella. Scale: 0.1 mm.

41. PLATYPALPUS FENESTELLA Kovalev, 1971
Figs. 105, 108, 113.

Platypalpus fenestella Kovalev, 1971: 204.

Resembling ruficornis but pleura dusted except for sternopleura, mid legs only slightly thickened with tibial spur smaller and blunt in ♂, and fore tarsi with sharp dark annulations in ♂. Mesonotum with a large polished patch in posterior half and another small one in front between humeri in both sexes.

♂. Head densely light grey dusted with face distinctly narrower than frons in front and cheeks shining. Antennae (Fig. 105) pale yellow with arista brownish and about 2.5 times as long as segment 3. Palpi (Fig. 108) pale yellow, large and broad but considerably smaller than in ruficornis, with 2 long whitish hairs at tip.

Thorax (Fig. 113) densely whitish to silvery-grey dusted even on pleura, but sternopleura largely polished; mesonotum with a large polished black patch in posterior third, broadly separated from scutellum, and another smaller patch in front between humeri; sometimes with a slight indication of a narrow polished line on acr; scutellum narrowly polished anteriorly. Acr narrowly irregularly biserial, dc uniserial with numerous hairs at sides and 2 prescutellar pairs longer; all pale, fine but distinct, disappearing on polished areas. Large bristles yellowish, rather a distinct curved humeral bristle.

Legs pale yellow including knees, fore tarsi with sharp blackish-brown annulations, posterior tarsi with only faint brown annulations or practically yellow. Fore femora thickened with long pale bristly-hairs in two rows beneath, tibiae very indistinctly spindle-shaped. Mid femora scarcely deeper than fore femora, the pale pv bristles not very long, and also with several shorter bristly-hairs anteroventrally; the ventral spines in two rows, somewhat yellowish, more distinct on apical half of femur. Mid tibia rather slender, not as thickened as in ruficornis, apical spur shorter, at most as long as tibia is deep, blackish, and blunt apically.

Wings clear or faintly pale yellowish, veins very pale; crossveins very narrowly separated, the vein closing anal cell slightly recurrent. Squamae and halteres whitish-yellow. Abdomen with anterior 3 tergites narrowly grey dusted at sides, genitalia not broader than end of abdomen.

♀. Antennal segment 3 orange-yellow to somewhat brownish, fore tarsi with rather faint brownish annulations, and apical tibial spur sharply pointed, even though scarcely as long as tibia is deep; acr irregularly almost quadriserial (in the Finnish specimen).

Length (for both sexes): body 1.8 - 2.6 mm, wing 2.3 - 2.6 mm (Kovalev, 1971).

Very rare. Finland: St, Reposaari, ♀ (Lauro). - Rare in European USSR (Leningrad, Moscow and Kaluga regions). - June-August. The Finnish specimen was taken at light.

Note. Closely related to ruficornis and particularly to P.ingenuus (Collin, 1926), which is known from England and C.Europe; both fenestella and inge-

<u>nuus</u> have the same structure and colour of legs, and the same thoracic pubes-
cence, but <u>ingenuus</u> is the only species of this complex with strikingly large vt
bristles, extensively darkened antennae even on basal segments, almost entirely polished pleura, and mesonotum polished right up to the front in both sexes;
<u>fenestella</u> undoubtedly has much in common with <u>pallidicornis</u>, with which it
was compared by Kovalev (1971).

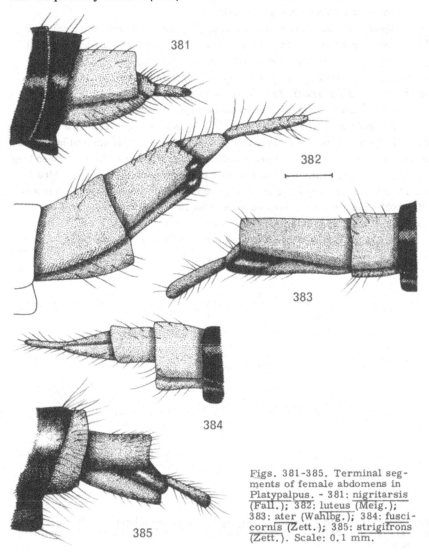

Figs. 381-385. Terminal segments of female abdomens in
Platypalpus. - 381: <u>nigritarsis</u>
(Fall.); 382: <u>luteus</u> (Meig.);
383: <u>ater</u> (Wahlbg.); 384: fuscicornis (Zett.); 385: <u>strigifrons</u>
(Zett.). Scale: 0.1 mm.

42. PLATYPALPUS ATER (Wahlberg, 1844)

Figs. 107, 203, 383, 386-388, 706.

Tachydromia atra Wahlberg, 1844: 106.

A larger polished black species with shining frons and vertex, short black antennae and legs uniformly blackish; tibial spur large and pointed. Mesonotum mostly polished with pale hairs and bristles.

♂. Head thinly dark grey dusted on occiput, but frons and vertex polished black; frons broad and parallel-sided, densely silvery-grey dusted face slightly narrower above. A pair of vt bristles rather close together, darkened. Antennae (Fig. 107) black, segment 3 slightly longer than deep, arista almost 3 times as long. Palpi small, dark brown with silver pile.

Thorax polished black on mesonotum except for humeri, narrow side margins and scutellum, pleura rather densely grey dusted but mesopleura and sternopleura largely polished. All hairs and bristles pale, biserial acr and uniserial dc fine but distinct, 2 pairs of prescutellar dc and a humeral bristle rather long.

Legs uniformly rather shining black but all tibiae and often tips of hind femora translucent brownish. Fore femora only slightly thickened, mid femora (Fig. 203) almost twice as deep with short black ventral spines and not very long pale pv bristles; tibial spur very strong and pointed.

Wings (Fig. 706) clear with dark veins, vein M slightly undulating before

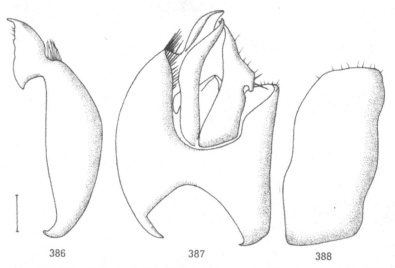

386   387   388

Figs. 386-388. Male genitalia of Platypalpus ater (Wahlbg.). - 386: right periandrial lamella; 387: periandrium with cerci; 388: left periandrial lamella. Scale: 0.1 mm.

tip, crossveins not widely separated and 2nd basal cell distinctly longer; the vein closing anal cell at right-angles to vein Cu, vein A abbreviated on basal half. Squamae and halteres yellowish. Abdomen polished black with scattered minute pale hairs; genitalia (Figs. 386-388) large and very long, polished black; cerci very strong and curved apically, overlapping lamellae.

Length: body 2.0 - 2.6 mm, wing 2.5 - 3.0 mm.

♀. Resembling male but lower part of frons and whole of occiput always more densely grey dusted. Abdomen shining but apical two segments (Fig. 383) and cerci densely grey dusted, sternite 8 polished black apically.

Length: body 2.3 - 3.2 mm, wing 2.5 - 3.1 mm.

Rather common in extreme north of Fennoscandia; in Norway in Fø, in Sweden S to P.Lpm., in Finland south rarely down to N; common in the Kola Peninsula. - N and NW of European USSR (S to the Leningrad region), ? Siberia. The record from the Alps (Engel, 1939, and followed by Frey, 1943 and Ringdahl, 1951) is obviously an error. - July. On leaves of Caltha in moist biotopes (Wahlberg, 1844).

43. PLATYPALPUS MINUTUS (Meigen, 1804)
Figs. 103, 206, 207, 389-391, 707.

Tachydromia minuta Meigen, 1804: 238.
Tachydromia exigua Meigen, 1822: 81.
Tachydromia pygmaea Macquart, 1823: 154.

Medium-sized species with mesonotum thinly greyish dusted, leaving a narrow median stripe polished black, pleura with silver pile. Legs yellowish with coxae and femora except for both tips black, fore tibiae with a rim-like projection at tip beneath, tibial spur large. Antennae black, a pair of dark vt bristles and a very fine humeral bristle.

♂. Head dark grey dusted, frons rather broad and widening above, face narrower and covered with dense silvery pile. A pair of dark widely separated vt bristles, antennae (Fig. 103) black with segment 3 about twice as long as deep, arista not very much longer. Palpi blackish with silver pile.

Thorax very thinly greyish dusted on mesonotum with a narrow median line (widened posteriorly) and sides of humeri polished black; pleura covered with silver pile, sternopleura largely polished. Acr narrowly biserial, dc uniserial with numerous hairs at side, all pale and rather small. Large bristles strikingly darkened including a strong pair of prescutellar dc, but humeral bristle small and very fine.

Legs yellow and black in colour, covered with rather long and densely-set

153

pale hairs; coxae and femora (Figs. 206, 207) mostly black but fore coxae at tip and all femora on both tips yellow, hind femora very broadly so on basal half; mid tibiae (Fig. 206) often brownish and all tarsi with distinct black annulations. Fore tibiae (Fig. 207) with a blackish rim-like projection at tip beneath; mid femora (Fig. 206) slightly deeper than fore femora, the short ventral black spines becoming pale and hair-like on basal third, a few long pv bristles almost whitish. Apical tibial spur (Fig. 206) very long and sharply pointed, blackish.

Wings (Fig. 707) clear with dark veins, crossveins widely separated, and the vein closing anal cell at right-angles to vein Cu. Squamae darkened, halteres pale. Abdomen entirely polished black with scattered pale hairs; genitalia (Figs. 389-391) rather smaller and shining except for dull grey short cerci.

Length: body 2.2 - 2.5 mm, wing 2.4 - 2.7 mm.

♀. Legs usually with shorter pale hairs, and somewhat brownish on yellow parts. Abdomen almost bare, apical two segments and cerci small, densely grey dusted.

Length: body 2.0 - 2.5 mm, wing 2.4 - 2.7 mm.

Very common throughout Denmark and Fennoscandia except for the extreme northern regions; in Norway north to TRy, in Sweden to Lu.Lpm. and Nb., in Finland to Lk; not yet recorded from the Kola Peninsula. - Very common and widespread in Europe but the records from the south need verification. - Mid May-mid October. Often in large numbers on bushes and vegetation, but quite frequently on the leaves of trees and on flowering shrubs, and also on flowers (Tuomikoski, 1952).

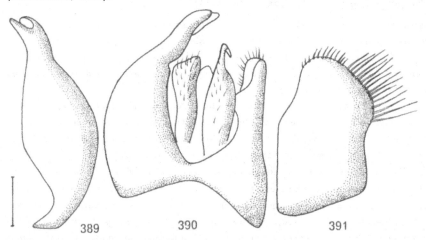

Figs. 389-391. Male genitalia of Platypalpus minutus (Meig.). - 389: right periandrial lamella; 390: periandrium with cerci; 391: left periandrial lamella. Scale: 0.1 mm.

154

Note. The very closely allied P.albifacies (Collin, 1926), described from England but fairly common in C.Europe, has the mesonotum entirely thinly dusted, femora entirely black and no rim-like projection on fore tibiae beneath. There are several species superficially resembling minutus, viz., P.montanus (Beck.), P.nigrinus (Meig.), P.tergestinus (Egger), P.apicalis (Beck.), P.argenteomicans (Beck.), P.eumelaenus (Mik), P.desertorum (Beck.), P.ostiorum (Beck.) and P.pedestris (Beck.), but they all have well-developed vt and humeral bristles, and the mesonotum is usually densely dusted. P.obscuripes (Strobl, 1899), described as a variety of minutus, is a distinct species, not a dark-legged form of minutus as stated by Collin (1961) or identical with albifacies as stated by Frey (1943).

44. PLATYPALPUS NIGER (Meigen, 1804)
Figs. 106, 204, 392-394, 708.

Tachydromia nigra Meigen, 1804: 238.
Tachydromia geniculata Fallén, 1815: 7.
Tachydromia femoralis Zetterstedt, 1842: 299.
Tachydromia exigua Meigen; Lundbeck, 1910: 310.
Platypalpus doormani Theowald, 1962: 192, syn.n.

A smaller species resembling minutus but femora only yellow broadly at base, tibiae and tarsi uniformly brownish, and no rim-like projection on fore tibiae. Mesonotum polished black or thinly dusted to a varying extent, pleura extensively polished.

♂. Head as in minutus but frons almost parallel and subshining above, a pair of widely separated vt bristles blackish. Antennae (Fig. 106) black with segment 3 small, slightly longer than deep, arista about twice as long.

Thorax entirely polished black on mesonotum including humeri or mesonotum uniformly thinly grey dusted (so-called "Continental" form after Collin (1961)), with all intermediate forms; pleura always extensively polished black. Acr and dc minute and inconspicuous, last prescutellar pair of dc only slightly longer; large bristles blackish.

Legs (Fig. 204) extensively blackish, but fore coxae except for base and broad bases on all femora yellowish, tibiae and tarsi brown to blackish-brown, latter not annulated. Fore tibiae without a rim-like projection at tip beneath; mid femora (Fig. 204) with the black ventral spines in two rows, short right up to base, pv bristles less numerous and very long.

Wings (Fig. 708) sometimes slightly brownish on costal half, veins dark;

veins R4+5 and M indistinctly diverging. Abdomen including genitalia (Figs. 392-
394) polished black and almost bare, genitalia small.

Length: body 1.4 - 1.9 mm, wing 1.8 - 2.1 mm.

♀. Resembling male; apical two abdominal segments very small, usually
concealed in the larger segment 6, and densely grey dusted like the long slender
cerci.

Length: body 1.4 - 2.0 mm, wing 2.0 - 2.2 mm.

Uncommon in Denmark and S.Sweden, Sk. - England, The Netherlands,
NW and C parts of European USSR, C.Europe. - June-mid August. In temperate
zones; rather common on vegetation, on large leaves of Petasites and Arctium,
and often on leaves of trees, particularly on fruit-trees.

Note. A variable species in the covering of greyish dust on mesonotum and
frons; var.doormani only represents an intermediate form between specimens
with the mesonotum entirely polished (Atlantic form) and evenly thinly dusted
specimens (Continental form). P.cinereovittatus (Strobl, 1899) is a closely related but distinct species.

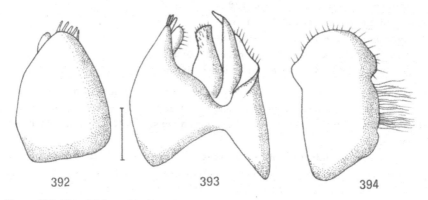

Figs. 392-394. Male genitalia of Platypalpus niger (Meig.). - 392: right periandrial lamella; 393: periandrium with cerci; 394: left periandrial lamella.
Scale: 0.1 mm.

IX. pallidiventris - cursitans - group

45. PLATYPALPUS AENEUS (Macquart, 1823)
Figs. 120, 208, 395-397, 709.

Tachydromia aenea Macquart, 1823: 153.
Tachydromia aeneicollis Zetterstedt, 1849: 3008.

Extensively black species, mesonotum subshining metallic black, large bristles yellowish. Frons very narrow; legs blackish-brown with anterior four femora and fore tibiae thickened, tibial spur long, slender.

♂. Head greyish dusted, frons and face conspicuously narrow, former as deep below as front ocellus, slightly widening above. A pair of vt bristles rather small and close together, yellowish-brown. Antennae (Fig. 120) uniformly blackish-brown, segment 3 almost twice as long as deep, arista more than twice as long. Palpi dark, very small.

Thorax subshining black on mesonotum, with a slight metallic tinge, pleura greyish dusted, sternopleura mainly polished. Biserial acr and uniserial dc (becoming longer posteriorly) rather long, pale. Large bristles pale yellowish-brown.

Legs blackish-brown with rather long pale hairs, knees somewhat yellowish, and tibiae and basal segment of tarsi translucent brownish. Fore tibiae spindle-shaped dilated, mid femora (Fig. 208) very thickened but not much deeper than rather thick fore femora, pv bristles pale, long. Tibial spur (Fig. 208) dark, slender and sharply pointed.

Wings (Fig. 709) often slightly yellowish-brown tinged, veins dark. Vein M slightly bowed at middle, crossveins widely separated , the vein closing anal cell recurrent and S-shaped. Squamae brown, knob of halteres pale yellow. Abdomen shining blackish-brown; genitalia (Figs. 395-397) small.

Length: body 2.1 - 2.4 mm, wing 2.4 - 2.6 mm.

♀. Resembling male, but fore tibiae less thickened and with shorter hairs. Abdomen polished with scattered and shorter pale hairs, apical two segments and cerci greyish dusted.

Length: body 2.2 - 2.5 mm, wing 2.4 - 2.6 mm.

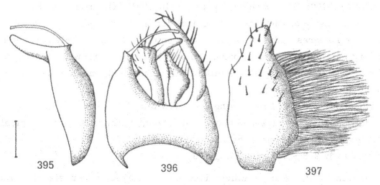

Figs. 395-397. Male genitalia of Platypalpus aeneus (Macq.) - 395: right periandrial lamella; 396: periandrium with cerci; 397: left periandrial lamella. Scale: 0.1 mm.

Rare. Sweden: Gtl., Näher, Gothem, 3 ♂ 3 ♀ (Boheman). - England, C. Europe, France; rare everywhere. - June-July.

46. PLATYPALPUS MACULIPES (Meigen, 1822)
Figs. 116, 209, 398-400, 710.

Tachydromia maculipes Meigen, 1822: 79.

A species with 2 pairs of dark vt bristles, rather narrower frons and face, antennae black with long slender segment 3 and subequal arista. Mesonotum rather thinly brownish dusted, legs yellow.

♂. Head rather densely light grey dusted; frons rather narrower, about as deep as antennal segment 2 and parallel; face still narrower, clypeus and narrow jowls black. 2 pairs of vt bristles darkened. Antennae (Fig. 116) black, segment 3 about 3 times as long as deep and conspicuously slender, arista subequal in length. Palpi small, dark.

Thorax rather thinly light brownish-grey dusted on mesonotum, pleura more silvery-grey, sternopleura with a bare polished patch anteriorly. Acr and dc brownish, inconspicuous; former irregularly 4-serial at least in front, posteriorly almost biserial in two widely separated rows, dc uniserial. Large bristles obviously pale.

Legs yellow, long and slender; base of posterior four coxae, knees and tips of basal two tarsal segments darkened, apical three tarsal segments almost blackish. Fore femora slightly thickened, with very long pale hairs beneath, fore tibiae somewhat dilated and with long hairs everywhere. Mid femora (Fig. 209) slightly deeper than fore femora, the ventral black spines absent on basal third, pv bristles pale, very long. Tibial spur (Fig. 209) almost black, long and pointed.

Wings (Fig. 710) clear with dark veins, vein M slightly undulating before tip, crossveins contiguous, the vein closing anal cell at right-angles to vein Cu. Squamae and halteres rather pale. Abdomen shining black, densely covered by rather long pale hairs, sides of basal two segments and narrow fore-margin of anterior 4 segments greyish. Genitalia (Figs. 398-400) rather small, polished, but cerci very dull grey.

Length: body 2.0 - 2.4 mm, wing 3.0 - 3.3 mm.

♀. Mid femora often darkened above, and abdomen almost bare with apical two segments and rather small cerci greyish dusted.

Length: body 2.3 - 2.7 mm, wing 3.0 - 3.3 mm.

Common in Denmark and southern Fennoscandia including Russian Carelia, rare towards the north; in Sweden to Ly. Lpm. and Nb., in Finland to Om. - Widespread in Europe, south to Italy and Spain. - Adults may be found in any month of the year, and at least some of them hibernate; mainly August-November but individually (even in Fennoscandia) also in February, March, June and July.

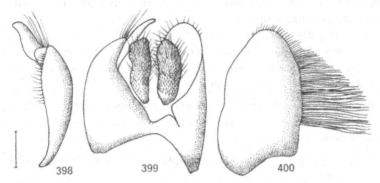

Figs. 398-400. Male genitalia of Platypalpus maculipes (Meig.). - 398: right periandrial lamella; 399: periandrium with cerci; 400: left periandrial lamella. Scale: 0.1 mm.

47. PLATYPALPUS RAPIDUS (Meigen, 1822)
   Figs. 121, 210, 401-403, 711.

Tachydromia rapida Meigen, 1822: 81.
Platypalpus mundus Walker, 1837: 228.
Tachydromia diversipes Strobl, 1910: 84, syn.n.

A medium-sized species with 2 pairs of black vt bristles resembling agilis but antennal segment 3 very broad, acr 4-serial, hypopleura partly polished, and legs very thickened on mid femora.

♂. Head darker grey dusted, frons rather broad and widening above, face as deep as frons in front; clypeus and narrow jowls subshining. 2 pairs of vt bristles black. Antennae (Fig. 121) blackish-brown, segment 3 very conspicuously broad, almost spherical, arista more than twice as long. Palpi small, brownish.

Thorax rather thinly dark grey dusted on mesonotum, more greyish on pleura but most of sternopleura and a small central area of hypopleura polished black. Acr 4-serial on a broad median stripe, dc biserial, all black like the large bristles.

Legs yellow, but black on posterior four coxae and trochanters, whole of mid femora (Fig. 210) except for narrow tip and on apical half of hind femora; apical tarsal segment darkened; fore tarsi with conspicuously enlarged black claws and pale pulvilli. Fore femora rather slender, mid femora (Fig. 210) very thickened, pv bristles long, black. Tibial spur (Fig. 210) large and pointed, fore and hind tibiae with dark bristly-hairs above.

159

Wings (Fig. 711) very weakly yellowish-brown clouded, veins brown; veins R4+5 and M almost parallel. Squamae brown, halteres whitish-yellow. Abdomen shining black with sparse fine pale hairs; genitalia (Figs. 401-403) small, polished black, lamellae with only fine pale hairs.

Length: body 2.0 - 2.2 mm, wing 2.5 mm.

♀. Abdomen with longer and more densely-set pale hairs, apical two segments small, usually hidden within segment 6, and greyish dusted like the small cerci.

Length: body 2.1 - 2.3 mm, wing 2.5 - 2.6 mm.

Rare. Finland: Ta, Forssa, 2 ♀ (Käpylä); Jokioinen, ♀ (Väre). - Widespread in Europe but rather rare everywhere; England, C.Europe, France. - April-early September.

Note. P.diversipes (Strobl, 1910), of which I have studied the type specimens, is identical with rapidus.

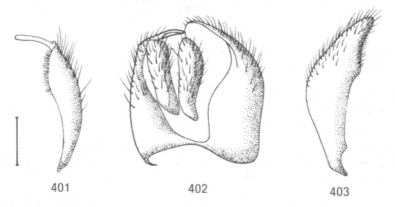

401   402   403

Figs. 401-403. Male genitalia of Platypalpus rapidus (Meig.). - 401: right periandrial lamella; 402: periandrium with cerci; 403: left periandrial lamella. Scale: 0.1 mm.

48. PLATYPALPUS PALLIDICOXA (Frey, 1913)
Figs. 117, 211, 404-406, 712.

Tachydromia flavipes var.pallidicoxa Frey, 1913: 80.
Tachydromia agilella Collin, 1926: 153.

A smaller species with 2 pairs of vt bristles and biserial acr resembling agilis, but legs yellow in both sexes with posterior four coxae brownish and all tarsi with faint annulations; hypopleura polished.

♂. Head grey dusted with rather broad parallel frons and much narrower face, clypeus dull grey but narrow jowls shining. 2 pairs of black vt bristles. Antennae (Fig. 117) black to blackish-brown, segment 3 rather narrow, ovate, almost twice as long as deep, arista twice as long.

Thorax thinly dark grey dusted on mesonotum with inner edge of humeri shining; pleura lighter grey dusted, subshining in central areas of meso- and pteropleura, sternopleura and hypopleura largely polished. Narrowly biserial acr and uniserial dc dark and hair-like, latter with 5 larger bristles in each row, last pair very large, other large bristles black.

Legs yellow with posterior four coxae brownish and all tarsi with faint brown annulations, apical two segments and knees darkened.. Fore femora thickened towards base and fore tibiae distinctly spindle-shaped dilated. Mid femora (Fig. 211) remarkably thickened, the black ventral spines equal in length through-out, long pv bristles pale; apical tibial spur (Fig. 211) sharply pointed, large, black towards tip.

Wings (Fig. 712) practically clear, crossveins rather separated, and the vein closing anal cell often slightly recurrent. Abdomen shining black; genitalia (Figs. 404-406) small with lamellae dull towards tip.

Length: body 1.7 - 2.2 mm, wing 2.0 - 2.3 mm.

♀. Resembling male, but fore tibiae somewhat less dilated. Abdomen almost bare with apical two segments and cerci dull grey.

Length: body 1.8 - 2.3 mm, wing 2.1 - 2.3 mm.

Uncommon in southern parts of Sweden and Finland to approximately 62° north; also in Russian Carelia. - Widespread in Europe but never common; Scotland, NW of European USSR (Leningrad region, Estonia) and C.Europe (Czechoslovakia, Austria). - End May-July.

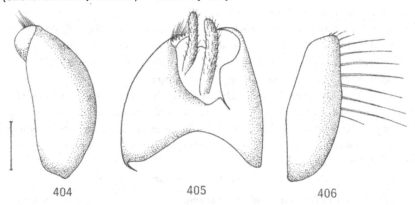

404   405   406

Figs. 404-406. Male genitalia of Platypalpus pallidicoxa (Frey). - 404: right periandrial lamella; 405: periandrium with cerci; 406: left periandrial lamel-la. Scale: 0.1 mm.

49. PLATYPALPUS AGILIS (Meigen, 1822)
Figs. 51, 118, 212, 407-409, 713.

Tachydromia agilis Meigen, 1822: 80.
Platypalpus dubius Walker, 1851: 132.
Tachydromia agilis var.hybrida Frey, 1907: 411.
Coryneta maculipes - nigrosetosa Strobl; Engel, 1939: 83.

A robust, black bristled species with 2 pairs of vt bristles and small blackish antennae. Legs yellow in ♂, with posterior four coxae black and annulated tarsi, extensively blackish in ♀.

♂. Head thinly dark grey dusted with rather broad frons and narrower face; 2 pairs of vt bristles black. Antennae (Fig. 118) small, dark; segment 3 narrowly ovate, pointed at tip, almost twice as long as deep, arista about 1.5 times as long.

Thorax very thinly dark grey dusted on mesonotum with humeri almost shining black, pleura lighter grey dusted with only sternopleura largely polished. All hairs and bristles black; acr closely biserial, small; dc uniserial and becoming longer posteriorly, last 2 pairs large.

Legs extensively yellow with rather short pale hairs, but posterior four coxae and trochanters black, tarsi with distinct dark annulations; fore coxae at base and mid femora dorsally often darkened. Fore femora and tibiae rather slender; mid femora (Fig. 212) very thickened, the pale pv bristles strikingly short; tibial spur (Fig. 212) large, black. In addition to dark bristly-hairs on fore and hind tibiae above, and to black bristles on apical third of mid femora anteriorly, there is a row of distinct dark anteroventral bristly-hairs on hind femora. Rarely with legs extensively blackish as in ♀.

Wings (Fig. 713) practically clear with dark veins, 2nd basal cell distinctly longer. Abdomen shining black but sides of basal two segments and narrow anterior margins dull grey. Genitalia (Figs. 407-409) large, polished black.

Length: body 2.4 - 2.9 mm, wing 2.7 - 3.2 mm.

♀. Legs extensively darkened to almost blackish, and the dark anteroventral ciliation on hind femora absent; pv bristles on mid femora darkened; females with extensively yellow legs and with the same bristling as in the ♂ (var. hybrida) are sometimes to be found. Apical two abdominal segments elongated, segment 8 often subshining black.

Length: body 2.6 - 3.8 mm, wing 2.8 - 3.4 mm.

Very common in Denmark and the extreme south of Fennoscandia, rare towards the north: in Sweden to Dlr., in Finland to Ta; not yet recorded from Norway. - Widespread and common in Europe except for southern areas.

162

May-July; a typical early spring species, but a male taken in September in Estonia may well represent a second generation. On ground-vegetation, even in sunny drier biotopes, often in large numbers.

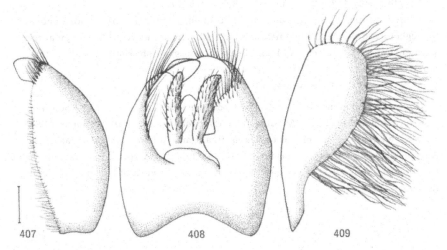

Figs. 407-409. Male genitalia of Platypalpus agilis (Meig.). - 407: right periandrial lamella; 408: periandrium with cerci; 409: left periandrial lamella. Scale: 0.1 mm.

50. PLATYPALPUS PSEUDORAPIDUS Kovalev, 1971
    Figs. 119, 213, 410-412, 714.

Platypalpus rapidus Meigen; Frey, 1943: 9, 16.
Platypalpus pseudorapidus Kovalev, 1971: 209.

Resembling agilis but smaller, legs yellow in both sexes with a constant black pattern (a dorsal stripe on mid femora and apical half of hind femora black); only fine hairs on hind femora beneath, and humeri dull grey.

♂. Head as in agilis with 2 pairs of black vt bristles but antennae (Fig. 119) with longer arista, at least twice as long as segment 3. Mesonotum thinly dark grey dusted even on humeri, and only last pair of prescutellar dc large.

Legs yellow with posterior four coxae and trochanters, an ovate patch on dorsum of mid femora (Fig. 213) (or covering the whole dorsum) and apical half of hind femora blackish; tarsi with distinct dark annulations. Fore femora with longer pale hairs beneath, mid femora (Fig. 213) much deeper than fore femora, anteriorly with a single black bristle on apical third and pale pv bristles longer than in agilis. Hind femora with finer anteroventral bristly-hairs, becoming black towards base.

163

Wings (Fig. 714) clear with brown veins, veins R4+5 and M slightly diverging just before tip. Abdomen usually entirely polished black with rather dense short pale hairs; genitalia (Figs. 410-412) rather small.

Length: body 1.7 - 2.3 mm, wing 2.1 - 2.6 mm.

♀. Resembling male, but anteroventral bristly-hairs on hind femora pale throughout and thinner. Abdomen shining with dorsum almost bare; apical two segments small, tergites and long slender cerci greyish dusted but sternites mostly polished.

Length: body 2.0 - 2.6 mm, wing 2.2 - 2.7 mm.

Common in Finland north to Ks, but not found west of the Gulf of Bothnia; also in Russian Carelia. - A North East European species, in European USSR (Leningrad region, Estonia) E as far as Ural, and in mountainous areas of C. Europe (Czechoslovakia, Austria). - May-early July. A spring species, on trees and bushes, also in grasses near the coast and on flowers of Sorbus and Prunus (Kovalev, 1971).

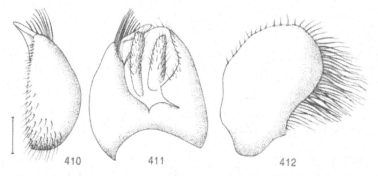

Figs. 410-412. Male genitalia of Platypalpus pseudorapidus Kov. - 410: right periandrial lamella; 411: periandrium with cerci; 412: left periandrial lamella. Scale: 0.1 mm.

51. PLATYPALPUS NIGROSETOSUS (Strobl, 1893)
    Figs. 124, 214, 413-415, 715.

Tachydromia nigrosetosa Strobl, 1893: 117.

Medium-sized black bristled species with 2 pairs of vt bristles and broadly separated biserial acr; palpi small, dark; antennae blackish. Legs yellow including posterior four coxae.

♂. Head rather densely grey dusted, frons narrower, about as deep as antennal segment 2, parallel; face much narrower than frons, clypeus black. 2 pairs of vt bristles black. Antennae (Fig. 124) black with segment 3 triangular in shape, slightly more than twice as long as deep, arista less than twice as long. Palpi very dark brown, small and narrow, covered with silver pile and a long dark bristle at tip.

Thorax with mesonotum more thinly dusted than pleura, sternopleura largely polished. All hairs and bristles black; acr and dc hair-like but fairly long; acr in two widely separated rows (as in interstinctus), the distance between the two rows of acr equalling that between acr and dc; latter uniserial with 2 prescutellar pairs longer.

Legs uniformly yellow to yellowish-brown, tarsi with narrow dark annulations, apical two segments and knees almost black. Fore femora thickened on basal two-thirds, with long pale hairs beneath that are also present on slender fore tibiae and metatarsi. Mid femora (Fig. 214) not much more thickened, pale pv bristles rather short, a distinct black bristle anteriorly on apical third; tibial spur (Fig. 214) strong, sharply pointed. Hind femora with long pale hairs anteroventrally.

Wings (Fig. 715) clear with brownish veins, veins R4+5 and M slightly bowed, converging towards tip, crossveins usually contiguous. Squamae light brown, halteres whitish-yellow. Abdomen shining black except for sides of segment I, covered with conspicuously long and dense pale hairs. Genitalia (Figs. 413-415) polished and rather large, lamellae fringed with long yellowish hairs.

Length: body 1.8 - 2.4 mm, wing 2.3 - 2.8 mm.

♀. Legs uniformly yellow, not brownish; mid femora more thickened; and front pair in particular with shorter pale hairs on tibiae. Abdomen with only scattered pale hairs, apical two segments and cerci dull.

Length: body 2.0 - 2.8 mm, wing 2.4 - 3.0 mm.

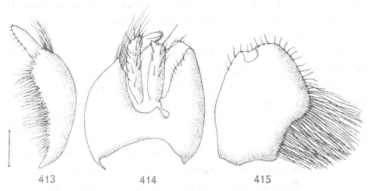

Figs. 413-415. Male genitalia of Platypalpus nigrosetosus Strobl. - 413: right periandrial lamella; 414: periandrium with cerci; 415: left periandrial lamella. la. Scale: 0.1 mm.

165

Rather rare. Norway: Ak, Toyen, ♀ (Siebke); Finland: Alandia, ♀ (Palmén); Ab, Pargas, ♂ (Ingelius), Hangö, ♀ (Lindberg); N, Esbo, ♀ (Hackman), Tvärminne, ♀ (Lindberg). - NW of European USSR (Ib, Leningrad region, Estonia) and C. Europe (Austria). - Mid June-August.

Note. P. balticus Kovalev, 1971, described recently from the Leningrad region and Estonia, and also known to me from Czechoslovakia, Austria and Switzerland, has the antennae smaller (Fig. 123), much broader frons, larger and pale yellow coloured palpi, large bristles on head and thorax usually paler, acr closely biserial, and legs clear yellow with more slender mid femora; the male genitalia of balticus (Kovalev,1971: 207) differ only slightly from those of nigrosetosus. P. nigrosetosus of Engel (1939: 83) and Frey (1943: 10) is only agilis.

52. PLATYPALPUS COTHURNATUS Macquart, 1827
Figs. 122, 215, 416-418, 716.

Platypalpus cothurnatus Macquart, 1827: 100.
Tachydromia socculata Zetterstedt, 1838: 550.

A smaller species with 1 pair of dark vt bristles and black antennae. Legs yellow with tibial spur short and blunt, about as long as tibia is deep, in ♂ with a tiny spine at tip.

♂. Head densely grey dusted, frons rather broader and parallel, face much narrower with clypeus black. A pair of dark vt bristles, more widely separated. Antennae (Fig. 122) black, segment 3 rather slender, about twice as long as deep, arista twice as long; palpi small, pale.

Thorax rather darker grey dusted on mesonotum, paler on pleura, sternopleura with a bare polished patch. Biserial acr somewhat diverging and, like the uniserial dc (except for last two pairs), numerous and equally short, yellowish-brown like large bristles.

Legs yellow with apical two tarsal segments very darkened, tarsi sometimes with very indistinct annulations. Mid femora (Fig. 215) very slightly thickened, anteriorly with a row of longer brown bristly-hairs, pv bristles pale, rather small, fine and densely-set. Fore and hind tibiae with several dark bristly-hairs on dorsum. Tibial spur (Fig. 215) blunt with a tiny curved spine at tip, almost as long as tibia is deep.

Wings (Fig. 716) slightly yellowish-brown tinged, veins R4+5 and M parallel, crossveins distant, 2nd basal cell longer; the vein closing anal cell almost at right-angles to vein Cu. Abdomen subshining black, basal two segments broadly grey at sides; pubescence pale, dense. Genitalia (Figs. 416-418) small, polished black, left lamella fringed with long pale hairs.

Length: body 1.7 - 2.1 mm, wing 2.3 - 2.4 mm.

♀. Apical tibial spur blunt and trowel-like, without a tiny apical spine. Abdomen almost bare and more shining, apical two segments and very long slender cerci densely brownish-grey dusted.

Length: body 2.0 - 2.5 mm, wing 2.3 - 2.5 mm.

Rather uncommon in Denmark and southern parts of Fennoscandia; in Norway north to Ak, in Sweden to Sdm., in Finland along the Baltic coast north to Ta. - Widespread in Europe including the south, but never common. - May-July. On ground-vegetation and on bushes.

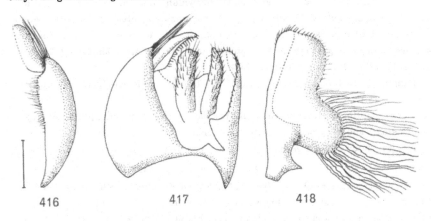

Figs. 416-418. Male genitalia of Platypalpus cothurnatus Macq. - 416: right periandrial lamella; 417: periandrium with cerci; 418: left periandrial lamella. Scale: 0.1 mm.

53. PLATYPALPUS CRYPTOSPINA (Frey, 1909)

Figs. 125, 216, 419-421, 717.

Tachydromia cryptospina Frey, 1909: 8.
Tachydromia tantula Collin, 1926: 158.

Resembling cothurnatus but antennae with smaller segment 3, dc longer and less numerous, and legs with dark annulated tarsi, very spindle-shaped fore tibiae and much shorter tibial spur.

♂. Head with narrower frons scarcely as deep as antennal segment 2, and still narrower and more silvery-grey face. A pair of dark widely separated vt bristles, antennae (Fig. 125) black with segment 3 shorter and more ovate, slightly longer than deep, arista more than twice as long. Palpi pale yellow, somewhat larger and ovate.

Thorax thinly brownish-grey dusted on mesonotum, slightly subshining, with small hairs brownish; acr small and closely biserial, dc uniserial, longer and bristle-like, only few in number (5-6) with last pair strong.

Legs yellow with posterior four coxae often slightly darkened as in cothurna-tus but tarsi with distinct dark annulations which are broader towards tip, anterior femora of the same thickness, and fore tibiae conspicuously spindle-shaped dilated. Mid femora (Fig. 216) with long pale pv bristles, anteriorly on apical third with a long pale spine-like bristle. Tibial spur (Fig. 216) very small, much shorter than tibia is deep.

Wings (Fig. 717) clear or very faintly brownish clouded, veins R4+5 and M inconspicuously but evenly diverging, crossveins separated by a distance equal to the length of mid crossvein. Abdomen shining black, segment 1 and sides of segment 2 densely greyish dusted, pubescence short, pale. Genitalia (Figs. 419-421) rather larger, polished black.

Length: body 1.0 - 1.3 mm, wing 1.5 - 1.6 mm (holotype body 1.2 mm, wing 1.6 mm).

♀. Resembling male, abdomen shorter haired at sides, and apical two segments small and greyish dusted like the cerci.

Length: body 1.2 - 1.6 mm, wing 1.6 - 1.7 mm.

Rare. Sweden: Öl., Långlöt, 2 ♂ (Hedström); Finland: Ab, Karislojo, ♂ (Frey, holotype). - England, NW and C parts of European USSR, Caucasus, Czechoslovakia, Austria, ? Jugoslavia. - June-July. On ground-vegetation; more common on short herbage than on leaves of bushes (Collin, 1961).

Note. The closely related P. aliterolamellatus Kovalev, 1971, known from the north of the USSR (Leningrad region) and Austria (Styria), has entirely yel-

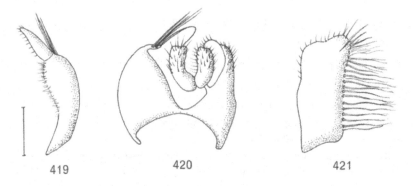

419   420   421

Figs. 419-421. Male genitalia of Platypalpus cryptospina (Frey). - 419: right periandrial lamella; 420: periandrium with cerci; 421: left periandrial lamella. Scale: 0.1 mm.

low tarsi without annulations with at most apical two segments slightly darkened; biserial acr wider apart; crossveins on wings more distant; and different hairing on left lamella of male genitalia (Kovalev, 1971: 201, Fig. 1). P.pseudomaculipes (Strobl, 1899) is another related species.

54. PLATYPALPUS OPTIVUS (Collin, 1926)
Figs. 126, 127, 217, 422-424, 718.

Tachydromia optiva Collin, 1926: 157.

A larger species with 1 pair of vt bristles, long black antennae and all bristles on head and thorax black. Legs yellow with annulated tarsi, tibial spur large but somewhat blunt; acr quadriserial.

♂. Head rather densely grey dusted with broader frons widening above, face scarcely narrower but more silvery-grey, clypeus polished black. A pair of rather close black vt bristles. Antennae (Fig. 126) black, segment 3 at least 3 times as long as deep, arista equal in length or slightly longer. Palpi broadly ovate, not very small, dark with a dense covering of yellowish pile and hairs.

Thorax dark grey dusted on mesonotum, pleura more silvery-grey, leaving sternopleura largely polished. Acr and dc fairly long and numerous, dark brownish; acr broadly 4-serial, dc uniserial with 2 prescutellar pairs large, other large bristles black.

Legs yellow but posterior four knees blackish and tarsi with black annulations; pubescence pale but fore and hind tibiae with distinct dark bristles on dorsum, and mid femora anteriorly with a row of about 6 very distinct dark bristles. Mid femora (Fig. 217) distinctly thickened, pv bristles pale; tibial spur (Fig. 217)

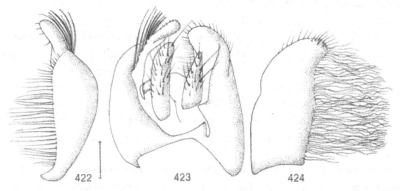

Figs. 422-424. Male genitalia of Platypalpus optivus (Coll.). - 422: right periandrial lamella; 423: periandrium with cerci; 424: left periandrial lamella. Scale: 0.1 mm.

large, black, but apically somewhat blunt and with a tiny curved hair at tip.

Wings (Fig. 718) faintly yellowish-brown clouded, veins R4+5 and M slightly bowed, crossveins separated for a short distance; the vein closing anal cell slightly recurrent and S-shaped. Abdomen polished black, covered with short pale hairs; genitalia (Figs. 422-424) somewhat larger with dull cerci, and left lamella densely fringed with long, apically tangled pale hairs on its outer margin.

Length: body 2.6 - 3.6 mm, wing 3.0 - 3.3 mm.

♀. Antennal segment 3 (Fig. 127) slightly shorter, scarcely 3 times as long as deep, arista longer than segment 3; tibial spur similarly somewhat blunt but without a recurved hair at tip. Last two abdominal segments long and densely grey dusted like the slender cerci.

Length: body 2.7 - 3.7 mm, wing 3.0 - 3.2 mm.

Rare. Denmark: LFM, Guldborg Storskov, ♀ (N.M.Andersen). - England, C.Europe (W.Germany, Czechoslovakia); never a common species. - May-September, but mostly in June and July. On ground vegetation and bushes.

Note. First record from Scandinavia. Both annulatus and melancholicus also possess 4-serial acr, but have shorter antennae, large thoracic bristles pale, sharply pointed tibial spur and extensively darkened legs. P.dalmatinus (Strobl, 1902) and P.flavicoxis (Becker, 1907) also belong here.

55. PLATYPALPUS ANNULATUS (Fallén, 1815)
    Figs. 128, 129, 218, 425-427, 719.

Tachydromia annulata Fallén, 1815: 7.
Tachydromia fascipes Meigen, 1822: 78.
Tachydromia fulvipes Meigen, 1822: 78

A medium-sized species with densely-set long pale hairs on legs and abdomen, antennae black. All thoracic hairs and bristles pale, acr 4-serial. Legs darkened on coxae and femora, tarsi annulated, and a large, sharply pointed tibial spur.

♂. Head rather densely grey dusted, frons scarcely as deep as antennal segment 2, parallel, face slightly narrower above. A pair of rather close light brownish vt bristles. Antennae (Fig. 128) black, basal segment not visible, unlike optivus, and segment 3 shorter, about twice as long as deep; arista almost twice as long. Palpi small, dark, covered with silvery pile and 2 long pale apical hairs.

Thorax darker grey dusted on mesonotum, paler grey on pleura, sternopleura narrowly polished at middle. Small thoracic hairs longer than usual, numerous and rather whitish; acr broadly 4-serial throughout, dc practically multiserial. Large bristles pale.

Legs yellowish with a variable dark brown pattern, usually base of fore coxae and all of the posterior coxae, and a more or less broad ring on all femora (Fig. 218) blackish-brown; tarsi with distinct annulations and tibiae often brownish towards tips. Legs with dense long pale hairs, particularly on femora, anterior four femora of about the same thickness; mid femora (Fig. 218) with very long pale pv bristles, tibial spur sharp.

Wings (Fig. 719) almost clear with brown veins, veins R4+5 and M convergent on apical half but quite parallel just before tip, crossveins separated, the vein closing anal cell at right-angles to vein Cu. Squamae and halteres dirty yellow. Abdomen polished black except for dusted sides of basal segment, rather densely covered with long pale hairs. Genitalia (Figs. 425-427) polished with dull long slender cerci, overlapping both lamellae; left lamella with shorter pale hairs on outer margin.

Length: body 2.3 - 2.8 mm, wing 3.0 - 3.5 mm.

♀. Resembling male, apical 2 abdominal segments and small slender cerci densely grey dusted.

Length: body 2.5 - 3.3 mm, wing 3.3 - 3.6 mm.

Common in Denmark and southern parts of Fennoscandia, rare towards the north; in Norway to Nsi, in Sweden to Nrk., in Finland to Ks. - Widespread and rather common in Europe, also in N.Africa, C.Asia and Siberia; ? N.America (Kovalev, 1969). - May-September. On ground-vegetation, both in woods and in open places.

Note. The leg-colour varies considerably, from extremely pale-legged specimens with only indications of black femoral rings (var. fascipes of Collin), to

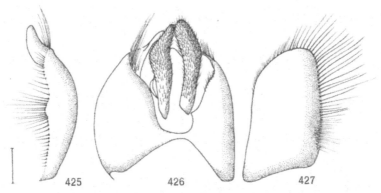

Figs. 425-427. Male genitalia of Platypalpus annulatus (Fáll.). - 425: right periandrial lamella; 426: periandrium with cerci; 427: left periandrial lamella. Scale: 0.1 mm.

171

specimens with extensively black legs somewhat resembling melancholicus. Fallén's annulata was confused with minuta by nearly all subsequent authors.

56. PLATYPALPUS MELANCHOLICUS (Collin, 1961)
Figs. 131, 219, 428-430, 720.

Coryneta atra Wahlberg; Engel, 1939: 59 (p.p.).
Coryneta nigrina Meigen; Engel, 1939: 87.
Tachydromia melancholica Collin, 1961: 147.

Resembling annulatus but generally larger and darker, wings brownish with dark veins. Large bristles brownish, acr smaller and darker, obviously only 4-serial anteriorly. Legs extensively blackish.

♂. Head as in annulatus, a pair of vt bristles almost blackish-brown and antennal arista (Fig. 131) longer, about 2.5 times as long as segment 3. Thorax darker brownish-grey dusted on mesonotum and the polished patch on sternopleura obviously larger. Acr and dc smaller and brownish, former 4-serial anteriorly, usually biserial at middle and posteriorly. Large bristles brownish.

Legs considerably darker than in annulatus but with similar dense long pale hairs. Practically all coxae, trochanters and femora (Fig. 219) except for a narrow tip blackish; tibiae rather pale brownish and tarsi with less distinct dark annulations.

Wings (Fig. 720) distinctly light brown clouded with blackish-brown veins, vein A distinct. Squamae brownish, halteres whitish-yellow with darker stalks. Abdomen with dense pale hairs even at sides, that are shorter than in annulatus;

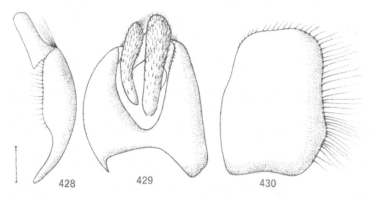

Figs. 428-430. Male genitalia of Platypalpus melancholicus (Coll.). - 428: right periandrial lamella; 429: periandrium with cerci; 430: left periandrial lamella. Scale: 0.1 mm.

genitalia (Fig. 428-430) distinctly smaller than end of abdomen, subshining, the dull cerci conspicuously long.

Length: body 2.9 - 3.6 mm, wing 3.6 - 4.1 mm.

♀. Resembling male but abdomen with less dense and shorter pale hairs, apical two segments and small cerci densely grey dusted.

Length: body 3.0 - 3.8 mm, wing 3.3 - 4.0 mm.

Rare. Finland: N, Lappvik, 2♀ (Frey); Ob, Uleåborg, ♂♀ (coll. Becker, Berlin); also in Russian Ingria borealis: Terijoki, ♂♀ (Krogerus) and Kivinebb, ♂ (Sahlberg). - Great Britain, The Netherlands, NW of European USSR (Leningrad region, Estonia) and C. Europe (W. Germany, Czechoslovakia, Austria, Hungary); rare everywhere. - May-July.

57. PLATYPALPUS NOTATUS (Meigen, 1822)
Figs. 130, 220, 431-433, 721.

Tachydromia notata Meigen, 1822: 78.

A smaller blackish species with a pair of more widely separated pale vt bristles, and black antennae with segment 3 and arista equal in length. Legs darkened at least on coxae and central area of femora, male genitalia with strikingly long and rather blunt cerci.

♂. Head darker grey dusted with narrower frons (scarcely as deep in front as antennal segment 2), face even narrower. A pair of pale widely separated vt bristles, at least twice as widely separated as frons is broad opposite front ocellus. Antennae (Fig. 130) black with segment 3 at least 2.5 times as long as deep, arista subequal. Palpi dark, small ovate.

Thorax rather thinly dark grey dusted on mesonotum, sternopleura with a large polished patch at middle. All thoracic hairs and bristles pale; acr biserial but the two rows not very close, dc uniserial with similar hairs at sides and a large bristle in front of scutellum.

Legs often extensively darkened on coxae, femora (Fig. 220) and tip of tibiae, but at least all coxae darkened at base and femora with blackish dorsal patch or complete ring at middle; tarsi annulated. Mid femora (Fig. 220) only slightly thickened, not very much deeper than slender fore femora, pv bristles long, pale; tibial spur (Fig. 220) long and pointed.

Wings (Fig. 721) almost clear with dark veins, veins R4+5 and M wider apart, parallel; crossveins usually only narrowly separated, lower crossvein very oblique; the vein closing anal cell almost at right-angles to vein Cu. Squamae light brown, halteres pale. Abdomen entirely polished black, with rather

173

dense fine pale hairs. Genitalia (Figs. 431-433) large, polished; cerci over-lapping both lamellae, and with blunt tips.

Length: body 2.0 - 2.5 mm, wing 2.5 - 3.0 mm.

♀. Resembling male; abdomen with only scattered minute pale hairs, api-cal two segments and short slender cerci grey dusted.

Length: body 1.9 - 2.7 mm, wing 2.2 - 3.1 mm.

Very common in Denmark and S.Sweden, but becoming rare towards the north; in Norway to MRy, in Sweden to Boh. and Ög., in Finland along the Bal-tic coast including Ib, but a single ♀ was taken in Ks. - Great Britain, NW of European USSR, C.Europe; probably widely distributed but often confused with other related species. - End May-early October. On ground-vegetation, mainly in sandy coastal biotopes or near water, but also in forest clearings.

Note. The closely related P.insperatus Kovalev, 1971, described from Es-tonia, has pale palpi, shining black clypeus and cheeks, longer antennal seg-ment 3 (3 times as long as deep), extensively yellow legs and abdomen light brown in ground-colour.

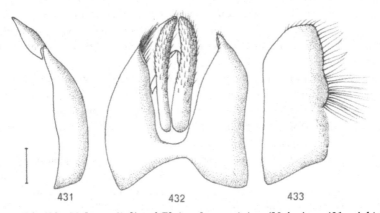

431   432   433

Figs. 431-433. Male genitalia of Platypalpus notatus (Meig.). - 431: right pe-riandrial lamella; 432: periandrium with cerci; 433: left periandrial lamella. Scale: 0.1 mm.

58. PLATYPALPUS STRIGIFRONS (Zetterstedt, 1849)
Figs. 132, 221, 385, 434-436, 722.

Tachydromia strigifrons Zetterstedt, 1849: 3005.

Resembling notatus but larger, legs extensively yellow, frons broader and face equally deep, dc more bristle-like; male genitalia with left cercus sclerotised apically and pointed.

♂. Head densely grey dusted with frons broader than antennal segment 2 is deep, and face almost as deep as frons in front, covered with silvery-grey pile even on clypeus. A pair of widely separated vt bristles as in <u>notatus</u> but brownish, and antennal segment 3 (Fig. 132) almost 3 times as long as deep, as long as or very slightly shorter than arista.

Thorax rather thinly dark brownish-grey dusted on mesonotum; all hairs and bristles pale but biserial acr rather close together and dc uniserial, longer than acr and rather bristle-like, becoming longer posteriorly with last pair very long.

Legs yellow including fore coxae, posterior four coxae slightly darker at base, tarsi annulated; sometimes posterior coxae extensively darkened and femora at about middle and hind femora also towards tip slightly clouded. Legs somewhat more slender and with longer pale hairs; mid femora (Fig. 221) with pale pv bristles shorter than in <u>notatus,</u> and anteriorly on apical third with 2 or 3 distinct dark bristles,

Wings (Fig. 722) with vein M slightly bowed, crossveins distinctly separated or almost contiguous; squamae pale. Abdomen polished black but segment 1 sometimes slightly dull at sides. Genitalia (Figs. 434-436) rather large, polished black with dull cerci overlapping lamellae; left cercus larger, sclerotised apically (black) and with a pointed hook at tip.

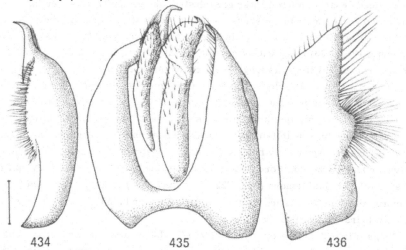

Figs. 434-436. Male genitalia of <u>Platypalpus strigifrons</u> (Zett.). - 434: right periandrial lamella; 435: periandrium with cerci; 436: left periandrial lamella. Scale: 0.1 mm.

175

Length: body 2.5 - 3.4 mm, wing 3.0 - 3.5 mm.

♀. Closely resembling male; abdomen with sparse hairs on dorsum, apical two segments (Fig. 385) and short cerci densely grey dusted but sternite 8 slightly produced and subshining above.

Length: body 2.5 - 3.5 mm, wing 3.0 - 3.5 mm.

Very common on coasts of Denmark and southern Fennoscandia; in Norway approximately to 60° north, in Sweden to Vrm., and on the Baltic coast of Finland. - Common on coasts of Great Britain, E to NW of European USSR; not found in C. Europe, and the records from the south very probably refer to P. approximatus (Becker, 1902), P. turgidus (Becker, 1907) or other related species. - End May-early October. On ground-vegetation in sandy dunes: a typical coastal species.

59. PLATYPALPUS INFECTUS (Collin, 1926)
    Figs. 133, 222, 437-440, 723.

Tachydromia infecta Collin, 1926: 157.

Resembling strigifrons but vt bristles closer together, arista distinctly longer than segment 3, and usually 3 notopleural bristles. Legs yellow with blackish knees and annulated tarsi, mid femora very thickened and fore tibiae very spindle-shaped dilated in ♂.

♂. Head densely rather lighter grey dusted, face slightly narrower than frons and silvery-grey including clypeus. A pair of vt bristles close together, the distance between them 1.5 times as long as depth of frons opposite front ocellus. Antennae (Fig. 133) black with segment 3 about 2.5 times as long as deep but arista at least 1.5 times as long. Palpi rather smaller, ovate.

Thorax with pale acr closely biserial and somewhat diverging, dc uniserial but more numerous in front, ending in 2 large bristles. Large bristles yellowish, usually 3 notopleural, the third anterodorsal (? posthumeral) as in pallidiventris.

Legs yellow, tarsi with faint dark brown annulations, last segment blackish, and all knees conspicuously black; sometimes posterior four coxae and femora more or less darkened. Fore femora thickened and fore tibiae distinctly spindle-shaped dilated. Mid femora (Fig. 222) very thickened, much stouter than fore femora, the pale pv bristles rather shorter. Fore and hind tibiae with dark bristly-hairs on dorsum.

Wings (Fig. 723) clear or indefinitely yellowish-brown clouded, crossveins separated by a distance equal to length of mid crossvein. Abdomen, unlike notatus and strigifrons, with basal two segments grey dusted at sides, and large genitalia (Figs. 437-440) with small blunt cerci enclosed within lamellae; dorsal process on right lamella stout, · and left lamella with a dense fringe of long yellow, apically tangled hairs on its outer margin.

Length: body 2.5 - 3.2 mm, wing 2.8 - 3.2 mm.

♀. Resembling male but fore tibiae somewhat less spindle-shaped dilated. Apical two abdominal segments and cerci densely grey dusted.

Length: body 2.9 - 3.4 mm, wing 2.5 - 3.0 mm.

Not uncommon in Denmark, but rather rare in southern Fennoscandia, in Sweden north to Öl., in Finland to Ta. - Great Britain, NW and C parts of European USSR and C. Europe (Czechoslovakia). - May-mid September.

Note. The closely related P. praecinctus (Collin, 1926) and P. carteri (Collin, 1926), known from Great Britain, and the C. European P. collini (Chvála, 1965), have distinct greyish patches at base of all tergites.

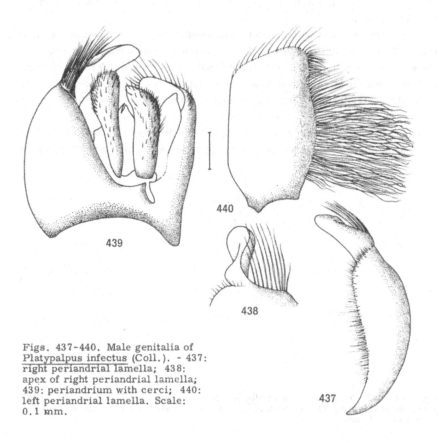

Figs. 437-440. Male genitalia of Platypalpus infectus (Coll.). - 437: right periandrial lamella; 438: apex of right periandrial lamella; 439: periandrium with cerci; 440: left periandrial lamella. Scale: 0.1 mm.

60. PLATYPALPUS INTERSTINCTUS (Collin, 1926)
   Figs. 134, 138, 223, 441-443, 724.

Tachydromia flavipes var. pseudofulvipes Frey, 1909: 9 (p.p.); Frey, 1913: 79.
Tachydromia interstincta Collin, 1926: 158.

A larger species, antennae black with short segment 3 and much longer arista, palpi brownish and strikingly narrow. Acr biserial and conspicuously wide apart.

♂. Head rather densely grey dusted, frons broad, parallel, face slightly narrower and including clypeus silvery-grey. A pair of rather close vt bristles, brownish. Antennae (Fig. 134) black, segment 3 short, scarcely more than 1.5 times as long as deep, arista almost 3 times as long. Palpi (Fig. 138) brown, rather small and narrow.

Thorax rather thinly brownish-grey dusted on mesonotum, acr and dc brownish and hair-like; acr in two widely separated rows, dc uniserial. Large bristles yellowish-brown.

Legs yellow, tarsi with distinct black annulations, and at least mid femora (Fig. 223) with a dark ring before tip, or all femora extensively darkened. Fore femora very thickened and with long pale hairs beneath; fore tibiae rather uniformly but slightly thickened, but not spindle-shaped. Mid femora (Fig. 223) scarcely stouter than fore femora, anteriorly with a row of brown bristles, pv bristles long, yellowish; tibial spur (Fig. 223) very sharp.

Wings (Fig. 724) faintly brownish tinged, vein M slightly undulating, crossveins separated. Squamae and halteres pale yellowish-brown. Abdomen polished

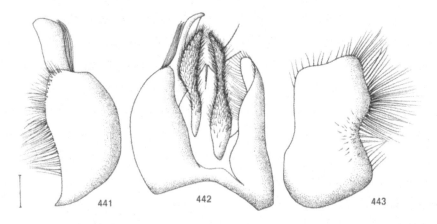

Figs. 441-443. Male genitalia of Platypalpus interstinctus (Coll.). - 441: right periandrial lamella; 442: periandrium with cerci; 443: left periandrial lamella. Scale: 0.1 mm.

black except for slightly dusted sides of basal segment; genitalia (Figs. 441-443) with long cerci, overlapping lamellae.

Length: body 2.4 - 3.4 mm, wing 3.3 - 4.0 mm.

♀. Resembling male; apical two abdominal segments and rather small cerci dull grey, but sternite 8 somewhat produced and subshining at sides.

Length: body 3.0 - 3.6 mm, wing 3.3 - 3.8 mm.

Very common in Denmark and southern parts of Fennoscandia but rare towards the north, approximately to 66° N; in Sweden to Nb., in Finland to Ks. - Widespread in Europe in cold and temperate zones, S to C. Europe and C parts of European USSR. - May -September. Mostly on ground-vegetation.

Note. The var. pseudofulvipes includes both interstinctus and coarctatus; however, the specimen labelled as "type" is coarctatus. Since the name pseudofulvipes has not appeared in the literature since 1913 (not mentioned by Frey in 1943), and to avoid further confusion in nomenclature, the I.C.Z.N. should be asked to suppress the name pseudofulvipes under plenary powers.

61. PLATYPALPUS COARCTATUS (Collin, 1926)
    Figs. 135, 139, 444-446, 725.

Tachydromia flavipes var. pseudofulvipes Frey, 1909: 9 (p.p.).
Tachydromia coarctata Collin, 1926: 158.

Resembling interstinctus but palpi larger and broadly ovate, acr bristles close together, legs with broadly annulated tarsi, and fore tibiae in ♂ very spindle-shaped dilated.

♂. Head with narrower frons which is scarcely as deep as antennal segment 2, and palpi (Fig. 139) rather light brownish, more broadly ovate than usual. Thorax as in interstinctus but the two rows of acr close together.

Legs mainly yellow including coxae, but anterior four femora often brownish above or legs extensively darkened even on coxae and at tip of tibiae; tarsi with broad dark annulations. Fore tibiae very spindle-shaped dilated, dorsally with several brown bristly-hairs. Mid femora unlike interstinctus stouter than fore femora and with only pale hairs anteriorly.

Wings (Fig. 725) somewhat yellowish with yellowish-brown veins, crossveins wider apart. Abdomen shining black, sides of anterior two segments greyish dusted, the following tergites sometimes also very narrowly greyish at sides. Genitalia (Figs. 444-446) rather large, polished black.

Length: body 2.4 - 3.3 mm, wing 3.0 - 3.3 mm.

♀. Legs extensively yellow except for the annulated tarsi, fore tibiae only slightly dilated and shorter haired beneath. Abdomen with at most anterior two tergites slightly grey at sides, apical two segments including the long and slender cerci densely grey dusted.

Length: body 2.6 - 3.8 mm (according to Kovalev (1969) up to 4.4 mm), wing 2.8 - 3.3 mm.

Rather common in Denmark and the southern parts of Fennoscandia; in Norway north to Ak, in Sweden to Upl., in Finland to Ta and Sa. - Widespread in Europe except for the south. - June-September. On ground-vegetation in forest clearings and in meadows, but also on bushes.

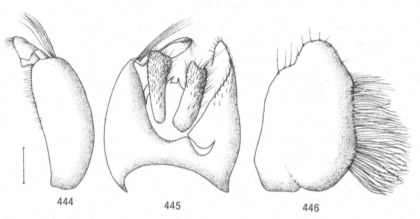

444   445   446

Figs. 444-446. Male genitalia of Platypalpus coarctatus (Coll.). - 444: right periandrial lamella; 445: periandrium with cerci; 446: left periandrial lamella. Scale: 0.1 mm.

62. PLATYPALPUS CLARANDUS (Collin, 1926)
Figs. 136, 224, 447-449, 726.

Tachydromia claranda Collin, 1926: 157.

Smaller than coarctatus with vt bristles more widely separated, clypeus polished black and mesonotum rather golden-yellow dusted. Abdomen somewhat brownish, at least on venter.

♂. Head densely grey dusted, with silvery-grey face distinctly narrower above than frons in front, clypeus bare. A pair of brownish vt bristles, more widely separated, and antennal segment 3 (Fig. 136) small, slightly longer than deep. Palpi dark, small-ovate.

Thorax rather golden dusted on mesonotum, pleura lighter grey, sternopleura with a large polished median patch. All hairs and bristles yellowish-brown; acr closely biserial, dc uniserial and somewhat longer.

Legs yellow including coxae, but tarsi with distinct dark annulations, sharper and blacker on front pair. Fore tibiae only slightly and uniformly thickened, mid femora (Fig. 224) slightly stouter than fore femora, pv bristles long and pale, and 2 or 3 short pale bristles on apical third in front.

Wings (Fig. 726) faintly yellowish with veins becoming darker towards tip; vein M slightly bowed, crossveins widely separated, squamae and halteres pale. Abdomen polished except for narrowly dusted sides of basal two segments, black on dorsum but somewhat yellowish-brown on venter and whole of apical two segments. Genitalia (Figs. 447-449) conspicuously polished black, rather large, short cerci enclosed within lamellae.

Length: body 1.5 - 2.1 mm, wing 2.3 - 2.6 mm.

♀. Acr and dc shorter and less conspicuous; abdomen shining dark brown on dorsum, venter rather yellowish but apical two segments and long slender cerci light grey dusted, sternite 8 subshining black at tip.

Length: body 1.8 - 2.6 mm, wing 2.3 - 2.6 mm.

Rare. Sweden: Sk., Kullaberg, ♂ (H. Andersson); Småland, 2♀ (Boheman). - Commoner in Great Britain, but no further data available; the record from Morocco, N. Africa (Frey, 1943), seems improbable. - June-August. I myself took this species when sweeping in grass on Hampstead Heath in London.

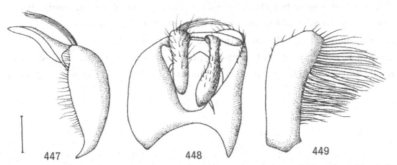

Figs. 447-449. Male genitalia of <u>Platypalpus clarandus</u> (Coll.). - 447: right periandrial lamella; 448: periandrium with cerci; 449: left periandrial lamella. Scale: 0.1 mm.

63. PLATYPALPUS ARTICULATUS Macquart, 1827
    Figs. 137, 225, 450-452, 727.

Platypalpus articulatus Macquart, 1827: 98.
Tachydromia maculimana Zetterstedt, 1842: 284.

A very small greyish dusted species with basal antennal segments yellow and a

181

very small, blunt spur on mid tibiae. Legs yellow, tarsi with dark annulations, male genitalia almost bare.

♂. Head densely grey dusted, frons not very deep below, widening out above, but face almost linear. A pair of long brownish vt bristles, widely separated. Antennae (Fig. 137) small, basal segments yellow, segment 3 blackish, about 1.5 times as long as deep, arista more than twice as long. Palpi small, dirty yellowish.

Thorax densely grey dusted, lighter grey on pleura, sternopleura polished except for posterior margin. Acr biserial, absent on posterior third of mesonotum, dc uniserial with last pair large, all pale.

Legs yellow, all tarsi with brownish annulations or rather evenly darkened towards tip, posterior four knees darkened. Mid femora (Fig. 225) scarcely stouter than fore femora, pv bristles yellowish, and another pale bristle on anterior third in front. Tibial spur (Fig. 225) very small, blunt.

Wings (Fig. 727) almost clear with brownish veins, veins R4+5 and M parallel, crossveins usually separated. Squamae and halteres pale yellow. Abdomen polished black, tergites usually broadly dusted grey at sides and on a narrow anterior margin. Genitalia (Figs. 450-452) small, polished, left lamella practically bare.

Length: body 1.2 - 2.1 mm, wing 1.7 - 2.2 mm.

♀. Resembling male but abdomen more shining even at sides, apical two segments very small and together with slender cerci grey dusted.

Length: body 1.5 - 2.3 mm, wing 2.0 - 2.1 mm.

Uncommon in Denmark and southern Fennoscandia, extending far into the north; in Norway to NTi, in Sweden to Nb. (1♀), in Finland to Om and Sb. -

450   451   452

Figs. 450-452. Male genitalia of <u>Platypalpus articulatus</u> Macq. - 450: right periandrial lamella; 451: periandrium with cerci; 452: left periandrial lamella. Scale: 0.1 mm.

Rare in England but rather common in SW., C. and E. Europe. - May-September.
On ground-vegetation and on the leaves of bushes.

Note. It is easily confused with articulatoides, and the genitalia illustrated by
Zusková (1966: Figs. 6, 10) are those of articulatoides. P. stigma (Collin, 1926),
known from England and C. Europe, is also related but is larger, with longer antennae, broader face and distinct brown stigma on wings.

64. PLATYPALPUS ARTICULATOIDES (Frey, 1918)
Figs. 140, 226, 453-455, 728.

Tachydromia articulatoides Frey, 1918: 11.

Closely resembling articulatus but fore tarsi with very striking black annulations,
and posterior four tarsi mostly yellow, darker towards tip. Male genitalia with
long pale hairs on outer margin of left lamella.

♂. Head as in articulatus but frons perhaps somewhat broader and more parallel even above, a pair of vt bristles not so conspicuously widely separated. Antennal segment 3 (Fig. 140) scarcely narrower or longer, as stated by Frey.
Thorax somewhat lighter grey on mesonotum.

Legs pale yellow with fore femora and tibiae perhaps more thickened, and
mid tibia (Fig. 226) with an even smaller apical spur; posterior knees at most
with indistinct brownish points. Fore tibiae with longer pale hairs above, and
the pale anterior bristle on apical third of mid femora absent. Fore tarsi with
very sharp black annulations, but posterior four tarsi mostly yellow with last
segment dark, or even segment 4 brownish at tip.

Wings (Fig. 728) clear with somewhat paler veins, veins R4+5 and M parallel

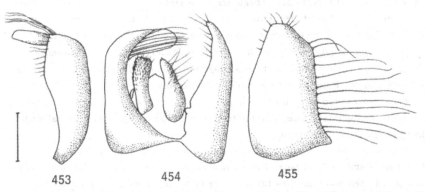

453   454   455

Figs. 453-455. Male genitalia of Platypalpus articulatoides (Frey). - 453: right
periandrial lamella; 454: periandrium with cerci; 455: left periandrial lamella. Scale: 0.1 mm.

but more evenly bowed upwards. Abdomen with all tergites at sides and narrow anterior margins on all sternites more or less lighter grey dusted. Genitalia (Figs. 453-455)larger, left lamella with several long whitish hairs on its outer margin.

Length: body 1.3 - 1.9 mm, wing 1.6 - 2.1 mm; holotype body 1.9 mm (not 2.25 mm as given by Frey, 1918), wing 1.9 mm.

♀. Resembling male but fore tarsi with distinct but rather blackish-brown annulations. Abdomen more evenly thinly grey dusted, rather subshining, apical two segments and slender cerci densely grey dusted.

Length: body 1.7 - 2.0 mm, wing 1.8 - 2.1 mm.

Rare.Denmark: NEZ, Bagsværd, ♂ (Lyneborg), Uggeløse Skov, ♂ (Chvála); Sweden: Gotland, ♂♀ (Boheman); Norway: On, Fron, ♂ (Siebke); Finland: Om, Jakobstad, ♀ (Frey). - N (Archangelsk) and NW (Leningrad region, Latvia) parts of European USSR, C.Europe (Czechoslovakia, Austria). - End May-July. On ground-vegetation and bushes.

65. PLATYPALPUS ANNULIPES (Meigen, 1822)
Figs. 145, 227, 456-458, 729.

Tachydromia annulipes Meigen, 1822: 77.
Tachydromia infuscata Meigen, 1822: 84.
Tachydromia coxata Zetterstedt, 1842: 281.

Basal antennal segments yellow, segment 3 dark, about 3 times as long as deep. Legs yellow or extensively darkened on coxae and femora, tarsi with dark annulations, very sharp on front pair; tibial spur long but blunt. Wings yellowish on costal half, acr irregularly 3-serial.

♂. Head grey dusted, frons narrower in front than depth of antennal segment 2, widening above; face almost linear above, produced below into polished black clypeus. A pair of brownish vt bristles, widely separated. Antennae (Fig. 145) yellow on basal segments, segment 3 blackish-brown, long, arista slightly longer. Palpi small and narrow, yellowish.

Thorax brownish-grey dusted on mesonotum, sternopleura largely polished. All hairs and bristles pale, acr bi- to triserial, dc irregularly biserial with last prescutellar pair strong.

Legs variable in colour from entirely yellow to extensively darkened, with long pale hairs on front pair, but tarsi always with distinct annulations, very sharp and deep black on fore tarsi, more faint on posterior two pairs. Mid femora (Fig. 227) considerably stouter and larger than fore femora, pale pv birstles short; tibial spur (Fig. 227) long but blunt-tipped.

Wings (Fig. 729) large and yellowish on costal half, darkened posteriorly, anterior veins yellow but veins R4+5, M and particularly Cu very dark. Crossveins separated, squamae and halteres pale. Abdomen shining black except for greyish dusted tergite 1 and narrow sides of tergite 2, densely covered with fairly long whitish hairs. Genitalia (Figs. 456-458) subshining with dull cerci enclosed within lamellae, hypandrium very convex.

Length: body 2.3 - 2.8 mm, wing 2.8 - 3.3 mm.

♀. Legs usually entirely yellow without any darkening and with longer pv bristles on mid femora, antennal segment 3 generally slightly shorter. Abdomen with shorter hairs, apical two segments and very long slender cerci densely light grey dusted.

Length: body 2.5 - 3.2 mm, wing 2.7 - 3.5 mm.

Common in Denmark and S. Sweden, rare towards the north; in Norway to SFi, in Sweden to Upl.; not yet known from Finland, but recorded from Estonia. - Great Britain, W. and C. Europe, NW of European USSR. - May-July, in the north until early September.

Note. A variable species, not only in colour of legs but also in shape of

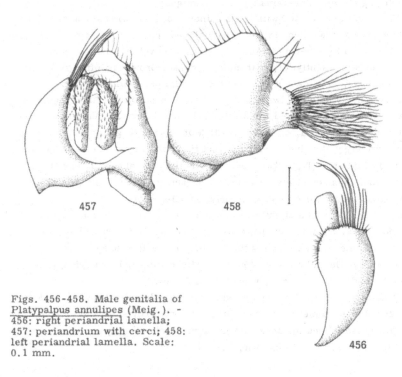

Figs. 456-458. Male genitalia of Platypalpus annulipes (Meig.). - 456: right periandrial lamella; 457: periandrium with cerci; 458: left periandrial lamella. Scale: 0.1 mm.

antennae; the W. and C. European populations have segment 3 almost 3.5 times as long as deep with arista scarcely longer, but it is sometimes hardly 2.5 times as long as deep with arista almost twice as long in E populations. The closely related P. subtilis (Collin, 1926), known from England and Estonia, is smaller, with tibial spur blunt in ♂ but sharply pointed in ♀, antennal segment 3 always 2.5 times as long as deep, mesonotum with biserial acr and only a single notopleural bristle, wings evenly tinged with yellow with pale veins, and male genitalia with hypandrium not so convex but otherwise rather similar.

66. PLATYPALPUS ECALCEATUS (Zetterstedt, 1838)
   Figs. 141, 228, 459-461, 730.

Tachydromia bicolor Meigen; Zetterstedt, 1838: 549, 1842: 276 (p.p.), Lundbeck, 1910: 298 (p.p.).
Tachydromia ecalceata Zetterstedt, 1838: 550.

Medium-sized light grey dusted species with legs entirely yellow including tarsi; antennae yellow on basal segments, segment 3 short, brown. Frons broad and diverging above, crossveins on wings contiguous.

   ♂. Head light grey dusted with frons rather broad and widening above, face slightly narrower than frons in front, with clypeus polished. A pair of very pale vt bristles, broadly separated. Antennae (Fig. 141) yellow on basal segments, segment 3 broad, slightly longer than deep, darker brown and often yellowish at extreme base, arista about twice as long. Palpi pale yellow, ovate.

   Thorax densely pale yellowish-grey dusted on mesonotum, silvery-grey on pleura, sternopleura polished black anteriorly. All hairs and bristles pale yellow, acr closely biserial, dc biserial with 3 or 4 pairs longer.

   Legs entirely yellow including tarsi, but tip of trochanters and a large sharply-pointed tibial spur (Fig. 228) black towards tip; apical tarsal segments on posterior four legs sometimes indefinitely darkened. Mid femora (Fig. 228) slightly stouter than fore femora, pv bristles pale, short.

   Wings (Fig. 730) often slightly yellowish tinged, veins very pale, crossveins contiguous. Squamae and halteres pale yellow. Abdomen polished black with sides of anterior two tergites thinly dull grey, with dense but not very long pale hairs. Genitalia (Figs. 459-461) polished black, cerci dull grey.
   Length: body 2.4 - 2.6 mm, wing 3.1 - 3.3 mm.

   ♀. Resembling male, fore femora with shorter pale hairs beneath. Abdomen with sparse hairs, sometimes all tergites narrowly greyish at sides; apical two segments small and together with slender cerci dull grey.
   Length: body 2.0 - 3.2 mm, wing 2.6 - 3.3 mm.

The commonest Platypalpus species in C and N parts of Fennoscandia, much less common in southern parts and in Denmark; males very rare, commoner in the extreme north. - Iceland, Great Britain, C. and E. Europe. - May-mid August. Females are common on ground-vegetation and also on leaves of bushes.

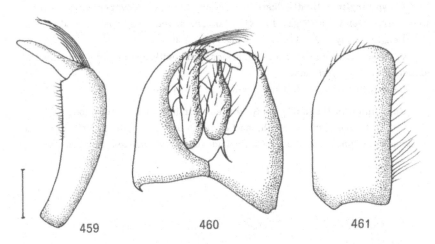

Figs. 459-461. Male genitalia of Platypalpus ecalceatus (Zett.). - 459: right periandrial lamella; 460: periandrium with cerci; 461: left periandrial lamella. Scale: 0.1 mm.

67. PLATYPALPUS CALCEATUS (Meigen, 1822)
Figs. 146, 229, 462-464, 731.

Tachydromia calceata Meigen, 1822: 87.

Resembling ecalceatus but smaller, legs with apical tarsal segments darkened and long pv bristles on mid femora, antennal segment 3 narrower and very dark. Frons narrower and crossveins on wings widely separated.

♂. Head with frons narrower and almost parallel below, broadly widening above, clypeus dusted. A pair of broadly separated vt bristles. Antennae (Fig. 146) small, basal segments yellow, but segment 3 narrower and uniformly blackish, unlike ecalceatus; arista about 1.5 times as long. Palpi smaller and narrow.

Thorax rather thinly brownish-grey dusted on mesonotum, all hairs and bristles pale; acr small, narrowly biserial, dc uniserial, less numerous and much longer, posterior two pairs large.

Legs rather small and slender, yellow, but last tarsal segment and segment 4 at tip darkened, sometimes all tarsi yellowish-brown. Anterior four femora of about the same thickness, mid femora (Fig. 229) with long pale pv bristles almost as long as femur is deep.

Wings (Fig. 731) clear with pale veins; crossveins widely separated, by a distance at least equal to length of mid crossvein. Abdomen shining black, anterior two tergites slightly dusted anteriorly at sides, covered with scattered pale hairs. Genitalia (Figs. 462-464) larger, at least as broad as end of abdomen.

Length: body 1.3 - 1.6 mm, wing 1.9 - 2.0 mm.

♀. Resembling male, abdomen almost bare with apical two segments and slender cerci greyish dusted.

Length: body 1.5 - 2.4 mm, wing 1.9 - 2.4 mm.

Uncommon in Denmark and S. Sweden, north to Sdm.; commoner in S. Finland, but rare towards north, as far as Ks and Lk. - Great Britain, C. and E. Europe. - June-September. On ground-vegetation and leaves of bushes.

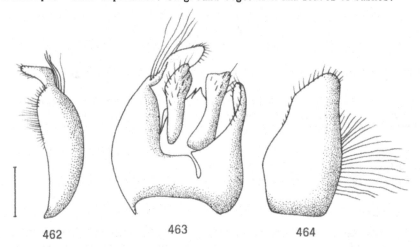

462   463   464

Figs. 462-464. Male genitalia of Platypalpus calceatus (Meig.). - 462: right periandrial lamella; 463: periandrium with cerci; 464: left periandrial lamella. Scale: 0.1 mm.

68. PLATYPALPUS STABILIS (Collin, 1961)

Figs. 142, 465-467, 732.

Tachydromia stabilis Collin, 1961: 188.

Larger species with narrow parallel frons and yellow antennae except for darkened apical half of segment 3 and arista. Crossveins on wing broadly separated, acr wider apart, and fore tibiae and femora in ♂ very thickened.

♂. Head densely grey dusted, frons narrow and parallel, face almost as deep; a pair of pale broadly-separated vt bristles. Antennae (Fig. 142) mainly yellow with apical half of segment 3 and arista brown to dark brown, latter not very much longer than segment 3. Palpi small, ovate, pale yellow.

Thorax rather brownish-grey dusted on mesonotum, all hairs and bristles pale yellow; acr biserial and wider apart, dc practically uniserial or in two alternating rows, last pair bristle-like.

Legs yellow, tarsi with faint brown annulations, apical 2 or 3 segments darker brown, and the sharply pointed tibial spur almost black. Anterior four femora equally thickened, fore femora with a double row of very long pale bristly-hairs beneath, mid femora with long pv bristles and usually 2 short pale bristles in front at tip; hind femora with only short hairs. Fore tibiae very spindle-shaped dilated and with several longer bristly-hairs on dorsum.

Wings (Fig. 732) clear with very pale veins, vein M distinctly bowed and crossveins broadly separated; the vein closing anal cell recurrent and somewhat S-shaped. Abdomen shining black with all tergites narrowly grey anteriorly at sides, anterior 2 or 3 segments more broadly so. Genitalia (Figs. 465-467) polished black, large, the grey cerci concealed within long lamellae.

Length: body 2.5 - 3.0 mm, wing 2.8 - 3.0 mm.

♀. Fore tibiae not spindle-shaped dilated and fore femora less thickened. Abdomen almost bare, with grey lateral dusting on anterior tergites only, apical two segments and long slender cerci light grey.

Length: body 2.6 - 3.2 mm, wing 2.6 - 3.3 mm.

Uncommon in Denmark and in S. Sweden north to Gtl., in Finland along the Baltic coast. - England, NW of European USSR (Estonia). - June-August.

Note. The closely related P.pictitarsis (Becker, 1902), known from Eng-

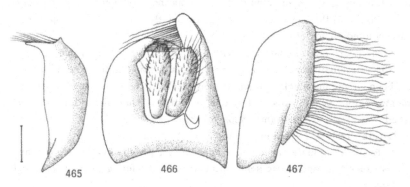

Figs. 465-467. Male genitalia of Platypalpus stabilis (Coll.). - 465: right periandrial lamella; 466: periandrium with cerci; 467: left periandrial lamella. Scale: 0.1 mm.

land, C.Europe and S to N.Africa, is generally a smaller species with more broadly diverging frons, darker palpi and extensively blackish antennal segment 3, acr very diverging and almost alternating in one row, and legs often darkened with slender fore tibiae, tarsi with blackish annulations, and hind femora fringed with long pale bristles beneath.

69. PLATYPALPUS PALLIDIVENTRIS (Meigen, 1822)
Figs. 147, 230, 468-470, 733.

Empis flavipes Fabricius, 1794: 406 (nec flavipes Scopoli, 1763).
Tachydromia pallidiventris Meigen, 1822: 82.
Tachydromia dichroa Meigen, 1822: 83.
Platypalpus robustus Walker, 1837: 228.

Medium-sized greyish species with anterior notopleural bristle, fore and hind tibiae with dark bristles above. Antennal segment 3 very darkened, and legs yellow with black annulated tarsi in both sexes.

♂. Head densely grey dusted, frons rather narrow in front, widening above, face as deep as frons in front; a pair of light brownish vt bristles, rather close together. Antennae (Fig. 147) yellow on basal segments, segment 3 entirely blackish, more than twice as long as deep, arista not very much longer. Palpi very small, pale.

Thorax somewhat golden-grey dusted on mesonotum, sternopleura with a large bare patch in anterior half. All hairs and bristles yellowish, acr and dc biserial, latter alternating in the outer row, 2 pairs large.

Legs yellow, tarsi with almost black annulations, fore tibiae very indistinctly thickened, these and hind tibiae with distinct dark bristles above. Mid femora (Fig. 230) thickened with long pale pv bristles, tibial spur large and sharply pointed, blackish.

Wings (Fig. 733) almost clear with pale veins, crossveins widely separated, the vein closing anal cell recurrent. Squamae and halteres whitish-yellow. Abdomen shining black with small grey lateral patches on anterior two segments, rather densely covered by long pale hairs. Genitalia (Figs. 468-470) polished, with left cercus slightly hooked at tip.

Length: body 2.2 - 2.8 mm, wing 2.5 - 2.9 mm.

♀. Resembling male, abdomen with indefinite grey lateral anterior margins on all tergites, apical two segments and cerci densely grey dusted; cerci very slender.

Length: body 2.4 - 3.1 mm, wing 2.6 - 3.3 mm.

Very common in Denmark and southern parts of Fennoscandia, rare towards north; in Sweden to Äng., in Finland to Om and Ks (1 ♀ in each district). - Widespread and common throughout Europe, but the records from N.Africa need verification. - Mid May-September. Mostly on leaves of bushes and trees.

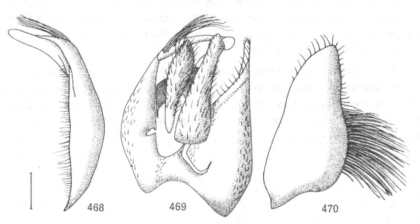

Figs. 468-470. Male genitalia of Platypalpus pallidiventris (Meig.). - 468: right periandrial lamella; 469: periandrium with cerci; 470: left periandrial lamella. Scale: 0.1 mm.

70. PLATYPALPUS LONGISETA (Zetterstedt, 1842)
Figs. 148, 231, 471-473, 734.

Tachydromia longiseta Zetterstedt, 1842: 278.
Tachydromia extricata Collin, 1926: 188.

Closely resembling pallidiventris but rather a robust species with antennal segment 3 yellowish at base, and legs in ♂ with only apical segment on fore tarsi black and fore tibiae thickened.

    ♂. Head as in pallidiventris but frons lighter grey dusted and antennal segment 3 (Fig. 148) more slender, at least 2.5 times as long as deep, and obviously yellowish at base; arista scarcely longer.

    Thorax lighter grey dusted, but acr somewhat closer together and smaller than dc in the inner row. Large bristles including a distinct anterior notopleural birstle (? posthumeral) as in pallidiventris. Fore tarsi yellow with only last segment black, posterior four tarsi with faint brown to dark brown annulations, last segment always blackish. Fore femora scarcely stouter but fore tibiae conspicuously spindle-shaped dilated, about as deep at middle as hind femora. Genitalia (Figs. 471-473) with long dull grey cerci, left cercus apically more distinctly pointed and with a claw at tip.

191

Length: body 2.3 - 3.1 mm, wing 2.8 - 3.3 mm.

♀. Can only be separated from pallidiventris with difficulty because of the annulated fore tarsi and only indistinctly thickened fore tibiae, but the larger size, paler antennal segment 3, lighter grey dusted thorax with smaller and closer acr, more distinct greyish fasciae on abdomen and not so sharply annulated tarsi with darker last segment may help in separating these two species.

Length: body 2.8 - 3.8 mm, wing 3.0 - 3.6 mm.

Almost the same pattern of distribution as in pallidiventris: common in southern and warmer areas up to 61 - 62 ° north, but only individually in Norway to NTi, in Sweden to Ly.Lpm. (1♂), and in Finland to Sa. - Widespread in Europe, ? Iran (Frey, 1943) . - Mid May-September. Mostly on bushes and trees together with pallidiventris.

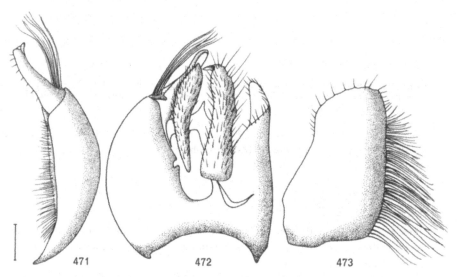

Figs. 471-473: Male genitalia of Platypalpus longiseta (Zett.). - 471: right periandrial lamella; 472: periandrium with cerci; 473: left periandrial lamella. Scale: 0.1 mm.

71. PLATYPALPUS LATICINCTUS Walker, 1851
Figs. 149, 151, 232, 474-476, 735.

Platypalpus laticinctus Walker, 1851: 127.
Tachydromia cursitans Fabricius; Lundbeck, 1910: 303.
Tachydromia candicans Fallén; Lundbeck, 1910: 303 (p.p., ♂).

A large, light grey dusted species with a bare polished patch on sternopleura;

192

antennae yellow with longer segment 3 dark, arista only slightly longer. Legs yellow even on tarsi, veins R4+5 and M slightly bowed.

♂. Head with not very broad frons slightly widening above, face as deep as frons in front, a pair of vt bristles close together. Antennae (Fig. 149) yellow on basal segments, segment 3 about 2.5 - 3 times as long as deep, blackish-brown but often yellowish near base, arista as long as or slightly longer. Palpi (Fig. 151) ovate, pale yellow, rather large but not as large as in major.

Thorax light grey with two indefinite darker stripes between acr and dc when viewed from above, sternopleura with a narrow polished black patch anteriorly. All hairs and bristles pale, acr and dc small and both narrowly biserial, 2 large prescutellar dc and 2 or 3 posterior notopleural, anterior notopleural not developed.

Legs yellow even on fore tarsi, posterior four tarsi with very indefinite brownish annulations, sometimes posterior femora (including coxae and base of hind tibiae) considerably darkened. Fore tibiae slender, mid femora (Fig. 232) thickened, almost twice as deep as fore femora; the pale pv bristles small, also with similar bristles anteroventrally; tibial spur (Fig. 232) large.

Wings (Fig. 735) faintly yellowish tinged, veins R4+5 and M evenly slightly bowed towards each other, crossveins separated by a short distance, and the vein closing anal cell only slightly recurrent. Abdomen shining black with anterior two tergites broadly grey dusted at sides, the following 2 or 3 tergites with a narrow grey lateral stripe in front. Genitalia (Figs. 474-476) polished black, closed, not broader than end of abdomen.

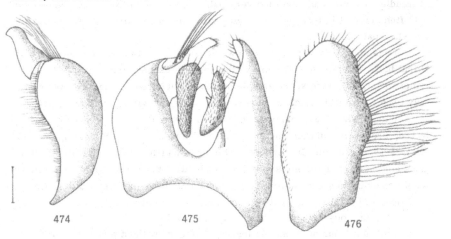

474   475   476

Figs. 474-476. Male genitalia of Platypalpus laticinctus Walk. - 474: right periandrial lamella; 475: periandrium with cerci; 476: left periandrial lamella. Scale: 0.1 mm.

Length: body 3.0 - 3.8 mm, wing 3.5 - 4.1 mm.

♀. Resembling male, apical two abdominal segments and very long slender cerci densely grey dusted, posterior margin of preceding segment (or two) dull.

Length: body 3.0 - 4.5 mm, wing 3.4 - 4.8 mm.

Uncommon in Denmark and S. Sweden (Sk.). - Great Britain and C. Europe; not found in European USSR. - May-early July. On leaves of bushes.

Note. Needs comparison only with fasciatus and cruralis: for details, see under major.

72. PLATYPALPUS ALBICORNIS (Zetterstedt, 1842)
Figs. 143, 152, 233, 477-479, 736.

Tachydromia albicornis Zetterstedt, 1842: 279 (p.p.).

A very light grey dusted species with antennae whitish-yellow and irregularly tri- to quadriserial acr, large palpi pointed at tips. Legs pale yellow with last tarsal segment darkened, tibial spur blunt in ♂, longer and more pointed in ♀.

♂. Head with broad parallel frons, face silvery-grey and much narrower. A pair of long pale yellow broadly separated vt bristles. Antennae (Fig. 143) pale whitish-yellow, segment 3 scarcely twice as long as deep, arista slightly longer and darker towards tip. Palpi (Fig. 152) large, pale, pointed at tips in contrast to the blunt palpi of flavicornis (Fig. 153) and pallidicornis (Fig. 154).

Thorax almost silvery-grey dusted, sternopleura with a large bare patch at middle. All hairs and bristles very pale, acr irregularly tri- to quadriserial but often biserial in front, dc uniserial with numerous hairs at side, last pre-scutellar pair long.

Legs uniformly very pale yellow but last segment of all tarsi blackish on apical half. Anterior four femora of the same thickness, tibiae slender, mid femora (Fig. 233) with short pale pv bristles. Tibial spur (Fig. 233) blunt, scarcely longer than tibia is deep, darkened apically and with a tiny hair at tip.

Wings (Fig. 736) clear with pale veins, veins R4+5 and M very indistinctly bowed, crossveins distant and 2nd basal cell longer; the vein closing anal cell at right-angles to vein Cu. Abdomen shining blackish-brown, anterior two tergites and almost all sternites yellowish; tergite 1 thinly dull grey, the following tergites sometimes very narrowly greyish on anterior margin at sides. Genitalia (Figs. 477-479) shining with small, grey pilose cerci.

Length: body 2.1 - 2.6 mm, wing 2.8 - 3.1 mm.

♀. Apical spur on mid tibia more pointed and without a tiny hair at tip; apical two abdominal segments and long slender cerci densely light grey dusted.

Length: body 2.2 - 3.2 mm, wing 2.8 - 3.6 mm.

194

Rather rare in Denmark, S.Sweden north to Ög., and SW.Finland (Al, Ab); in Norway only known from NTi. - England, W. and C.Europe, and NW and C. parts of European USSR. - May-July; a typical spring species.

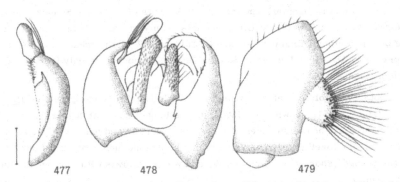

Figs. 477-479. Male genitalia of Platypalpus albicornis (Zett.). - 474: right periandrial lamella; 478: periandrium with cerci; 479: left periandrial lamella. Scale: 0.1 mm.

73. PLATYPALPUS FLAVICORNIS (Meigen, 1822)

Figs. 150, 153, 234, 480-482, 737.

Tachydromia flavicornis Meigen, 1822: 83.

Light grey to yellowish-grey dusted species resembling albicornis, but tarsi with sharp black annulations, and tibial spur pointed in both sexes; abdomen broadly grey dusted at sides and palpi ovate.

♂. Head as in albicornis but frons rather yellowish-grey dusted and face silvery-grey with narrow polished black cheeks. Antennae (Fig. 150) pale yellow with dark arista almost twice as long as segment 3. Palpi (Fig. 153) broadly ovate, distinctly blunt-tipped.

Thorax rather yellowish-grey dusted on mesonotum, silvery-grey on pleura, sternopleura with a smaller black patch at middle. Acr narrowly biserial, dc uniserial, all small and fine. Legs pale yellow, all tarsi with sharp blackish-brown to black annulations, tarsal segments at least on front pair slightly dilated apically. Tibial spur (Fig. 234) sharply pointed, blackish towards tip. Wings (Fig. 737) as in albicornis.

Abdomen very broadly light grey dusted at sides of all tergites, leaving often only broad median triangles polished black; venter uniformly yellowish or light brown. Genitalia (Figs. 480-482) rather shining, blackish-brown, with longer dull grey cerci.

Length: body 2.2 - 2.7 mm, wing 2.6 - 3.0 mm.

195

♀. Antennae deeper yellow in colour and segment 3 often slightly darkened at extreme tip. Otherwise as in the male.

Length: body 2.2 - 2.6 mm, wing 2.6 - 2.8 mm.

Uncommon in Denmark but rare in other parts of southern Fennoscandia, in Norway to Ak, in Sweden to Dlr., in Finland only in Ta. - Widespread in Europe but uncommon; England, W. and C.Europe, NW of European USSR; ? S. Europe, ? N.Africa. - April-mid September, in Fennoscandia only from June. Often on bushes.

Note. Very often confused with verralli (Coll.). The closely related P.divisus Walker, 1851, known from England, has longer antennal segment 3, which is yellow in ♂ but brownish except for base in ♀, longer and narrower palpi, vt bristles closer together, extensively shining black abdomen, only last tarsal segment darkened, shorter blunt spur in ♀, and wings yellowish with brownish clouding at tip.

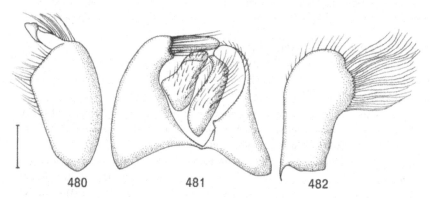

480   481   482

Figs. 480-482. Male genitalia of Platypalpus flavicornis (Meig.). - 480: right periandrial lamella; 481: periandrium with cerci; 482: left periandrial lamella. Scale: 0.1 mm.

74. PLATYPALPUS PALLIDICORNIS (Collin, 1926)
Figs. 144, 154, 235, 483-485, 738.

Tachydromia albicornis Zetterstedt, 1842: 279 (p.p.); Lundbeck, 1910: 308.
Tachydromia pallidicornis Collin, 1926: 186.

Closely resembling flavicornis but frons narrower and face rather yellowish, vt bristles closer together, abdomen more extensively shining and legs with a small blunt spur in ♂, longer and pointed in ♀.

♂. Head silvery-grey dusted on frons and upper part of occiput, former narrower, scarcely as deep as antennal segment 2; face usually conspicuously yellowish-grey dusted. A pair of vt bristles close together. Thorax silvery-grey dusted even on mesonotum,and the bare median patch on sternopleura larger, acr irregularly bi- to triserial.

Legs with more slender fore femora and tarsi with dark brown annulations, less distinct than in flavicornis. Mid tibiae (Fig. 235) with a small blunt apical spur which is much shorter than tibia is deep, somewhat pointed when viewed from the side. Wings (Fig. 738) clear but smaller and crossveins less separated.

Abdomen subshining black, tergite 1 extensively light grey dusted but the following tergites with less conspicuous grey lateral patches. Genitalia (Figs. 483-485) rather large, polished, with grey pilose cerci.

Length: body 2.0 - 2.5 mm, wing 2.2 - 2.6 mm.

♀. Antennal segment 3 often darkened towards tip, the black ventral spines on mid femora all black even at base, and tibial spur long, pointed. Abdomen usually with only anterior three tergites greyish dusted at sides.

Length: body 2.2 - 2.6 mm, wing 2.3 - 2.5 mm.

Rather common in Denmark and southern Fennoscandia; in Sweden north to Nrk. and Sdm., in Finland to Ta; not yet recorded from Norway. - England, NW and C parts of European USSR, C.Europe. - End May-August but a single ♀ was taken by Zetterstedt at Lund as early as 17 March. Common in fields of wheat.

Note. P.ochrocerus (Collin, 1961), known from England and NW of European USSR (Leningrad region), has the same yellow antennae and small tibial spur, but is generally darker grey dusted and smaller (body about 1.5 mm in

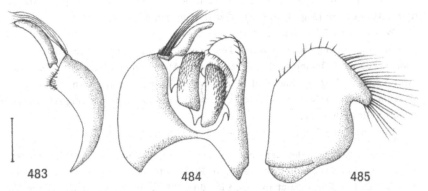

Figs. 483-485. Male genitalia of Platypalpus pallidicornis (Coll.). - 483: right periandrial lamella; 484: periandrium with cerci; 485: left periandrial lamella. Scale: 0.1 mm.

length), with smaller palpi, 6 - 7 stronger pairs of dc, and tarsi with only last segment darkened. P.vegetus Frey, 1943 and P.villeneuvei (Becker, 1910) are also related species.

75. PLATYPALPUS MAJOR (Zetterstedt, 1842)
Figs. 157, 160, 236, 739.

Tachydromia major Zetterstedt, 1842: 287.
Tachydromia major var.minor Strobl, 1899: 79, syn.n.

A very large grey species with yellow antennae and a bare patch on sternopleura, palpi large, pale, and vein M very conspicuously bowed. Legs yellow even on tarsi, mid femora very thickened.

♂. Unknown.

♀. Head densely light grey dusted, face almost as deep as the not very broad frons, very prominent and often yellowish; the narrow jowls polished. A pair of pale vt bristles rather close together. Antennae (Fig. 160) yellow, extreme tip of segment 3 often darker, or faintly brownish on apical half; arista at least twice as long, almost blackish. Palpi (Fig. 157) pale, blunt, very large.

Thorax light grey dusted, sternopleura with a long-ovate, polished black patch at middle. Mesonotum with an indication of two darker stripes between acr and dc, these setae rather inconspicuous and narrowly biserial; two pre-scutellar pairs of dc longer, all pale.

Legs pale yellow including tarsi, sometimes apical tarsal segment slightly darkened, or with very indistinct light brown annulations on all segments. Fore femora slightly thickened, fore tibiae only very indistinctly and uniformly thickened; mid femora (Fig. 236) very thickened, at least twice as deep as fore femora, tibial spur (Fig. 236) large and sharply pointed.

Wings (Fig. 739) clear or faintly yellowish, vein R4+5 straight but vein M very bowed, ending close to vein R4+5; crossveins separated for a short distance, the vein closing anal cell recurrent. Abdomen shining black, anterior two tergites light grey dusted at sides, the following tergites with only an indication of lateral dusting anteriorly; apical two segments and long slender cerci dull yellowish-grey.

Length: body 2.8 (var.minor) - 5.5 mm, wing 3.5 - 5.5 mm.

Very common in Denmark and southern parts of Fennoscandia; in Norway north to On, in Sweden to Jmt. (♀♀), in Finland to Ta; also in the south of Russian Carelia. - Throughout Europe including the south. - May-mid August. Common on ground-vegetation in moist and shaded places, often on Urtica, but also on bushes.

Note. P.fasciatus (Meigen, 1822), known from Great Britain and practically throughout Europe, but uncommon everywhere, has antennal segment 3 extensively darkened, fore femora and tibiae very thickened, and abdominal tergites broadly greyish dusted at sides. The very closely related P.cruralis (Collin, 1961) has frons broader above, smaller palpi, antennal segment 3 more brownish, fore tibiae spindle-shaped, abdomen extensively grey dusted at sides, and vein M, even if strongly bowed, more widely separated from vein R4+5 at tip.

P.major is still known only from the female sex: Lundbeck (1910) described males by mistake, and there are only females in his collection; the males recorded from C.Europe by Zusková (1969) all belong to cruralis.

P.cruralis (Collin, 1969), described from material collected by Dr. O. Ringdahl at Col du Lautaret, France (see also Ringdahl, 1957), is a primary homonym (in Tachydromia) of P.cruralis (Collin, 1961); for the former species (with 2 pairs of vt bristles and allied to P.agilis) the new name Platypalpus ringdahli nom.n. (= Tachydromia cruralis Collin, 1969, nec Collin, 1961) is proposed.

76. PLATYPALPUS ANALIS (Meigen, 1830)
   Figs. 161, 237, 486-488, 740.

Tachydromia analis Meigen, 1830: 343.
Tachydromia varia Walker; Lundbeck, 1910: 302.

A large species with yellowish abdomen, a very small polished patch on sternopleura, irregularly 4-serial acr and distinct anterior notopleural bristle. Antennae with segment 3 blackish; tarsi with black annulations, and dark bristles on fore and hind tibiae above.

♂. Head as in major but face including cheeks silvery-grey, a pair of yellowish-brown vt bristles close together, antennal segment 3 (Fig. 161) blackish (sometimes yellowish at base) and palpi small.

Thorax densely grey dusted, more yellowish-grey on mesonotum, sternopleura with a small, indistinct polished patch just above mid coxae. All hairs and bristles pale, acr irregularly tri- to quadriserial, dc irregularly biserial with 2 prescutellar pairs large; a distinct third anterior notopleural bristle developed.

Legs yellow, tarsi with distinct blackish annulations, metatarsi only brownish. Fore tibiae slightly thickened, with 4 to 5 large blackish bristles above. Mid femora (Fig. 237) deeper than fore femora, the pale pv bristles rather short, anteroventrally with several longer pale hairs and anteriorly on apical third with 2 or 3 dark bristles; tibial spur (Fig. 237) large, sharply pointed.

Hind femora slender, anteroventrally with a row of densely-placed light brown spine-like bristles, hind tibiae with about 8 distinct black bristles antero- and posterodorsally.

Wings (Fig. 740) clear or inconspicuously yellowish, veins R4+5 and M slightly bowed but almost parallel before tip, crossveins widely separated. Abdomen yellowish, tergites sometimes darkened to almost blackish-brown. Genitalia (Figs. 486-488) large, polished black, much darker compared with the pale abdomen; cerci brownish, left cercus sclerotised apically.

Length: body 3.4 - 3.9 mm, wing 3.8 - 4.6 mm.

♀. Resembling male but hind femora with only fine pale hairs beneath. Abdomen more uniformly yellowish-brown, apical two segments and very long slender cerci densely light grey dusted.

Length: body 3.5 - 4.2 mm, wing 4.0 - 4.8 mm.

Rather rare in Denmark and S.Sweden (Sk.). - England (Slindon, Sx., ♂, 17 July 1951 (G.E.Shewell) - first record) and along the Baltic coast S to C. Europe. - July-August. On leaves of bushes.

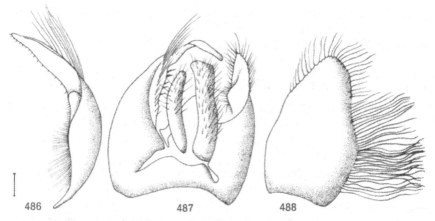

Figs. 486-488. Male genitalia of Platypalpus analis (Meig.). - 486: right periandrial lamella; 487: periandrium with cerci; 488: left periandrial lamella. Scale: 0.1 mm.

77. PLATYPALPUS CANDICANS (Fallén, 1815)
Figs. 159, 163, 238, 489-491, 741.

Tachydromia candicans Fallén, 1815: 10 (p.p., only ♀).
Tachydromia ventralis Meigen, 1822: 85.
Platypalpus varius Walker, 1851: 126.

Tachydromia oedicnema Strobl, 1898: 211, syn.n.

Tachydromia candicans var. flaviventris Strobl, 1898: 211, syn.n.

A light grey species with dusted sternopleura, very narrow frons, broader face and broad greyish dusted jowls. Antennal segment 3 blackish, vein M strongly but evenly bowed, fore tibiae in ♂ very spindle-shaped, and abdomen extensively yellowish at least on venter.

♂. Head densely light grey dusted, with very narrow frons which is scarcely broader than front ocellus is deep; face broader, jowls (Fig. 159) rather deep and grey dusted. A pair of pale vt bristles, close together. Antennae (Fig. 163) yellow on basal segments, segment 3 rather small and narrow, extensively blackish-brown even at base. Palpi (Fig. 159) pale yellow, large and pointed.

Thorax densely light grey dusted including sternopleura; acr small, pale, narrowly biserial and diverging; dc uniserial with 2 prescutellar pairs bristle-like; 2 posterior notopleural bristles.

Legs yellow including tarsi, or tarsi with very indefinite and faint annulations. Fore femora rather slender, fore tibiae very conspicuously spindle-shaped dilated. Mid femora (Fig. 238) very thickened, almost 3 times as deep as fore femora, the pale pv bristles rather shorter; tibial spur (Fig. 238) very large and sharply pointed, black towards tip.

Wings (Fig. 741) clear with vein R4+5 practically straight, but vein M very

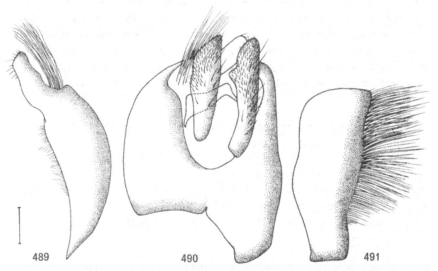

489   490   491

Figs. 489-491. Male genitalia of Platypalpus candicans (Fall.). - 489: right periandrial lamella; 490: periandrium with cerci; 491: left periandrial lamella; (oedicnema Strobl, Paralectotype). Scale: 0.1 mm.

strongly and evenly bowed towards tip; crossveins usually contiguous and the vein closing anal cell recurrent and S-shaped. Abdomen extensively yellowish at least on venter, tergites 1 - 5 often largely polished brown at sides, leaving a dull median stripe yellow; extreme sides of tergites slightly grey dusted anteriorly. Genitalia (Figs. 489-491) black.

Length: body 2.8 - 3.5 mm, wing 3.7 mm.

♀. Fore tibiae only slightly but uniformly thickened, not spindle-shaped dilated, narrower than fore femora; vein M not so strongly bowed, and tarsi often with very indistinct and faint annulations. Abdomen with tergites polished blackish-brown with all variants to entirely yellow abdomen (var.flaviventris), apical two segments and cerci always densely light grey dusted.

Length: body 2.8 - 4.5 mm, wing 3.3 - 4.3 mm.

Very common (♀ only) in Denmark and southern Fennoscandia, less common towards the north; in Norway up to TR, in Sweden to Jmt., in Finland to Om and Sb. - Females are very common throughout Europe including N.Africa, but males are so far known only from Austria, Czechoslovakia and Italy. - May-August. On ground-vegetation, preferably in shaded and humid places, often on patches of nettles together with ♀ of cursitans and major, but also on tree-trunks and the leaves of bushes and trees.

Note. The closely related P.nigrimanus (Strobl, 1880) (= fasciatus Meigen sensu Frey) has thorax rather golden-grey dusted, golden-yellow coloured palpi, broader frons, narrow polished jowls, extensively darkened legs and abdomen broadly grey dusted at sides. The male sex of candicans was previously unknown.- Zetterstedt confused it with longiseta and stabilis, and Lundbeck with laticinctus; the new synonymy oedicnema = candicans is based on a type revision, and the male sex is fully described and illustrated here for the first time.

78. PLATYPALPUS CURSITANS (Fabricius, 1775)
   Figs. 158, 164, 239, 492-494, 742.

Musca cursitans Fabricius, 1775: 782.
Tachydromia bicolor Meigen, 1804: 237.
Tachydromia cursitans var.denominata Frey, 1907: 409.
Tachydromia fasciata Meigen; Lundbeck, 1910: 305 (only ♀); Engel, 1939: 71
   (as Coryneta).

Resembling candicans but frons broader, and narrower jowls black, antennal segment 3 yellowish at base, tarsi annulated, abdomen with small grey lateral patches on all tergites, and different venation.

202

♂. Head with broader frons which is as deep below as antennal segment 2, face as deep and rather silvery-grey; jowls narrower, black. Antennal segment 3 (Fig. 164) somewhat broader, brownish, yellowish at base.

Thorax as in candicans but acr and dc even smaller and very inconspicuous. Tarsi with faint brown annulations, broader on apical two segments. Fore femora considerably thickened and fore tibiae slightly spindle-shaped, mid femora (Fig. 239) not very much deeper than fore femora. Wings (Fig. 742) with the vein M curved up before tip and ending close to vein R4+5 for some distance.

Abdomen polished black, all tergites with light grey dusting at sides forming rather broad lateral triangular patches. Genitalia (Figs. 492-494) not very large, polished black, but broad cerci dull grey and right lamella conspicuously wrinkled and somewhat subshining.

Length: body 2.9 - 3.2 mm, wing 3.2 - 4.1 mm.

♀. Fore tibiae rather slender and with shorter pale hairs beneath; abdomen with the same grey dusting, apical two segments and cerci greyish dusted, segment 7 rather yellowish-grey dusted.

Length: body 2.3 (var. denominata) - 4.8 mm, wing 3.3 - 4.6 mm.

Very common (♀), with a similar pattern of distribution to candicans, but extending further northwards; in Norway to Fn, in Sweden to Lu.Lpm. and Nb., in Finland to ObS; also from the Kola Peninsula. Males are very rare in the north: in Norway in N, in Sweden Ly.Lpm., Ög. and Sk., in Finland Sb; also in Lr. - Widespread and common in Europe, but records from the southern parts need verification. - May-August. Common together with candicans and major.

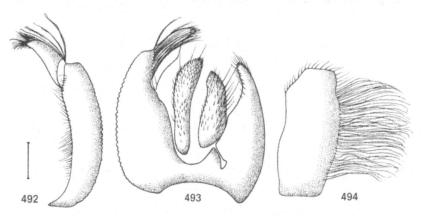

Figs. 492-494. Male genitalia of Platypalpus cursitans (Fabr.). - 492: right periandrial lamella; 493: periandrium with cerci; 494: left periandrial lamella. Scale: 0.1 mm.

203

Note. A very variable species, but the var. hispanica of Strobl (1899) is a distinct species with a polished patch on sternopleura, and var. chrysonata of Strobl (1899, 1906) is very probably a species close to flavicornis.

79. PLATYPALPUS VERRALLI (Collin, 1926)
    Figs. 162, 240, 495-497, 743.

Tachydromia Verralli Collin, 1926: 185.

A smaller light grey dusted species with dusted sternopleura and yellow antennae, segment 3 darkened in ♀. About 5 pairs of dc large and bristle-like, veins R4+5 and M almost parallel.

♂. Head light grey dusted, with narrower frons somewhat yellowish-grey and widening above, silvery-white face as deep. A pair of pale vt bristles rather close together, antennae (Fig. 162) yellow with segment 3 not much longer than deep, darkened at extreme tip, blackish arista almost 3 times as long. Palpi large, pale, broadly ovate.

Thorax entirely densely light grey to brassy-yellow dusted even on sternopleura, all hairs and bristles pale; acr small, closely biserial; dc uniserial with about 5 pairs large and bristle-like.

Legs yellow except for dark annulated tarsi, last segment almost black. Anterior four femora of about the same thickness, fore tibiae slightly but gradually thickened, not spindle-shaped. Mid femora (Fig. 240) with about six long pale pv bristles, the double ventral spines rather fine, longer and paler towards base; tibial spur (Fig. 240) large, sharply pointed.

Wings (Fig. 743) clear or slightly yellowish anteriorly, veins R4+5 and M

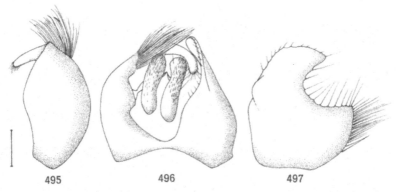

495   496   497

Figs. 495-497. Male genitalia of Platypalpus verralli (Coll.). - 495: right periandrial lamella; 496: periandrium with cerci; 407: left periandrial lamella. Scale: 0.1 mm.

almost parallel, crossveins contiguous and the vein closing anal cell indistinctly recurrent. Abdomen polished black with anterior margins of all tergites
grey dusted, the grey fasciae strongly broadened at sides. Venter uniformly
grey. Genitalia (Figs. 495-497) rather large with small dull grey cerci and
polished black lamellae.

Length: body 2.1 - 2.6 mm, wing 2.7 - 3.1 mm.

♀. Antennal segment 3 extensively darkened leaving a narrow base yellowish. Abdomen more extensively polished, the grey pattern restricted to triangular lateral patches; apical two segments and cerci grey dusted.

Length: body 2.3 - 3.1 mm, wing 2.5 - 3.3 mm.

Uncommon but widespread in Fennoscandia far towards the north; in Norway to NTi, in Sweden to Nb., in Finland to Ks. - Great Britain, forest zone
of European USSR, C.Europe. - (May) June-July. On ground-vegetation, often
in deciduous forests.

X. brevicornis - group

80. PLATYPALPUS BREVICORNIS (Zetterstedt, 1842)
Figs. 155, 165, 241, 498-500, 744.

Tachydromia brevicornis Zetterstedt, 1842: 293.
Tachydromia brevicornis var.subbrevis Frey, 1913: 76.(p.p.).

A robust, lighter grey dusted species with 2 pairs of blackish vt bristles and
irregularly 5-serial acr, antennae pale on basal segments, segment 3 dark.
Legs dark brown in ♂, yellow in ♀, tibial spur very small.

♂. Head (Fig. 155) lighter grey dusted with very broad frons and face, 2
pairs of vt bristles blackish, and considerably broad jowls below eyes. Antennae (Fig. 165) yellowish-brown on basal segments; segment 3 blackish and almost bare, slightly longer than deep; arista not very much longer. Palpi yellowish, small-ovate, apical hairs pale.

Thorax with sternopleura largely polished black. Acr small, dark brown,
irregularly 5-serial on a broad median stripe, dc multiserial with 2 or 3 inner
prescutellar pairs longer. Large bristles blackish.

Legs yellowish with coxae, basal two-thirds of femora and tarsi brownish,
or legs extensively brown to blackish-brown. Mid femora (Fig. 241) scarcely
deeper than fore femora, pv bristles black; tibial spur (Fig. 241) very small
and blunt, much shorter than tibia is deep.

Wings (Fig. 744) clear with brown veins, veins R4+5 and M almost parallel, crossveins contiguous; the vein closing anal cell recurrent. Abdomen polished black but basal two segments slightly dull grey at sides. Genitalia (Figs. 498-500) polished black and rather smaller.

Length: body 2.0 - 2.2 mm, wing 2.3 - 2.5 mm.

♀. Generally paler, with basal antennal segments yellow, legs yellow except for slightly brownish base of posterior coxae and apical 2 or 3 tarsal segments; halteres paler, yellowish-brown. Abdomen polished with last segment (at least posteriorly) and cerci greyish.

Length: body 2.3 - 2.7 mm, wing 2.2 - 2.7 mm.

Uncommon but widespread in eastern Fennoscandia, in Sweden from Sk. to T.Lpm., in Finland from N to Li; also in Russian Carelia but not known from Denmark, Norway or the Kola Peninsula. - NW of European USSR (Leningrad region, Estonia) and in the mountains of C.Europe (Krkonoše Mts., Czechoslovakia). - End April-August. On tree-trunks and logs in the Krkonoše Mts.

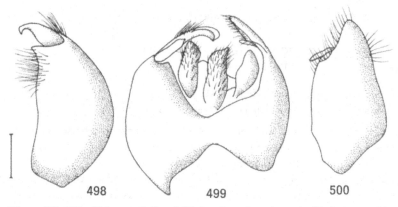

Figs. 498-500. Male genitalia of Platypalpus brevicornis (Zett.). - 498: right periandrial lamella; 499: periandrium with cerci; 500: left periandrial lamella. Scale: 0.1 mm.

81. PLATYPALPUS SORDIDUS (Zetterstedt, 1838)
Figs. 166, 242, 501-503, 745.

Tachydromia sordida Zetterstedt, 1838: 552.

Resembling brevicornis but antennal segment 3 longer and arista equal in length, palpi dark, and legs darkened on coxae and femora, mid femora very thickened and tibial spur as long as tibia is deep.

♂. Head as in brevicornis with 2 pairs of blackish vt bristles, but antennae (Fig. 166) almost blackish-brown, basal segments rather brownish, with larger, triangular-shaped segment 3 quite twice as long as its basal depth, arista equal in length. Palpi dark with apical hairs brown.

Thorax rather dark grey dusted but otherwise as in brevicornis. Legs blackish-brown on coxae and femora except for tips, tibiae and tarsi mostly yellowish, leaving hind tibiae on apical half and apical tarsal segments darkened. Mid femora (Fig. 242) very conspicuously thickened, almost twice as deep as fore femora and larger, tibial spur (Fig. 242) blunt but longer, about as long as tibia is deep. Abdomen polished blackish-brown, slightly dull at base; genitalia (Figs. 501-503) rather large, polished black and practically bare including the broad cerci.

Length: body 2.0 - 2.7 mm, wing 2.5 - 3.0 mm.

♀. Distinctly larger, abdomen polished except for greyish dusted apical two segments and slender cerci.

Length: body 3.2 - 3.4 mm, wing 3.0 - 3.3 mm.

Rather rare in northern Fennoscandia but not known from Norway, extending south in Sweden as far as Sm.; also from the Kola Peninsula. - Also in Estonia, according to Kovalev (in litt.) - July-August.

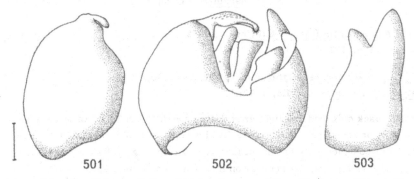

Figs. 501-503. Male genitalia of Platypalpus sordidus (Zett.). - 501: right periandrial lamella; 502: periandrium with cerci; 503: left periandrial lamella. Scale: 0.1 mm.

82. PLATYPALPUS SUBBREVIS (Frey, 1913)
   Figs. 168, 243, 746.

Tachydromia brevicornis var. subbrevis Frey, 1913: 76 (p.p.).

Much more closely allied to sordidus than to brevicornis, having similar but darker antennae and dark palpi, but mid femora more slender and a very short apical spur on mid tibia.

♂. Unknown.

♀. Head black in colour and rather thinly dark grey dusted with all characters as in sordidus but antennae (Fig. 168) uniformly blackish-brown in colour, basal segments scarcely paler. Thorax dark grey dusted with the same hairing and bristling.

Legs uniformly dark brown to blackish-brown, sometimes apical third of fore femora and whole of anterior tibiae paler. Fore femora not very thickened but mid femora (Fig. 243) much more slender, only slightly deeper than fore femora (not twice as deep as in sordidus); mid tibiae with apical spur (Fig. 243) somewhat pointed but very small, much shorter than tibia is deep. Wings (Fig. 746) and abdomen as in sordidus but halteres paler, pale yellow with darker stalks, not brown.

Length: body 2.0 - 2.5 mm, wing 2.2 - 2.4 mm.

Rare. Finland: Le, Enontekis, 2 ♀ (Sahlberg); Ks, Kuusamo, ♀ (Sahlberg). - No further data available.

XI. hackmani - group

83. PLATYPALPUS HACKMANI Chvála, 1972
    Figs. 156, 167, 244, 504-506, 747.

Platypalpus brevicornis Zetterstedt; Frey, 1943: 15, 17.
Platypalpus hackmani Chvála, 1972: 1.

A small black coloured and light grey dusted species with 1 pair of broadly separated vt bristles, prominent humeri and dusted sternopleura. Antennae yellow, frons conspicuously broad but face very linear, legs yellow with very thickened fore femora; pv bristles on mid femora and tibial spur absent.

♂. Head (Fig. 156) densely light grey dusted, with conspicuously broad frons that widens above, face very linear throughout. A pair of very pale, long, widely separated vt bristles. Antennae (Fig. 167) yellow, segment 3 often orange-yellow or extreme tip darkened, broad and almost spherical, blackish arista about twice as long. Palpi pale, long and slender, proboscis yellowish.

Thorax light grey dusted even on sternopleura, humeri very large, globose. All hairs and bristles pale yellow; acr irregularly narrowly biserial and very minute, dc uniserial with 3 pairs bristle-like; a long humeral, a posthumeral (!) and 2 posterior notopleural bristles.

Legs yellow except for very darkened last tarsal segment. Fore femora very thickened, mid femora (Fig. 244) not so thick as fore femora, with the usual ventral spines rather fine, more distinct in the posterior row towards base, no pv bristles; tibial spur (Fig. 244) quite absent.

Wings (Fig. 747) clear with pale veins, veins R4+5 and M almost parallel, crossveins usually contiguous and the vein closing anal cell indistinctly recurrent. Abdomen black to blackish-brown but uniformly thinly light grey dusted. Genitalia (Figs. 504-506) with a large and almost globular polished black right lamella, cerci very small.

Length: body 1.4 - 2.0 mm, wing 2.0 - 2.4 mm.

♀. Antennal segment 3 rather yellowish-brown and mid femora more slender with longer ventral spines. Abdomen including rather short cerci uniformly thinly grey dusted, often brownish in ground-colour.

Length: body 1.6 - 2.6 mm, wing 2.2 - 2.4 mm.

Rather rare, and only known from Sk., S. Sweden and from Finnish Lapland (Li, Le, LkW); also from the Kola Peninsula. - N and NW (Leningrad region) parts of European USSR. - June-early July. Frey (1943) collected adults on birch and conifer tree-trunks, under exactly the same conditions as de Meijere (1907) took his Chersodromia brevicornis (= P. nanus).

Note. The very closely related P. varicolor (Becker, 1908), P. minutissimus (Strobl, 1899) and P. nanus (Oldenberg, 1924) (syn. Chersodromia brevicornis de Meijere, 1907) form, together with hackmani, a very distinct group which represents an intermediate link between Platypalpus and Dysaletria.

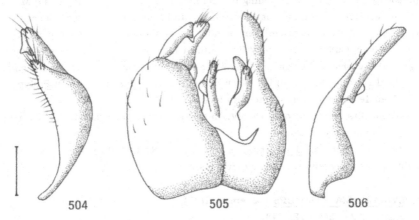

504   505   506

Figs. 504-506. Male genitalia of Platypalpus hackmani Chv. - 504: right periandrial lamella; 505: periandrium with cerci; 506: left periandrial lamella. Scale: 0.1 mm.

Genus *Dysaletria* Loew, 1864

Dysaletria Loew, 1864, Z.Ent., 14(1860): 30.

Type-species: Dysaletria melanocephala Loew, 1864 = unjustif. emend.
(originat. as lapsus calami) of Tachypeza atriceps Boheman (mon.).

A tiny yellow species, body about 2 mm in length and entirely thinly dusted.
Head (Fig. 5) deeper than either long or broad, eyes bare with facets equal in
size, broadly separated on frons but meeting below antennae. A pair of ocellar
and a pair of widely separated vertical bristles subequal, long and bristle-like.
Antennae (Fig. 509) inserted below middle of head in profile, basal segment invisible, apical two segments almost equal in size, terminal arista long. Palpi
small-ovate, rather flattened as in Platypalpus species and covered with fine
hairs. Proboscis slender, not half as long as head is high. Thorax rather narrow, slightly longer than deep and hardly convex above, humeri very large,
rounded and distinctly convex. Acr and dc practically absent except for last prescutellar pair of bristle-like dc, other large bristles rather thin: 1 humeral,
2 notopleural, 1 postalar and apical pair of scutellars. Legs with only short
hairs, without distinct bristles. Fore femora (Fig. 508) very thickened, almost
twice as deep as mid femora (Fig. 507), latter almost as slender as hind femora and with only fine small hairs in two rows beneath which become longer and
somewhat bristle-like towards base. Mid tibiae ventrally with a row of small
close-set bristly-hairs, no apical spur. Wings (Figs. 25, 748) clear and conspicuously narrow, vein R1 ending beyond middle of wing, veins R4+5 and M
almost parallel and rather close together. Both basal cells long but 2nd basal
cell slightly shorter, anal cell very small, the vein closing it almost at right-
angles to vein Cu but vein A very fine, almost invisible. Abdomen consisting of
eight fully sclerotised segments, male genitalia with conspicuously convex right
epandrial lamella (Fig. 510) but otherwise of the same structure as in Platypalpus. Female abdomen telescopic with long slender cerci.

The adults are found only rarely, in grasses or on bushes, and the life-hi-
story is completely unknown.

Only one species, D.atriceps, is known from North and Central Europe,
and it agrees fully with the generic diagnosis.

84. DYSALETRIA ATRICEPS (Boheman, 1851)
Figs. 5, 25, 507-512, 548.

Tachypeza atriceps Boheman, 1851: 190.

Entirely yellow species except for the blackish coloured and grey dusted head, and darkened last tarsal segment on all legs.

♂. Head (Fig. 5) with frons and occiput black in ground-colour but rather densely light grey dusted, a pair of ocellar and a pair of vt bristles pale, long but rather fine. Frons sometimes somewhat translucent yellowish below ocelli, almost as deep in front as antennal segment 3, widening strongly above, at least twice as broad at level of hind ocelli. Antennae (Fig. 509) whitish-yellow, segment 3 almost spherical but pointed at tip; arista thin, 3 times as long as segment 3, darker towards tip. Palpi very pale with longer pale hairs at tip; proboscis yellow, brownish towards tip.

Thorax yellow, uniformly thinly grey dusted including sternopleura. Large

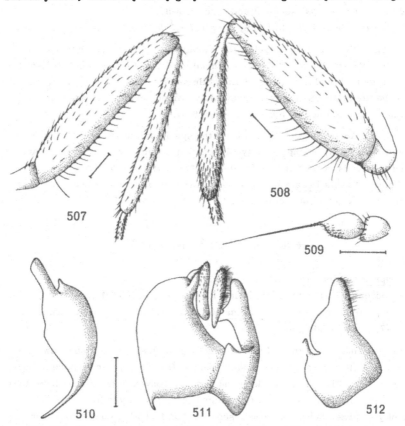

Figs. 507-512. _Dysaletria atriceps_ (Boh.), ♂ - 507: mid leg in posterior view; 508: fore leg in posterior view; 509: antenna; 510: right periandrial lamella; 511: periandrium with cerci; 512: left periandrial lamella. Scale: 0.1 mm.

bristles pale, rather fine. Legs (Figs. 507-508) yellow with last tarsal segment darkened, all hairs pale. Wings (Figs. 25, 548) clear with pale veins.

Thorax yellow, uniformly thinly grey dusted including sternopleura. Large bristles pale, rather fine. Legs (Figs. 507-508) yellow with last tarsal segment darkened, all hairs pale. Wings (Figs. 25, 548) clear with pale veins.

Abdomen yellow, thinly grey dusted and covered with sparse minute pale hairs, last segment with long bristly-hairs posteriorly. Genitalia (Figs. 510-512) subshining brown, contrasting with the pale abdomen and almost bare; large, at least as broad as tip of abdomen.

Length: body 1.6 - 1.8 mm, wing 1.7 - 1.9 mm.

♀. Resembling male but generally larger and abdomen very pointed with only short hairs, last segment and cerci brown.

Length: body 1.9 - 2.2 mm, wing 1.7 - 1.9 mm.

Rare. Sweden: Sk., Degeberga, ♀ (holotype); Öland, ♀; Östergötland, 4 ♂ 6 ♀; Södermanland, ♂ ♀ (all Boheman). - NW of European USSR (Leningrad region, Kovalev in litt.), Polish Silesia (Wroclaw)(Loew, 1864), Bavaria (Haunstetten)(Mus. Stuttgart). - June-July. Among grasses in dry biotopes (Boheman, 1851) and on bushes (Loew, 1864).

Note. According to Engel (1938), the specimens taken by Scholz at Wroclaw (Loew, 1864) belong to Platypalpus leucocephalus (von Roser), but according to Loew's redescription the single ♀ taken by Loew at Wroclaw undoubtedly belongs to this Dysaletria species. D. atriceps is a poorly known species, and the male sex is described here for the first time.

Genus *Tachypeza* Meigen, 1830

Tachypeza Meigen, 1830, Syst. Beschr., 6: 341.
Cormodromia Zetterstedt, 1838, Ins. Lapp. Dipt., p. 545 (MS).
Type-species: Tachydromia nervosa Meigen, 1822 (design. by Rondani, 1856:
147) = nubila (Meigen).

Uniformly medium-sized, slender, long-legged black to blackish-brown coloured and often densely grey dusted species. (Fig. 513), wings more or less brown. Head (Fig. 6) rather deep and narrow, convex in front and above neck, distinctly concave on lower half of occiput in profile. Eyes bare with facets slightly enlarged below, narrowly separated on frons but contiguous below antennae, with deep paraantennal excisions. Ocellar triangle placed rather lower on frons, latter again narrowed above. No ocellar and vt bristles, often a pair of small postvertical bristles and with some stronger bristles above neck, lower part of occiput with densely-set distinct whitish setae. Antennae (Fig. 16) small, basal segment indistinct, segment 3 uniformly short-ovate with very long, thin, bare terminal arista. Palpi small and narrow, apically with one or several long bristles; proboscis rather long and strong, somewhat recurved. Thorax (Figs. 519-521) rather long and narrow with distinct elongated humeri; mesonotum practically bare, usually with biserial acr and uniserial dc hair-like and scarcely visible, and with a few large bristles: 1 or 2 (occasionally 3) notopleural bristles, apical pair of scutellars, and sometimes a pair of strong prescutellar dc. Legs (Figs. 514-518) long and rather slender, fore femora the thickest, sometimes fore tibiae also thickened; no special raptorial adaptations. Wings (Figs. 26, 749-754) long and narrow with weakly developed anal lobe, more or less brownish clouded. Vein C ending at tip of vein M, vein R1 ending at or beyond middle of wing, vein R2+3 near tip of wing. Basal cells long but 2nd basal cell always slightly longer, anal cell and vein A absent, but the vein closing anal cell distinct, slightly recurrent. Abdomen consisting of eight fully sclerotised segments, but basal segment small and segment 8 in male usually completely hidden. Male genitalia generally of the same structure

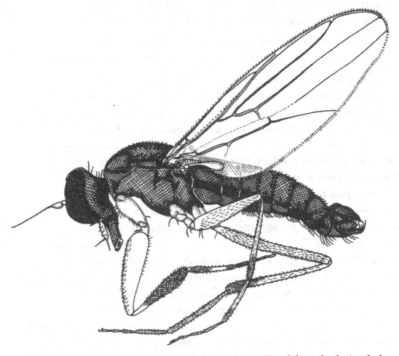

Fig. 513. Male of <u>Tachypeza nubila</u> (Meig.). Total length: 2.4 - 3.3 mm.

213

as in Platypalpus but varying greatly in shape among the individual species. Female abdomen telescopic with long slender cerci.

The adults are commonly found rapidly running on tree-trunks, and also on logs, stumps, palings or stones, but not on vegetation. They do not fly very much, and then only very rapidly and for a short distance. Like Tachydromia species, they carry their prey impaled on the proboscis. Nothing is known of the immature stages, but adults of nubila and fuscipennis have been bred several times from rotting wood or from the débris in a hollow tree.

The genus is mainly Holarctic in distribution, and is known throughout the Palaearctic region as far as Japan, and from North America; Brunetti (1920) recorded 2 species from India and I have seen an undescribed species from Laos. 10 species are known to me from Europe (3 of them still await description) and 18 species are described from the Nearctic region; of these only one, winthemi, is known to be Holarctic in distribution.

Key to species of Tachypeza

1 Thoracic pleura densely grey dusted, not shining (Fig. 513). Scutellar bristles large and widely separated (Figs. 519, 520)...........2

- Thoracic pleura extensively polished black; in doubtful cases (fuscipennis) scutellar bristles small and close together (Fig.521).......3

2(1) No distinct dc bristles (Fig.520). Fore coxae (Fig.514) with a strong black spine anteroventrally, and palpi with dark terminal bristles. ♂: Fore femora (Fig. 514) yellow, not maculated. ♀: Mid tibiae (Fig.518) with a long black apical bristly-hair anteriorly.................................... 85. nubila (Meig.)

- A pair of spine-like black dc bristles in front of scutellum (Fig. 519). Fore coxae (Fig. 515) without black spine, and palpi with pale terminal bristles. ♂: Fore femora (Fig.515) with black maculations anteriorly and with a cluster of long dark hairs near tip. ♀: Mid tibiae without a black apical bristle .... 86. truncorum (Fall.)

3(2) Scutellar bristles (Fig. 521) small and close together. Palpi with a very long strong black apical bristle, longer than palpus. The two rows of punctures from which the acr arise close together. Mid femora in ♀ with longer black bristles at base beneath ....87. fuscipennis (Fall.)

- Scutellar bristles large and widely separated. Palpi with smaller, usually pale apical bristles. The two rows of punctures

214

from which the acr arise widely separated. No longer bristles
or hairs (whitish in _sericeipalpis_) at base of mid femora
beneath in ♀. 4

4(3) Legs extensively yellow, fore femora in ♂ (Figs. 516, 517)
with distinct black maculations anteriorly. No distinct black
dc bristles; pleura polished black except for dusted hypopleura.
Wings (Figs. 752, 753) only faintly brownish. 5

- Legs blackish-brown. A pair of distinct black spine-like dc
bristles in front of scutellum; pleura entirely polished black
including hypopleura. Wings (Fig. 754) darker brown 6

5(4) Proboscis blackish; veins R4+5 and M parallel (Fig. 752).
♂: Fore femora (Fig. 516) anteriorly without a cluster of long
pale hairs; mid tibiae with a rather deep excision before tip
beneath; no tubercle on mid femora at base beneath, only a

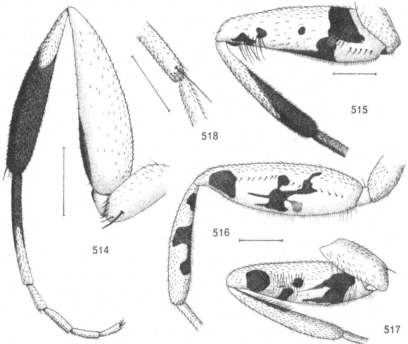

Figs. 514-517. Fore legs in anterior view of Tachypeza. - 514: _nubila_ (Meig.),
♂; 515: _truncorum_ (Fall.), ♂; 516: _heeri_ Zett., ♂; 517: _fennica_ Tuomik., ♂.

Fig. 518: Tip of mid tibia in posterior view of _Tachypeza nubila_ (Meig.), ♀.
Scale: 0.3 mm.

pair of fine long hairs .... 88. <u>heeri</u> Zett.

- Proboscis reddish-yellow; veins R4+5 and M slightly convergent before tip (Fig. 753). ♂: Fore femora (Fig. 517) anteriorly at middle with a cluster of long pale hairs; mid tibiae with a
shallow excision before tip beneath; mid femora with a large
tubercle near base beneath which is armed with several short
but strong bristles .... 89. <u>fennica</u> Tuomik.

6(4) Palpi small, dark; proboscis brownish, apically black; halteres dark. Mid tibiae in ♀ without longer hairs at base
beneath .... 90. <u>winthemi</u> Zett.

- Palpi large, pale yellow; proboscis entirely blackish; halteres
pale. Mid tibiae in ♀ with several long whitish hairs at base
beneath.. 91. <u>sericeipalpis</u> Frey

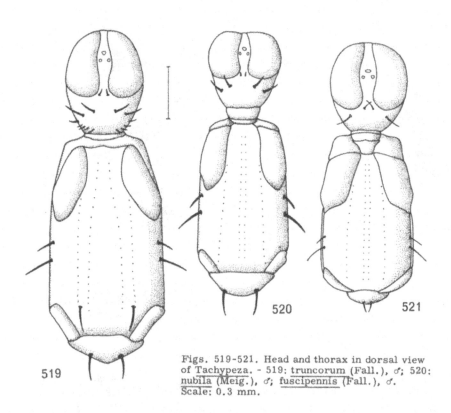

Figs. 519-521. Head and thorax in dorsal view
of Tachypeza. - 519: truncorum (Fall.), ♂; 520:
nubila (Meig.), ♂; fuscipennis (Fall.), ♂.
Scale: 0.3 mm.

519

520

521

85. <u>TACHYPEZA NUBILA</u> (Meigen, 1804)
   Figs. 6, 16, 26, 513, 514, 518, 520, 522-524, 749.

Tachydromia <u>nubila</u> Meigen, 1804: 239.
Tachydromia <u>nervosa</u> Meigen, 1822: 72.
Tachydromia <u>tibialis</u> Macquart, 1827: 91.

Rather a slender species with entirely grey dusted thorax and a pair of large
widely separated scutellar bristles; no distinct dc. Palpi with dark apical
bristles, fore coxae with a black spine on the inner side.

   ♂. Head (Fig. 6) densely grey dusted, upper part of occiput with 2 pairs
of strong black bristles, no pvt bristles. Antennae (Fig. 16) yellow with segment 3 darkened, arista blackish. Palpi small, whitish, with blackish bristles
at tip; proboscis blackish-brown.

   Thorax (Fig.520) entirely dull grey including pleura, the two rows of punctures from which the acr arise close together; 2 strong black notopleural bristles and 2 widely separated scutellars. Legs yellow with very spindle-shaped fore
tibiae (Fig. 514), base of fore metatarsus and tip of hind tibiae black, tarsi
brownish and fore femora (Fig. 514) with a blackish streak beneath. Mid femora
with several longer black bristles at base beneath, and fore coxae (Fig. 514)
with a strong black spine on the inner side.

   Wings (Figs. 26, 749) slightly brownish, more distinctly so along veins
R4+5 and Cu, crossveins widely separated. Abdomen black, brownish towards
base, entirely thinly grey dusted. Genitalia (Figs. 522-524) elongated, subshining.

   Length: body 2.4 - 3.3 mm, wing 2.5 - 3.3 mm.

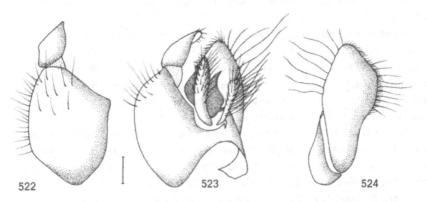

Figs. 522-524. Male genitalia of <u>Tachypeza nubila</u> (Meig.). - 522: right periandrial lamella; 523: periandrium with cerci; 524: left periandrial lamella.
Scale: 0.1 mm.

♀. Fore tibiae less thickened, and generally legs darker; however, fore metatarsus not black on basal part and no blackish streak on fore femora beneath; mid tibiae (Fig. 518) with a very long black bristle anteriorly at tip. Abdomen subshining, segment 8 and cerci very slender.

Length: body 3.0 - 3.7 mm, wing 2.6 - 3.2 mm.

Very common in Denmark and in S and C parts of Fennoscandia, rare towards the north; in Norway north to Nsi, in Sweden to Ly.Lpm., in Finland to Ks. - Throughout Europe, S to the Caucasus, France (Corsica) and Spain. - May-early November.

86. TACHYPEZA TRUNCORUM (Fallén, 1815)
     Figs. 515, 519, 525, 526, 750.

Tachydromia truncorum Fallén, 1815: 14.

A robust species with grey dusted thorax including pleura, and a pair of strong black prescutellar dc. Palpi whitish with mainly pale apical bristles.

♂. Head dull grey but ocellar tubercle shining, 1 or 2 pairs of small pvt bristles, behind them 2 pairs of long black occipital bristles. Antennae yellowish-brown, segment 3 darkened and arista almost black. Palpi pale, broader than in nubila, apically with several pale and a single dark bristles. Proboscis blackish-brown.

Thorax (Fig. 519) entirely grey dusted, more thinly on pleura; the two rows of punctures from which acr arise widely separated. Large bristles strong, black: 1 notopleural, 1 pair of prescutellar dc, 1 pair of widely separated scutellars.

Legs more thickened than in nubila, extensively darkened on posterior four femora, on hind tibiae at tip, and practically on all anterior tibiae. Fore femora (Fig. 515) thickened, with black maculations on the inner side and with a cluster of long black bristles on a black patch before tip, fore tibiae spindle-shaped. Mid femora with a tubercle near base beneath, and mid tibiae with a shallow excision covered with black spines before tip.

Wings (Fig. 750) long, rather uniformly but faintly brown clouded except for base, veins R4+5 and M slightly convergent just before tip, crossveins widely separated. Abdomen subshining black, often brownish anteriorly; genitalia (Figs. 525-526) large and globose, shining black.

Length: body 3.0 - 3.9 mm, wing 3.3 - 4.2 mm.

♀. Legs extensively darkened even on fore femora, the black maculations and dark bristly-hairs absent; mid femora with 2 long black bristles at base beneath, otherwise legs simple. Abdomen pointed, more densely grey dusted but apical two segments extensively shining black.

Length: body 3.4 - 4.5 mm, wing 3.6 - 4.3 mm.

Common throughout Fennoscandia including the extreme north, southwards in Sweden to Sk. but not yet recorded from Denmark. - Rare in Scotland, E to N and C parts of European USSR, Siberia, and mountains of C. Europe. - June-early September.

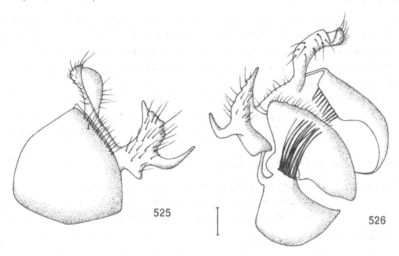

Figs. 525-526. Male genitalia of Tachypeza truncorum (Fall.). - 525: right periandrial lamella with cerci; 526: periandrium with cerci (left lamella anteriorly). Scale: 0.1 mm.

87. TACHYPEZA FUSCIPENNIS (Fallén, 1815)
Figs. 521, 527-529, 751.

Tachydromia fuscipennis Fallén, 1815: 14.

Smaller and more slender species with dark brown clouded wings, subshining pleura, and a pair of scutellars very small and close together. Palpi with a very long, strong black apical bristle.

♂. Head grey dusted, a pair of rather strong black, closely-set pvt bristles just behind the hind eye-corners, and another pair of slightly longer bristles on the upper half of occiput. Palpi pale, rather narrow, apically with a small dark bristle directed backwards, and another strong black bristle directed forwards which is distinctly longer than palpus. Proboscis brownish, black at tip.

Thorax (Fig. 521) thinly dark grey dusted, somewhat subshining on both mesonotum and pleura, but mesopleura always largely polished black. The two rows of punctures from which the acr arise close together; 2 long black noto-pleural bristles and a pair of very small, closely-set scutellar bristles.

Legs extensively yellow but hind femora except for basal quarter, practically all posterior tibiae and all tarsi darkened. Legs simple, fore femora strongly thickened, but mid femora and fore tibiae rather slender. Mid coxae with a strong black bristle anteriorly and mid femora with 2 long bristly-hairs at base beneath.

Wings (Fig. 751) somewhat shorter and narrower; dark brown clouded except for base and apex, veins R4+5 and M almost parallel, crossveins only narrowly separated. Abdomen blackish-brown, paler towards base, and entirely but very thinly grey dusted. Genitalia (Figs. 527-529) black, very large and globose, lamellae greyish towards tip.

Length: body 2.3 - 3.0 mm, wing 2.2 - 2.6 mm.

♀. Resembling male, legs with the same bristles on mid pair. Last abdominal segment long and stout, mostly shining black, sternite ending in two slender lateral processes.

Length: body 2.5 - 3.9 mm, wing 2.4 - 2.6 mm.

Uncommon in Denmark and the southern parts of Fennoscandia; in Norway north to On, in Sweden to Gstr. (a male in coll. Zetterstedt from Ly. Lpm.), in Finland to Tb. - England, C. and E. Europe, commoner on the Continent. - May-early August.

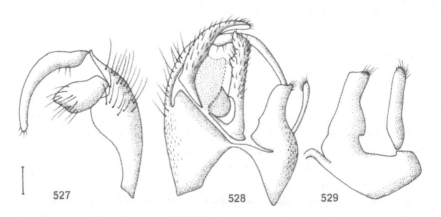

Figs. 527-529. Male genitalia of Tachypeza fuscipennis (Fall.). - 527: right periandrial lamella; 528: periandrium with cerci; 529: left periandrial lamella. Scale: 0.1 mm.

    Figs. 516, 530-532, 752.

Tachypeza heeri Zetterstedt, 1838: 547.
Tachypeza punctipes Zetterstedt, 1838: 547 (nom. nudum).

Larger species with extensively yellow legs and pleura polished black except for hypopleura, proboscis blackish. Strong prescutellar dc absent, fore femora bare in ♂ but black maculated anteriorly.

    ♂. Head grey dusted but ocellar tubercle shining, no pvt bristles, upper part of occiput with 2 pairs of strong black bristles. Antennae yellowish, segment 3 scarcely darker at tip, arista dark. Palpi pale and rather stout with arched anterior margin and several pale bristly-hairs at tip. Proboscis blackish.

    Thorax not very densely dark grey dusted on dorsum, pleura polished except for dull hypopleura. The two rows of punctures from which the hair-like acr arise widely separated. 1 strong black notopleural bristle and a subequal pair of widely separated scutellars.

    Legs yellowish but posterior four femora darker brown, tip of hind tibiae including metatarsus and last segment of all tarsi almost black. Fore femora (Fig. 516) thickened, on the inner side with black maculations but quite bare except for a row of dark tiny bristles; fore tibiae (Fig. 516) strongly thickened and arched in front, anteriorly with three large black patches. Mid femora covered with whitish pile beneath and usually with 2 long pale bristly-hairs at base, mid tibiae with a shallow excision on apical third beneath, which is black in colour and bears short black spinules.

    Wings (Fig. 752) rather faintly brownish clouded on costal half; veins R

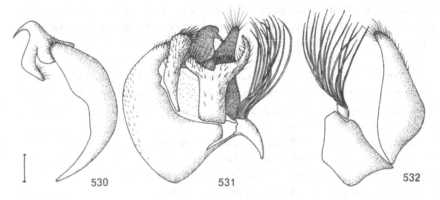

Figs. 530-532. Male genitalia of Tachypeza heeri Zett. - 530: right periandrial lamella; 531: periandrium with cerci; 532: left periandrial lamella. Scale: 0.1 mm.

4+5 and M practically parallel, even if somewhat upturned before tip; cross-veins widely separated. Squamae and halteres pale. Abdomen dull grey, genitalia (Figs. 530-532) large and globose, mainly shining black.

Length: body 3.3 - 4.0 mm, wing 3.6 - 4.0 mm.

♀. Legs mainly yellow except for brownish hind femora, darker tarsi and black tip of hind tibiae; fore femora and tibiae not maculated, no long bristles on mid femora at base, and mid tibiae simple. Apical two abdominal segments elongated and polished black.

Length: body 4.2 - 4.8 mm, wing 3.6 - 4.2 mm.

Rather uncommon in northern Fennoscandia; in Norway from HOy and Bø north to TRy, in Sweden from Boh. to T.Lpm., and throughout Finland; also from the Kola Peninsula, but not recorded from Denmark. - Scotland, NW of European USSR, mountains of C.Europe, Siberia, Japan. - June-September.

89. TACHYPEZA FENNICA Tuomikoski, 1932
   Figs. 517, 533-535, 753.

Tachypeza fennica Tuomikoski, 1932: 47.

Very closely resembling heeri but proboscis reddish-yellow and veins R4+5 and M slightly convergent before tip. Fore femora in ♂ with a cluster of long pale hairs anteriorly and mid femora with a tubercle at base beneath.

♂. Head as in heeri but proboscis reddish-yellow with only extreme tip somewhat darkened. Thorax with the same characters but fore femora (Fig. 517) with different black maculations on the inner side, with a cluster of long pale bristly-hairs at middle and with similar smaller hairs also near base. Fore tibia (Fig. 517) with only a black streak anteroventrally on basal half; mid femora with a distinct round tubercle at base beneath which is armed with 2 longer black spine-like bristles; apical excision on mid tibiae smaller.

Wings (Fig. 753) with the brownish clouding perhaps less distinct, and the entire wing rather greyish-brown tinged; veins R4+5 and M slightly convergent before tip. Abdomen with similar genitalia (Figs. 533-535) in general, but with slight structural differences.

Length: body 2.8 - 3.3 mm, wing 3.3 - 3.6 mm.

♀. Very similar to that of heeri, the paler proboscis and apically converging veins R4+5 and M being the only good differential features.

Length: body 3.4 - 4.0 mm, wing 3.3 - 3.6 mm.

Rather uncommon; in Sweden north to Ås.Lpm., rare in T.Lpm., in Finland to Ks; also from the Kola Peninsula but not yet recorded from Norway and

Denmark. - N and C parts of European USSR, mountainous regions of C. Europe south to Romanian Carpathians, Japan. - End May-August.

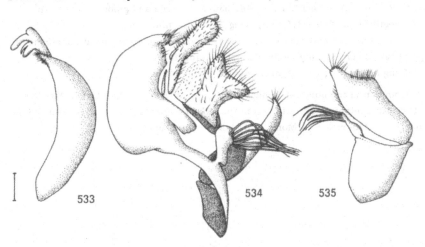

Figs. 533-535. Male genitalia of Tachypeza fennica Tuomik. - 533: right periandrial lamella; 534: periandrium with cerci; 535: left periandrial lamella. Scale: 0.1 mm.

90. TACHYPEZA WINTHEMI Zetterstedt, 1838.
   Figs. 536-538, 754.

Tachypeza winthemi Zetterstedt, 1838: 548.

Blackish species with pleura extensively polished black, palpi and halteres dark. Legs uniformly blackish-brown.

♂. Head grey dusted, a small pair of closely-set black pvt bristles behind eye-corners, upper part of occiput with 2 pairs of strong black bristles and numerous smaller bristles in a row behind postocular margin. Antennae uniformly dark brown; palpi dark, covered with pale hairs and with a black terminal bristle. Proboscis shining brown, tip blackish.

Thorax rather densely dark grey dusted on mesonotum, more silvery-grey on prothorax, pleura polished black including hypopleura. The two rows of punctures from which the acr arise widely separated. A pair of strong black prescutellar dc, and a pair of subequal widely separated scutellars.

Legs uniformly blackish-brown and simple; mid femora beneath with fine silver pile and 2 short black bristles at base; mid tibiae slightly thickened on apical quarter and anteroventrally with densely-set short black spiny-bristles.

Wings (Fig. 754) rather dark brown clouded except for a semi-hyaline poste-

223

rior margin, veins R4+5 and M very slightly convergent at tip, crossveins widely separated. Squamae and halteres dark brown. Abdomen subshining dark brown
and practically bare, genitalia (Figs. 536-538) large and globose, blackish.

Length: body 2.9 - 3.5 mm, wing 2.8 - 3.2 mm.

♀. Resembling male but mid femora without longer bristles at base beneath.
Apical two abdominal segments elongated, blackish.

Length: body 3.3 - 3.8 mm, wing 2.8 - 3.1 mm.

Common in the north of Fennoscandia, in Norway south to On, in Sweden
to Dlr., in Finland to Sb; also from the Kola Peninsula. - N of European USSR,
Siberia; N. America S to Mont. and Mass. - June-early August.

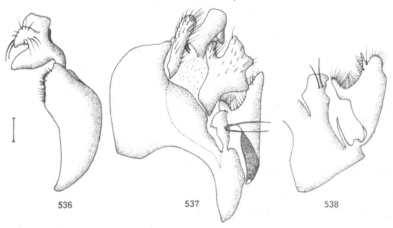

Figs. 536-538. Male genitalia of Tachypeza winthemi Zett. - 536: right periandrial lamella; 537: periandrium with cerci; 538: left periandrial lamella.
Scale: 0.1 mm.

91. TACHYPEZA SERICEIPALPIS Frey, 1913

Tachypeza sericeipalpis Frey, 1913 a: 10.

Smaller black species with extensively polished pleura, resembling winthemi
but palpi and halteres pale, proboscis uniformly blackish and all metatarsi paler.

♂. Unknown to me, but Frey (1915) described it as having large polished
black genitalia but otherwise resembling female.

♀. Resembling winthemi but basal antennal segments light brown, paler than
the dark segment 3. Palpi larger and pale, covered with densely-set pale hairs,
proboscis uniformly blackish. Thorax rather densely silvery-grey dusted on mesonotum and prothorax, otherwise pleura including hypopleura polished black.

224

Mesonotum with a distinct pair of black prescutellar dc, mentioned by Frey (1913 a) probably as postalar bristles.

Legs blackish as in <u>winthemi</u> but all metatarsi considerably paler, yellowish-brown, and mid femora at base beneath with several longer pale bristly-hairs. Wings brownish clouded, particularly on costal half and more distinctly along veins. Squamae dark, halteres yellowish. Abdomen very thinly grey dusted.

Length: body 2.5 - 2.7 mm, wing 2.5 mm.

Rare, known only from the Kola Peninsula: Lr, Ponoj, ♀ (Hellén). - North of European USSR, N. Siberia. - June-early August.

Note. The var. <u>dilutata</u> ♂, described by Frey (1913a) from Siberia, should differ from <u>sericeipalpis</u> by the less clouded wings, in which the clouding along the longitudinal veins in particular is not developed.

Genus *Tachydromia* Meigen, 1803

<u>Sicus</u> Latreille, 1796, Préc. car. gen. Ins., p. 158 (nec <u>Sicus</u> Scopoli, 1763: 369).

<u>Coryneta</u> Meigen, 1800, Nuov. classif. Dipt., p. 27 (suppressed by I. C. Z. N.,
    1963: 339).

<u>Tachydromia</u> Meigen, 1803, Illig. Mag., 2: 269.

<u>Sicodus</u> Rafinesque, 1815, Analyse de la Nature, p. 130 (nom. n. for <u>Sicus</u> La-
    treille, nec Scopoli).

<u>Phoneutisca</u> Loew, 1863, Berl. ent. Z., 7: 19.

<u>Tachista</u> Loew, 1864, Z. Ent., 14 (1860): 15.

Type-species: <u>Musca</u> <u>cimicoides</u> Fabricius, 1781 (design. by Curtis, 1833:
    pl. 477; misident.) = <u>connexa</u> Meigen.

Small to medium-sized polished black or greyish dusted species (Fig. 539), body 1 to 3.5 mm in length, wings (Figs. 27, 755-764) usually brown banded or clouded. Generally resembling species of <u>Tachypeza</u>, but ocelli placed higher up on frons and the latter distinctly widening towards vertex (Fig. 7). Usually a small, hair-like pair of ocellar bristles, a pair of fine postvertical bristles, and only fine bristly-hairs on upper half of occiput, without conspicuous whitish setae below neck. Antennae (Fig. 17) small as in <u>Tachypeza</u> but arista varying considerably in length. Thorax (Figs. 540, 541) very elongated with distinct long humeri; dc usually uniserial and scarcely visible, acr absent; 1 or 2 notopleural bristles, 1 postalar, 1 pair (or more) of scutellars, and sometimes 1 or several pairs of distinct prescutellar dc, all long and bristle-like to very strong and almost spine-like. Legs with thickened fore femora (Fig. 543), fore tibiae often spindle-shaped; mid femora (Figs. 544-546) rather slender, sometimes with ventral tubercles or excavations near base in ♂, mid tibiae in ♂ often with a projecting tooth at tip beneath. Wings (Figs. 27, 755-764) long and rather slender, rarely abbreviated, usually brown banded or clouded, rarely entirely clear. Venation as in Tachypeza, but first basal cell often very shortened and both crossveins very widely separated, and the lower branch of vein Cu, that usually closes anal cell, completely absent. Male genitalia of general tachydromiine structure, with the left lamella smaller but cerci considerably varying in size. Female abdomen pointed with rather shorter ovate cerci.

The adults run very rapidly searching for their prey on tree-trunks, palings and walls (thus on vertical surfaces), or on the ground, stones and ground-vegetation, often on large leaves of Petasites or Arctium (on horizontal surfaces); some of them are typical sand or coastal species. Unlike Tachypeza species,

Fig. 539. Male of Tachydromia umbrarum Hal. Total length: 2.3 - 2.8 mm.

the behaviour and ecology is usually clearly determined specifically. They rarely take flight but can fly very rapidly for short distances. Like the related Tachypeza species, they are not so highly predacious as for instance Platypalpus or Chersodromia species are. Nothing is known of the immature stages.

The genus is mainly Holarctic in distribution but a few species are also known from Mexico, Central America, South Africa, India and Formosa. 13 species are known from North America and 45 species from the Palaearctic region, but none of these is known to be Holarctic in distribution; the Palaearctic species have been recently revised (Chvála, 1970). 37 species are known from Europe of which only 11 have been found in Fennoscandia and Denmark, but the occurrence of a few others may be expected.

Seven natural groups of species have been established for the Palaearctic fauna (Chvála, 1970), of which three (ornatipes-, interrupta- and calcanea-groups) are mainly Central European and are not represented in North Europe; the other groups (terricola-, arrogans-, connexa- and annulimana-groups) occur here, and a further one, punctifera-group, is here established for the first time: it includes 2 distinctive northern species, previously known to the present author only from the existing inadequate descriptions of the female sex, and they have their closest relatives in the Nearctic fauna. A short diagnosis of each group is given in the following key to species.

Key to species of Tachydromia

1   Vein R2+3 almost straight and broadly separated from costa, cells R1 and R3 almost equal in width. Prothoracic episternum between humeri and fore coxae, or at least prosternum, silvery-grey dusted 2

-      Vein R2+3 arched towards costa, cell R1 much narrower than cell R3 (cf. Fig. 17). Prothorax polished black. Wings with two well-separated brown cross-bands (V. annulimana-group)....... 10

2(1)   Palpi unicolourous pale or dark, with a distinct terminal bristly-hair. Thorax silvery-grey dusted on prothorax between humeri and fore coxae, and on metapleura above hind coxae.....3

-      Palpi (Fig. 542) bicoloured pale and black in male, uniformly pale in female, without a terminal bristle. Thorax polished between humeri and fore coxae, the silvery-grey dusting confined to prosternum and to a small patch above hind coxae. Wing-pattern rather indistinct (IV. punctifera-group) 9

3(2)   Palpi pale yellow. Wings (Figs. 755, 756) more or less

brownish clouded, at least along veins, the brown clouding
reaching tip of wing; no cross-bands. Arista about 2.5
times as long as antenna, 1 notopleural bristle (I. terrico-
la-group).. 4

- Palpi dark in ground-colour. Wings (cf. Fig. 759) with two
brown cross-bands, sometimes broadly connected anteriorly;
tip of wing clear (except for halterata, see p. 235).................. 5

4(3) Wings (Fig. 755) only faintly brownish clouded along veins
and at tip (in cell R3), mainly clear. Mesonotum polished
black, all dc minute. Legs extensively yellow, with tarsi

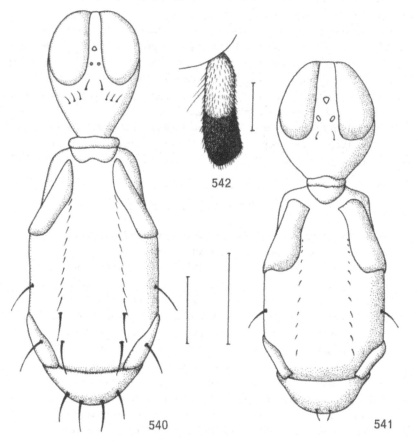

542

540 541

Figs. 540-541. Head and thorax in dorsal view of Tachydromia. - 540: umbrarum Hal., ♂; 541: woodi (Coll.), ♂. Scale: 0.3 mm.
Fig. 542: Palpus of Tachydromia incompleta (Beck.), ♂. Scale: 0.1 mm.

darkened at tip; mid femora (Fig. 544) in ♂ with a deep excision near base .... 92. terricola Zett.

-   Wings (Fig. 756) brown on apical two-thirds, base milky-white. Mesonotum thinly grey dusted, posterior 3 - 4 pairs of dc long, black. Legs yellow with distinct black median rings on femora (Fig. 543), or legs extensively darkened........
    .... 93. sabulosa Meig.

5(3) 2 notopleural bristles; arista short, at most twice as long as antenna. Wings (Figs. 757-758) rather broad and blunt-tipped, the brown cross-bands broadly connected anteriorly. Palpi small, covered with sparse black hairs. Male genitalia large, globular (II. connexa-group) 6

-   1 notopleural bristle; arista long, more than 3 times as long as antenna. Wings (Figs. 759-760) narrowed and somewhat pointed at tip, the brown cross-bands separated along their entire length. Palpi long and slender, covered with silvery-grey hairs. Male genitalia small, narrow (III. arrogans-group) 7

6(5) Legs (Fig. 545) extensively yellowish, hind femora yellow at base and fore femora with pale hairs beneath. Segment 2 of mid tarsi (Fig. 545) slightly more than half length of metatarsus ....94. connexa Meig.

-   Legs (Fig. 546) uniformly blackish-brown, hind femora entirely black, fore femora with short black bristles beneath. Segment 2 of mid tarsi (Fig. 546) almost as long as metatarsus .... 95. morio (Zett.)

7(5) Legs uniformly blackish-brown, anterior four femora with rather strong black bristles beneath. Vein R2+3 with a short appendix before tip, and the brown cross-bands disappearing beyond vein M. Larger, body about 3 mm in length ....
    .... 96. lundstroemi (Frey)

-   Legs bicoloured yellow and blackish-brown, anterior four femora with tiny black bristly-hairs beneath. No appendix to vein R2+3, and the brown cross-bands rather distinct even below. Smaller, body generally about 2 - 2.5 mm in length 8

8(7) Wings (Fig. 759) longer and blunter at tip, vein R2+3 sharply upturned towards costa. Hind femora all black, and posterior coxae yellow at tip. Occiput grey dusted along postocular margins....................................... 97. arrogans (L.)

-   Wings (Fig. 760) smaller and very pointed, vein R2+3 more

evenly upturned towards costa. Hind femora yellow at base
and posterior coxae mostly yellow, darker at base. Occiput
polished along postocular margins .... 98. <u>aemula</u> (Loew)

9(2)   Legs uniformly blackish-brown; all bristles on head and

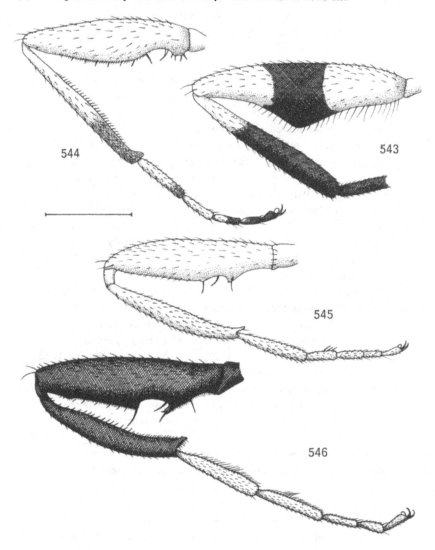

Fig. 543. Front leg in anterior view of <u>Tachydromia sabulosa</u> Meig., ♂.
Figs. 544-546. Mid legs in posterior view of <u>Tachydromia</u>. - 544: <u>terricola</u>
Zett., ♂; 545: <u>connexa</u> Meig., ♂; 546: <u>morio</u> (Zett.), ♂. Scale: 0.3 mm.

thorax whitish. Wings (Fig. 761) milky-white with conspicuous dark veins, the wing-pattern confined to a brown patch
at tip of vein R2+3, in ♂ also with a smaller patch at tip of vein
R1 and faint clouding along veins .... 99. punctifera (Beck.)
Legs partly yellowish, fore coxae always yellow; bristles
on head and thorax blackish. Wings (Fig. 762) clear, with two
brown cross-bands that disappear at posterior wing-margin....
.... 100. incompleta (Beck.)

10(1) Larger species; mesonotum (Fig. 540) with numerous strong
black spine-like bristles, at least 2 pairs of large prescutellar dc and 2 pairs of scutellars. Legs mostly blackish;
thorax with long yellowish bristles between posterior four
coxae; pvt bristles long, black .... 101. umbrarum Hal.

-      Smaller species; mesonotum (Fig. 541) with 1 pair of black
bristle-like scutellars, and all dc only minute. Legs mostly
yellowish-brown; no yellowish bristles between posterior four
coxae; pvt bristles dark and very small ....102. woodi (Coll.)

I. terricola-group

92. TACHYDROMIA TERRICOLA Zetterstedt, 1819
Figs. 544, 547-549, 755.

Tachydromia terricola Zetterstedt, 1819: 81.
Tachypeza apicata Zetterstedt, 1849: 3011.

Shining black species with wings faintly brownish clouded along veins and at
tip, antennae and legs mostly yellow.

♂. Head shining black with a pair of long black widely separated pvt bristles; antennae yellow, dark arista about 2.5 times as long as antenna. Palpi
yellowish, a long terminal bristle dark. Thorax shining black with whole of
pro- and metathorax silvery-grey dusted; 1 notopleural bristle long, black.

Legs yellow with tip of hind femora and all tibiae except for basal third
brownish, apical two tarsal segments very dark. Mid femora (Fig. 544) with a
distinct excision at base beneath, mid tibiae (Fig. 544) with only a short apical
projection.

Wings (Fig. 755) with faint brownish clouding at tip of wing in cell R3 and
along veins M and Cu, vein R2+3 straight. Abdomen shining black with large
genitalia (Figs. 547-549) of very complicated structure.

Length: body 2.0 - 2.3 mm, wing 2.0 - 2.1 mm.

♀. Resembling male but mid femora without the ventral excision at base beneath; apical two abdominal segments and cerci greyish dusted.

Length: body 2.2 - 2.5 mm, wing 2.1 - 2.2 mm.

Uncommon; known from Denmark, in Sweden north to Sm. and Gtl., in Finland to St and Ta; also in S of Russian Carelia. - From the Netherlands, E to NW of European USSR, S to its C parts and to C. Europe. - May-early September. In sandy coastal biotopes, also in grasses (Loew, 1864).

Note. The two related Siberian species, T.fuscinervis (Frey, 1915) ♀ and T.minima (Becker, 1900) ♀, are generally smaller and with extensively darkened legs.

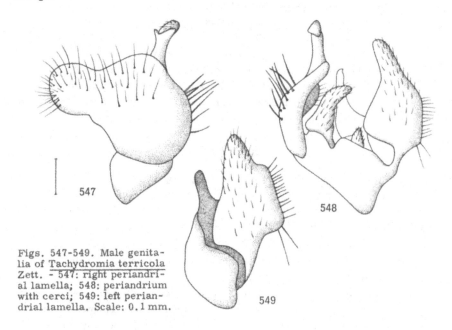

Figs. 547-549. Male genitalia of Tachydromia terricola Zett. - 547: right periandrial lamella; 548: periandrium with cerci; 549: left periandrial lamella. Scale: 0.1 mm.

93. TACHYDROMIA SABULOSA Meigen, 1830
Figs. 543, 550, 551, 756.

Tachydromia sabulosa Meigen, 1830: 342.
Tachypeza fenestrata Zetterstedt, 1842: 318.

Extensively grey dusted on thorax; wings milky-white on basal third, brown clouded towards and including apex. Legs yellow or darkened with black median ring on femora.

♂. Head shining black but faintly grey dusted on vertex and occiput above; a pair of close pvt bristles black. Antennae yellow on basal segments, segment 3 dark with arista about 2.5 times as long as antenna. Palpi pale with short black terminal bristle. Thorax entirely faintly grey dusted, more densely and silvery on prothorax but mesopleura polished. Mesonotum with numerous black bristles including 1 notopleural and 3 or 4 pairs of prescutellar dc.

Legs variable in colour but femora (Fig. 543) usually yellow with a black median ring, narrower on front pair but occupying almost all of hind femora except for tips. Fore femora (Fig. 543) with a large swelling at middle beneath, no apical projection on mid tibiae.

Wings (Fig. 756) milky-white on basal third or slightly more, rest of wing including apex brown clouded; vein R2+3 straight but placed rather closer to costa. Abdomen faintly grey dusted on dorsum, venter almost shining. Genitalia (Figs. 550, 551) large and globular, polished black, right lamella with distinct black spines.

Length: body 2.0 - 2.4 mm, wing 2.0 - 2.2 mm.

♀. Very like the male but the thickened fore femora without the median swelling beneath.

Length: body 2.0 - 2.5 mm, wing 2.1 - 2.3 mm.

Rather uncommon in Denmark and S. Sweden, in Norway in Ve, in Finland on the Baltic coast to Russian Ib, and along the Gulf of Bothnia north to Om. - Along the Baltic coast (Germany, Poland) E to Estonia, and along large rivers to C. Europe. - (May) June-July. A coastal species in sandy biotopes, and on sandy banks along large rivers, but also on drift-sand in Hungary.

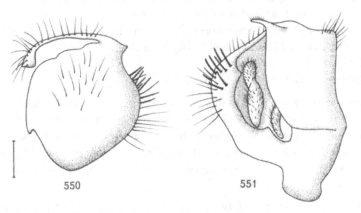

550   551

Figs. 550-551. Male genitalia of Tachydromia sabulosa Meig. - 550: right periandrial lamella; 551: periandrium with cerci (left lamella anteriorly). Scale: 0.1 mm.

94. TACHYDROMIA CONNEXA Meigen, 1822
Figs. 545, 552-554, 757.

Tachydromia connexa Meigen, 1822: 70.

Polished black species with short arista, small dark palpi and 2 notopleural bristles. Wings blunt-tipped, with brown bands broadly connected anteriorly; legs extensively yellow, with short segment 2 on mid tarsi, and hind femora pale at base.

♂. Head shining black but occiput subshining, slightly dull grey, a pair of long black widely separated pvt bristles. Antennae yellowish on basal segments, segment 3 brown; dark arista scarcely twice as long as antenna. Palpi very small, blackish, terminal bristle black. Thorax shining black except for largely silvery dusted pro- and metathorax, 2 black notopleural bristles.

Legs yellow, but hind femora except for base and entire hind tibiae black-ish-brown, anterior four femora with a brown streak above. Fore femora with fine pale hairs beneath, mid femora (Fig. 545) with a shallow excision on basal third beneath, armed with a yellow tooth and 3 black spines. Mid tibia (Fig. 545) rather short, with a spade-like apical projection, mid metatarsus almost twice as long as segment 2.

Wings (Fig. 757) rather broad and blunt-tipped, with two brown cross-bands broadly connected anteriorly, leaving a clear patch in cell R3. Vein R2+3 straight, cells R1 and R3 equally wide. Abdomen shining black with large globular geni-talia (Figs. 552-554) the dorsal process of right lamella slender.

Length: body 1.8 - 2.3 mm, wing 1.9 - 2.2 mm.

♀. Tibiae and tarsi rather brownish, fore tibiae and basal two tarsal seg-ments with shorter hairs anteroventrally, mid legs simple. Apical three abdo-minal segments greyish dusted.

Length: body 1.8 - 2.3 mm, wing 2.0 - 2.2 mm.

Very rare. Sweden: Sk., Hälsingborg, Pálsjo, ♀ (Ringdahl); Sdm., Stock-holm, ♂♀ (coll. Becker). - Widespread in Europe, including England and NW of European USSR (Leningrad region, Estonia); rather common in C.Europe, S to Romania and France. - April-early August, in Sweden in June. A typical springtime species, on ground-vegetation, but also on logs, stones and bushes along brooks (Strobl, 1893).

Note. There are at least two related species which may occur in southern Fennoscandia: T.costalis (von Roser, 1840) (syn. Sicodus submorio Collin, 1961), known from England and C. and S.Europe, has legs coloured as in con-

.nexa and shape of mid tarsi of morio, but the clear median stripe on wing only
reaches as far as lower part of cell R5; T.halterata (Collin, 1926), known from
England and eastwards as far as the Urals, has wings brown clouded on apical
two-thirds up to and including apex, and dark halteres.

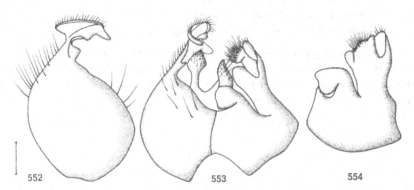

Figs. 552-554. Male genitalia of Tachydromia connexa Meig. - 552: right periandrial lamella; 553: periandrium with cerci; 554: left periandrial lamella.
Scale: 0.1 mm.

95. TACHYDROMIA MORIO (Zetterstedt, 1838)
Figs. 546, 555-557, 758.

Tachypeza morio Zetterstedt, 1838: 546.

Closely resembling connexa but legs extensively darkened with hind femora
blackish right up to base, segment 2 of mid tarsi almost as long as metatarsus,
and fore femora with fine black bristles beneath.

♂. Head and thorax as in connexa but antennae darker brown. Legs blackish-
brown, fore femora with short black bristles in anteroventral row and fore tibiae with predominantly black hairs beneath. Mid legs (Fig. 546) with the ventral femoral excision and tibial projection as in connexa but tarsal segment 2
longer, almost as long as metatarsus; hind femora not yellowish at base and
hind tibiae more curved towards tip.

Abdomen shining dark brown, tergite 6 with a brush of long black hairs at
hind margin at middle. Genitalia (Figs. 555-557) large and globular with the
apical process on right lamella apically broadened at sides.

Length: body 1.7 - 2.6 mm, wing 1.9 - 2.5 mm.

♀. Closely resembling male but mid legs simple, mid femora with 2 long
black hairs at base beneath, and hind tibiae only slightly curved at tip. Abdomen
subshining, apical three abdominal segments greyish dusted.

235

Length: body 1.8 - 2.7 mm, wing 2.0 - 2.4 mm.

Common and widely distributed in Denmark and Fennoscandia including the extreme north, less common towards south; also from the Kola Peninsula. - Great Britain, N and NW parts of European USSR; ? C. Europe (1 ♀, Styria). - June-early August. Mainly a coastal species and on sandy banks of lakes; usually on stones, or on sand under sea-weed and wood thrown up from the sea, but also on dunes and the adjacent ground-vegetation.

Note. T. acklandi Chvála, 1973a, recently described from Scotland, has head densely grey dusted, large yellowish-brown palpi with several long brown bristles, abdomen extensively grey dusted, and wings with the two brown bands broadly connected anteriorly even in cell R3.

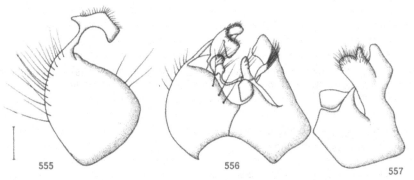

Figs. 555-557. Male genitalia of Tachydromia morio (Zett.). - 555: right periandrial lamella; 556: periandrium with cerci; 557: left periandrial lamella. Scale: 0.1 mm.

III. arrogans-group

96. TACHYDROMIA LUNDSTROEMI (Frey, 1913)
Figs. 558-560.

Tachista Lundströmi Frey, 1913: 73.

A species resembling arrogans but larger, legs uniformly blackish-brown with stronger black bristles on anterior four femora beneath, and the brown bands on wing disappearing below; vein R2+3 with a short appendix before tip.

♂. Head shining black but occiput behind vertex greyish dusted, a pair of more widely separated black pvt bristles. Antennae blackish-brown, arista about 3 times as long as antenna. Palpi dark brown, terminal bristle short, black. Thorax polished black, pro-and metathorax silvery-grey dusted; meso-

236

nótum almost bare, dc absent, and only 1 notopleural bristle.

Legs long and slender, uniformly blackish-brown, leaving knees and metatarsi slightly yellowish; anterior four femora with short black bristles in two rows beneath, and only a very small apical projection on mid tibia.

Wings with 2 brown cross-bands, widely separated throughout but only indistinct posteriorly; vein R2+3 almost straight and with a short downwardly pointed subapical appendix. Abdomen shining blackish-brown, genitalia (Figs. 558-560) small and slender, cerci short.

Length: body 2.8 - 3.2 mm, wing 2.5 - 3.0 mm.

♀. Resembling male, apical two or three abdominal segments dull grey.

Length: body 3.0 - 3.3 mm, wing 3.0 mm.

Rather rare in central parts of Fennoscandia; in Sweden in Ög. and Sdm., in Finland from the Baltic coast north to Ks; also in Russian Carelia. - NW of European USSR, E to Ural. - July-September. On tree-trunks (Frey, 1913).

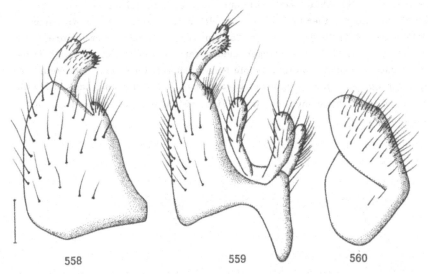

Figs. 558-560. Male genitalia of Tachydromia lundstroemi (Frey). - 558: right periandrial lamella; 559: periandrium with cerci; 560: left periandrial lamella. Scale: 0.1 mm.

97. TACHYDROMIA ARROGANS (Linné, 1761)

Figs. 7, 561-563, 759.

Musca arrogans Linné, 1761: 457.
Musca cimicoides Fabricius, 1781: 447.

Empis bifasciata Olivier, 1791: 390.

Tachista Fabricii Loew, 1864: 22.

Tachysta connexa var. c Strobl, 1893: 124.

A somewhat smaller slender species with yellowish and blackish-brown coloured legs, wings with 2 distinct brown bands wholly separated along their
entire length; vein R2+3 straight and sharply upturned towards costa, hind
femora entirely dark.

♂. Head (Fig. 7) shining black but upper part of occiput and postocular
margins at sides silvery-grey dusted, dark pvt bristles widely separated. Antennae yellow with segment 3 dark, arista about 3 times as long as antenna.
Palpi long, dark, covered with silvery hairs, terminal bristle black. Thorax
shining black except for silvery dusted pro- and metathorax, dc pale and minute, large bristles including 1 notopleural black but rather fine.

Legs yellowish but posterior four coxae except for tip, anterior four femora on dorsum, all tibiae and hind femora, and apical 3 tarsal segments, blackish-brown. Legs rather slender and simple, anterior four femora with short
black bristly-hairs beneath, at most on apical two-thirds on fore femora.

Wings (Fig. 759) rather longer and not very pointed at tip, the two brown
cross-bands separated on their whole length, often paler in cell R5. Vein
R2+3 straight and sharply upturned towards costa. Abdomen shining black with
small and somewhat conical genitalia (Figs. 561-563), cerci at least as long
as lamellae.

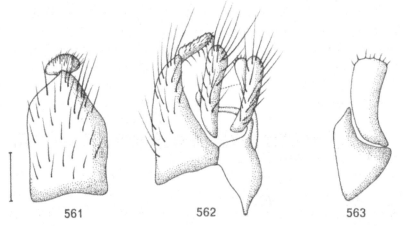

561   562   563

Figs. 561-563. Male genitalia of Tachydromia arrogans (L.). - 561: right periandrial lamella; 562: periandrium with cerci; 563: left periandrial lamella.
Scale: 0.1 mm.

Length: body 1.9 - 2.5 mm, wing 1.9 - 2.3 mm.

♀. Resembling male, apical two abdominal segments dull brownish-grey. Length: body 2.0 - 2.7 mm, wing 2.2 - 2.6 mm.

Common in Denmark, in Sweden north to Upl., in Finland to Sb; also in southern Russian Carelia but not yet recorded from Norway. - Common and widespread in Europe including the south, Palestine, Syria, ? N.Africa. - March-October. On logs, stumps, stones or tree-trunks, only rarely on adjacent vegetation; mostly in shaded and rather humid biotopes, more frequently in hilly areas.

Note. T.halidayi (Collin, 1926), known from England, is a smaller species, body about 1.5 mm in length, with pale palpi, dull grey occiput, brown crossbands on wing only narrowly separated at middle, legs extensively darkened, and anterior four femora without distinct bristles beneath.

98. TACHYDROMIA AEMULA (Loew, 1864)
Figs. 564-566, 760.

Tachypeza arrogans Linné; Zetterstedt, 1838: 546, 1842: 312 (p.p.); Lundbeck,
    1910: 267 (p.p.) (as Tachista).
Tachista aemula Loew, 1864: 22.

Closely resembling arrogans but occiput polished along postocular margins, hind femora yellow at base, wings smaller and very pointed with vein R2+3 only slightly curved towards costa.

♂. Head as in arrogans but occiput only very indistinctly and thinly dull grey above at middle, polished black along postocular margins at sides, and antennal arista even longer. Legs generally paler with posterior four coxae mainly yellow, dark at base only, and hind femora yellow at least on basal quarter, usually on the whole basal half.

Wings (Fig. 760) smaller and very narrow, pointed apically; wing-pattern as in arrogans but vein R2+3 closer to costa and not so sharply upturned before tip. Abdomen and genitalia (Figs. 564-566) practically as in arrogans.

Length: body 1.9 - 2.3 mm, wing 1.6 - 2.0 mm.

♀. With the main differential characters of the male.

Length: body 2.2 - 3.1 mm, wing 1.6 - 2.2 mm.

Common in Denmark, in Norway to Ry, in Sweden north to Dlr., in Finland along the Baltic coast north to Om; not yet recorded from Russian Carelia. - Widespread and very common in Europe, S to France and Jugoslavia. - June-October. On ground-vegetation, often on large leaves of Petasites and Arctium, only rarely on logs or stones. The adults prefer open sunny biotopes both in lowlands and in mountains.

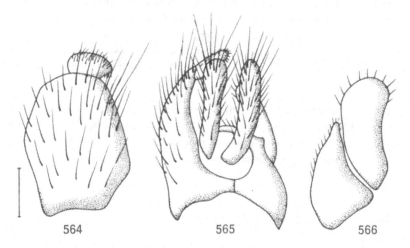

Figs. 564-566. Male genitalia of Tachydromia aemula (Loew). - 564: right peri-andrial lamella; 565: periandrium with cerci; 566: left periandrial lamella. Scale: 0.1 mm.

IV. punctifera-group

99. TACHYDROMIA PUNCTIFERA (Becker, 1900)
Figs. 571-573, 761.

Tachista punctifera Becker, 1900: 32.

Palpi long, pale, apically black in ♂, no terminal bristle. Legs blackish-brown, all bristles on head and thorax whitish. Wings somewhat milky-white with distinct dark veins and a conspicuous brown patch at tip of vein R2+3.

♂. Head polished black on frons but vertex and occiput densely light grey dusted, a pair of ocellar and a pair of widely separated pvt bristles whitish, long and fine, latter hardly distinguishable from other similar occipital hairs. Antennae blackish-brown, arista at most twice as long as antenna. Palpi longer than proboscis, whitish-yellow and narrow on basal third to half, apically flattened and black.

Thorax polished black but prosternum and a small patch between humeri and fore coxae silvery-grey dusted. All hairs and bristles whitish, dc uniserial and minute, 2 prescutellar pairs longer, as long as 1 or 2 notopleural bristles.

240

Legs including coxae uniformly blackish-brown, knees brownish, all hairs whitish except for a row of minute blackish bristly-hairs on mid tibiae beneath. Mid legs simple, no tibial projection; fore tibiae spindle-shaped.

Wings (Fig. 761) somewhat milky-white with very thick blackish veins, the brown pattern restricted to two small dark brown patches at costal margin and faint clouding along veins. Cell R1 broader than cell R3, crossveins widely separated.

Abdomen shining black with sparse minute whitish hairs on venter, genitalia (Figs. 571-573) small and slender, cerci short.

Length: body 2.2 - 2.3 mm, wing 2.2 mm.

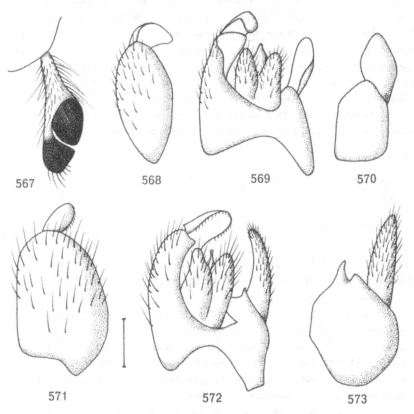

Fig. 567. Male palpus of Tachydromia bimaculata (Loew).
Figs. 568-570. Male genitalia of Tachydromia bimaculata (Loew). - 568: right periandrial lamella; 569: periandrium with cerci; 570: left periandrial lamella· Figs. 571-573. Male genitalia of Tachydromia punctifera (Beck.). - 571: right periandrial lamella; 572: periandrium with cerci; 573: left periandrial lamella. Scale: 0.1 mm.

♀. Palpi narrower even towards tip and entirely pale yellow, terminal bristle absent as in male. Wing-pattern restricted to the more distinct brown patch at end of vein R2+3.

Length: body 2.4 - 2.6 mm, wing 2.4 - 2.5 mm.

Rare. Norway: On, Fogstuen, 2♂ (Siebke); Sweden: T. Lpm., Jukkasjärvi, ♂, Kärkevagge, 2♀ (Hedström), Nuolja, ♂ 4♀, Tjuonajaure, ♀ (Ringdahl); Finland: Le, Enontekis, ♀ (Frey); also from the Kola Peninsula: Lr, Ponoj, ♂ 2♀ (Frey). - Siberia. - July. Hedström (in litt.) collected adults by sweeping ground-vegetation in marshy biotopes.

Note. The male sex is described here for the first time. The North American T. bimaculata (Loew, 1863) (syn. maculipennis Walker sensu Coquillett, 1903) is not identical with punctifera as was suggested by Ringdahl (1951). T. bimaculata has occiput polished along postocular margin and below neck, basal antennal segments yellowish, palpi in ♂ conspicuously bilobed apically (Fig. 567), and the whitish hairs becoming stouter and blackish on apical half of mid femora beneath; the wing-pattern is similar but the wings are faintly brownish clouded (not whitish) and vein R2+3 sharply upturned, almost at right-angles to costa; and the male genitalia show considerable differences (Figs. 568-570).

100. TACHYDROMIA INCOMPLETA (Becker, 1900)
Figs. 574-576, 762.

Tachista incompleta Becker, 1900: 33.
Tachydromia anderssoni Chvála, 1970: 494, syn. n.

Resembling punctifera but generally larger, legs paler, at least fore coxae yellow and wings clear with two indefinite brown cross-bands. All bristles on head and thorax blackish.

♂. Head including palpi as in punctifera but occiput polished black below neck, ocellar bristles minute and a pair of pvt bristles dark and rather close together; basal antennal segments yellowish and arista longer, about 2.5 times as long as antenna. Thorax without a small greyish patch between humeri and fore coxae, dc pale but last prescutellar pair longer and dark, other large bristles including 1 notopleural bristle black.

Legs yellow on fore coxae, at tip of posterior coxae and on trochanters, and on anterior four femora except for tip and dorsum; otherwise legs blackish-brown, fore tibiae very spindle-shaped. Wings (Fig. 762) with two broadly separated brown cross-bands which disappear below; vein R2+3 very indistinctly undulating but broadly separated from costa, cell R1 scarcely nar-

242

rower than cell R3. Abdomen shining black with scattered fine brownish hairs
on venter; genitalia (Figs. 574-576) small, superficially resembling those of
punctifera but cerci more slender and apical processes on both lamellae much
longer.

Length: body 2.2 - 2.6 mm, wing 2.4 - 2.6 mm.

♀. Palpi uniformly pale yellow and not widened apically, no terminal bristle. Legs darkened, rather brownish on yellow parts but fore coxae always yellow.

Length: body 2.4 - 3.0 mm, wing 2.5 - 2.6 mm.

Rather rare in northern parts of Fennoscandia including extreme north;
in Norway from On to Fn, in Sweden in P. and T.Lpm., in Finland in Le and
Li; also from the Kola Peninsula. - Widespread eastwards through Siberia
as far as Transbaicalia and Mongolia. - End June-early August. H.Andersson
(in litt.) collected this species when sweeping in a damp field of grass.

Note. Known only from the female sex until now, but when studying the
type-material I found it to be conspecific with anderssoni; the latter name becomes a junior synonym of incompleta.

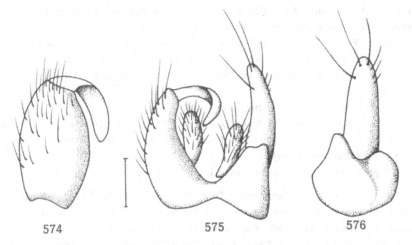

574   575   576

Figs. 574-576. Male genitalia of Tachydromia incompleta (Beck.). - 574: right
periandrial lamella; 575: periandrium with cerci; 576: left periandrial lamella.
Scale: 0.1 mm.

101. TACHYDROMIA UMBRARUM Haliday, 1833

Figs. 17, 27, 539, 540, 577-579, 763.

Tachydromia umbrarum Haliday, 1833: 161.

Tachypeza arrogans var.d. albitarsis Zetterstedt, 1838: 546.
? Tachypeza albitarsis var. nova Siebke, 1877: 24 (no material located).
Tachista annulimana Meigen; Lundbeck, 1910: 269; Frey, 1913: 75; Collin,
    1961: 90 (as Sicodus).

Larger polished black species without a silvery patch above fore coxae, mesonotum with numerous spine-like black bristles. Wings with two brown cross-bands and very narrow cell R1.

♂. Head shining black except for grey dusted vertex and upper part of occiput, black pvt bristles long. Antennae (Fig. 17) yellowish with segment 3 darker, arista more than 3 times as long as antenna. Palpi dark but covered with long silvery hairs, terminal bristle long, black. Thorax (Fig. 540) shining black including prothorax, mesonotum with numerous strong spine-like black bristles: 1 notopleural, 1 postalar, 2 - 3 pairs of prescutellar dc and 2 (or up to 4) pairs of scutellars; sternites between posterior four coxae with very long yellowish bristles.

Legs mostly blackish but fore coxae at least on apical half, base of all tibiae and basal two tarsal segments yellow, anterior four femora partly yellowish and with long pale bristles beneath. Mid tibia with a long spade-like projection at tip, mid metatarsus almost as long as tibia.

Wings (Figs. 27, 763) with 2 brown completely separated cross-bands, vein R2+3 arched towards costa, cell R1 much narrower than cell R3; veins R4+5 and M converging towards tip. Abdomen shining black, genitalia (Figs. 577-579) small.

Length: body 2.3 - 2.8 mm, wing 2.3 - 2.7 mm.

♀. Resembling male, but fore coxae except for narrow base and basal third of fore femora yellow; anterior four femora with shorter pale bristles beneath; mid tibia without apical projection.

Length: body 2.4 - 3.3 mm, wing 2.2 - 2.8 mm.

Very common and widespread in Denmark and Fennoscandia; in Norway to NTi, in Sweden to T.Lpm. and Nb., in Finland to Li; also in Russian Carelia but not yet recorded from Kola Peninsula. - Widespread in N parts of Europe, from Great Britain E to Ural, S to C.Europe (Austria, Czechoslovakia) and C parts of European USSR. - May-September. Very common on tree-trunks, also on palings, walls, guard-stones or telegraph-poles.

Note. There are several very closely related species, but the very common, rather Central European, T.annulimana Meigen, 1822 has very large globular genitalia in male, shorter legs with different pattern, and mesonotal bristles less numerous even if strong.

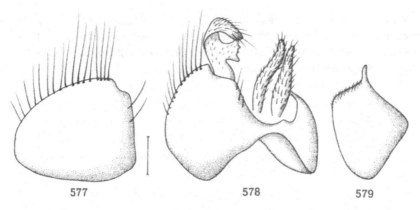

Figs. 577-579. Male genitalia of Tachydromia umbrarum Hal. - 577: right periandrial lamella; 578: periandrium with cerci; 579: left periandrial lamella.
Scale: 0.1 mm.

102. TACHYDROMIA WOODI (Collin, 1926)

 Figs. 541, 580-582, 764.

Tachista Woodi Collin, 1926: 151.

Resembling umbrarum but smaller, mesonotum with less numerous and only
fine dark bristles, 1 pair of scutellars and no distinct dc. Legs yellowish-brown,
mid metatarsus much shorter than tibia, and anterior femora with mainly short
hairs beneath.

 ♂. Head as in umbrarum but postocular margins also polished black, pvt
bristles very small and antennae paler with segment 3 light brownish. Thorax
(Fig. 541) with mesonotal bristles small and few in number, all dc only minute
and only 1 pair of scutellars; no yellowish bristles between posterior four
coxae.

 Legs mostly yellowish-brown, fore femora with a dark ring or at least a
small anterior patch before tip, with short hairs ventrally; apical projection
on mid tibia shorter and somewhat pointed, mid metatarsus about half as long
as tibia. Wings (Fig. 764) as in umbrarum but veins R4+5 and M parallel. Abdomen shining black, genitalia (Figs. 580-582) small with dorsal process on
right lamella ovate, not curved.

 Length: body 2.0 - 2.5 mm, wing 2.0 - 2.4 mm.

 ♀. Resembling male, but the dark subapical ring on fore femora usually
quite absent and no apical projection on mid tibia.

 Length: body 2.2 - 2.6 mm, wing 2.1 - 2.3 mm.

Very rare. Denmark: F, Fåborg, 2♀ (Schlick). - England, NW (Leningrad region, Estonia) and C parts of European USSR, C.Europe. - June-July. Biology unknown, but the adults of the very closely related C.European T.carpathica Chvála, 1966 live on ground-vegetation.

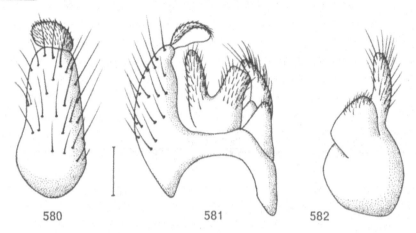

Figs. 580-582. Male genitalia of Tachydromia woodi (Coll.). - 580: right periandrial lamella; 581: periandrium with cerci; 582: left periandrial lamella. Scale: 0.1 mm.

TRIBE DRAPETINI

Genus *Drapetis* Meigen, 1822

Drapetis Meigen, 1822, Syst.Beschr., 3: 91.
Type-species: Drapetis exilis Meigen, 1822 (mon.).

Small, mainly polished black or yellow species (Fig. 583) with head close-set upon thorax, very narrow jowls below eyes and humeri not differentiated. Head (Fig. 9) deeper than long, convex both in front and behind, eyes microscopically pubescent. Frons very narrow in front, widening above, face very linear. Jowls narrow below eyes, not produced as in Crossopalpus. Bristles on head numerous but not very long, a pair of vt bristles usually with a similar pair of postocular hairs at sides, and 2 pairs of smaller ocellar bristles. Antennae (Fig. 18) inserted at or slightly above middle of head in profile, small; segment 2 at most with several longer fine bristles beneath as long as the segment, segment 3 short and pointed with long terminal arista. Proboscis stout, directed downwards; palpi flat, ovate. Thorax rather short and broad, humeri not differentiated, mesonotum uniformly covered with numerous small hairs, large bristles never very prominent. Metathorax well-developed, simulating the basal abdominal segment. Legs rather short and stout, fore and hind femora stouter, hind tibiae without a produced tooth posteriorly at tip. Wings (Figs. 28, 765-774) clear or uniformly slightly clouded, rather short and broad with developed axillary angle, the common stem of veins R2+3 and R4+5 rather long; vein R1 ending at middle of wing, vein R2+3 not far beyond middle of wing; 2nd basal cell much longer than 1st basal cell, vein A in the form of a fine fold, anal cell absent. Abdomen in ♂ short-conical, basal segment membraneous and hidden beneath enlarged metathorax, segment 8 ring-like and mainly hidden beneath segment 7. Segment 4 the longest, this and the narrow segment 5 with conspi-

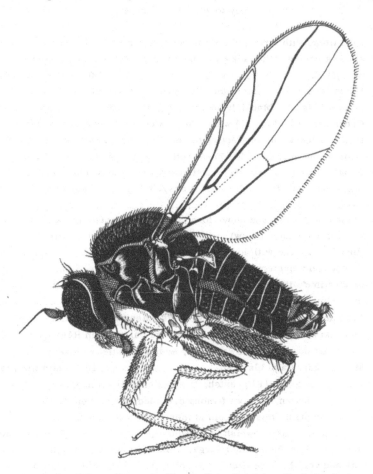

Fig. 583. Male of <u>Drapetis assimilis</u> (Fall.). Total length: 1.2 - 1.4 mm.

cuous apically flattened setae at sides, and their shape and structure are important specific features. Genitalia (Fig. 33) twisted around 180°, with periandrial lamellae quite separated; right lamella (placed on the left owing the twisting) large, the left one (on the right) much smaller but usually with a large dorsal process. Abdomen in ♀ broader and more flattened, short-pointed at tip, all segments almost equal in length.

The genus is worldwide in distribution and is represented in Europe by two easily distinguishable subgenera, Elaphropeza Macq. and Drapetis s.str.

Key to species of Drapetis

1   Hind tibiae with strong black anterodorsal bristles. European
    species yellow with black pattern on thorax (subgen. Elaphropeza
    Macq.) .... 103. ephippiata (Fall.)
-   Hind tibiae without anterodorsal bristles. European species
    polished black to blackish-brown (subgen. Drapetis s.str.) 2
2(1) Crossveins (cf. Fig. 28) widely separated for a distance greater
    than half length of last section of vein Cu; mid crossvein (r-m)
    at or before middle of 2nd basal cell (I. assimilis-group)............3
-   Crossveins closer together, distance apart less than half length
    of last section of vein Cu; mid crossvein beyond middle of 2nd
    basal cell (II. exilis-group) 6
3(2) Prothoracic episterna above fore coxae bare (except for se-
    veral minute hairs below), without a long pale upturned hair.
    Hind femora in ♂ with long pale posteroventral hairs 4
-   Prothoracic episterna with a long pale upturned hair. Hind
    femora in ♂ with shorter pale posteroventral hairs, shorter
    than femur is deep. 5
4(3) Face and clypeus polished black. Legs extensively blackish
    to blackish-brown, at most anterior legs paler. Vein R4+5
    somewhat downcurved towards tip, ending parallel with vein
    M (Fig. 28). Palpi blackish .... 104. assimilis (Fall.)
-   Face and clypeus dull greyish. Legs extensively yellow to
    yellowish-brown, at most femora darkened above. Vein R4+5
    almost straight, this and vein M more evenly diverging
    (Fig. 767). Palpi brownish to brownish-yellow .... 105. ingrica Kov.
5(3) Antennae (Fig. 587) rather large. Vein R4+5 almost straight,
    this and vein M more evenly diverging (Fig. 768). Hind femora
    in ♂ with longer pale posteroventral hairs .... 106. arcuata Loew

-   Antennae (Fig. 588) smaller. Vein R4+5 somewhat downcurved towards tip, ending parallel with vein M. Hind femora in ♂ with only short hairs .... 107. simulans Coll.

6(2) Wings clear. Antennal segment 2 (Fig. 589) with very short bristles beneath. Small species, body at most 1.5 mm in length 7

-   Wings more or less darkened. Bristles beneath antennal segment 2 more than half as long as the segment (Fig. 590); if short, then larger species with vein R4+5 abbreviated at tip 9

7(6) Vein R2+3 evenly curved towards costa, cell R1 rather narrower; vein M slightly undulating towards tip (Fig. 770). Legs extensively blackish even on tarsi. Antennal segment 3 (Fig. 589) small. ♂: fore femora with a recurved bristle at tip in front, tergite 4 dull. ♀: hind tibiae with minute short hairs behind... 108. pusilla Loew

-   Vein R2+3 more sharply upcurved towards costa, cell R1 broad;

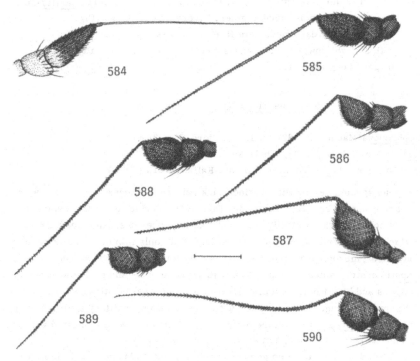

Figs. 584-590. Antennae of Drapetis. - 584: ephippiata (Fall.), ♂; 585: assimilis (Fall.), ♂; 586: ingrica Kov., ♂; 587: arcuata Loew, ♂; 588: simulans Coll., ♂; 589: pusilla Loew, ♂; 590: parilis Coll., ♂. Scale: 0.1 mm.

249

last section of vein M straight after a downcurved basal part
(Figs. 771, 772). Legs blackish leaving fore tibiae and all tarsi
yellowish. Antennal segment 3 larger. ♂: fore femora without a
recurved bristle at tip, tergite 4 shining 8

8(7)   Crossveins on wing rather closer together (Fig. 771). ♂:
right cercus slightly longer than left cercus and curved before
tip (Fig. 606). ♀: hind tibiae with conspicuous long pale hairs
behind .... 109. exilis Meig.

-       Crossveins slightly wider apart (Fig. 772). ♂: right cercus
much longer than the left one, almost straight (Fig. 607).
♀: hind tibiae without long hairs behind .... 110. infitialis Coll.

9(6)   Bristles beneath antennal segment 2 (Fig. 590) long, distinctly
more than half as long as the segment. Wings (Fig. 773) faintly
dark clouded, venation as in pusilla, vein R4+5 decidedly downcurved towards tip and distinct throughout. Smaller, body about
1.2 - 1.7 mm in length .... 111. parilis Coll.

-       Bristles beneath antennal segment 2 very short. Wings (Fig.
774) obviously darkened, vein R4+5 almost straight, apically
with a slight upward curve and abbreviated at tip. Generally
larger, body about 1.5 - 2.0 mm in length .... 112. incompleta Coll.

Subgenus Elaphropeza Macquart, 1827

Elaphropeza Macquart, 1827, Ins. Dipt. Nord Fr., 3: 86.
Ctenodrapetis Bezzi, 1904, Annls Mus. nat. hung., 2: 355.
Type-species: Tachydromia ephippiata Fallén, 1815 (mon.)

European species yellow with distinct dark pattern on thorax, hind tibiae with
long anterodorsal bristles. Antennae (Fig. 584) directed forwards, segment
3 more conical with lower edge straight like the upper one. Anterior pair of
ocellar bristles well-developed but posterior pair only minute. Thorax somewhat longer, mesopleura bare. Legs generally longer and more slender than in
Drapetis s. str., anterior four tibiae with small antero- and posteroventral apical spurs and hind tibiae with distinct long anterodorsal bristles. Wings (Fig.
765) somewhat longer and narrower with less developed axillary angle. Abdomen
of the typical Drapetis structure but male genitalia somewhat elongated with left
periandrial lamella larger, not so very small as in Drapetis s. str.

The adults are found on ground-vegetation and on bushes, and the larva
of ephippiata has been reared from woodland soil (Smith, 1969).

The subgenus is mainly tropical and subtropical in distribution; it is best

250

represented in the Oriental region, but is also present in South America, Africa and Australia. Only a few species are known from the cold and temperate zones of the northern hemisphere: 3 species are known from Europe, of which only ephippiata penetrates to its northern parts, whilst the other two (hutsoni Smith, 1967 a and boergei Chvála, 1971 a) are southern species; 1 species is known from the south of North America (Stone et al., 1965), 1 from Japan (Saigusa, 1964) and 6 species were described by Smith (1965) from Nepal.

103. DRAPETIS (ELAPHROPEZA) EPHIPPIATA (Fallén, 1815)
Figs. 584, 591-593, 765.

Tachydromia ephippiata Fallén, 1815: 11.
Tachydromia nigromaculata von Roser, 1840: 54.

Yellow species, thorax with four black ovate patches and hind tibiae with 2 long anterodorsal bristles at middle.

♂. Head black, thinly grey dusted except for polished frons, large bristles brownish. Antennae (Fig. 584) pale yellow on basal segments and base of segment 3, darkened at tip including arista. Palpi pale. Thorax shining yellow, a large oblong black patch near root of wing and another round one on hypopleura; scutellum and metanotum black. Small mesonotal hairs pale, large bristles blackish-brown; prothoracic episterna with a long pale upwardly directed bristle.

Legs yellow, apical tarsal segments brownish. Hind tibiae with longer

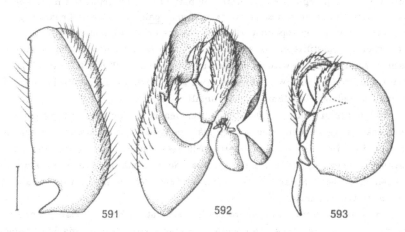

Figs. 591-593. Male genitalia of Drapetis (Elaphropeza) ephippiata (Fall.). - 591: right periandrial lamella; 592: periandrium with cerci; 593: left periandrial lamella with cerci. Scale: 0.1 mm.

pale hairs beneath and 2 large dark anterodorsal bristles at middle. Wings (Fig. 765) very faintly yellowish tinged, veins R4+5 and M evenly diverging and almost straight, crossveins widely separated as in Drapetis assimilis-group; 1st basal cell narrow, less than half length of 2nd basal cell. Halteres pale. Abdomen with large segment 4 and narrow segment 5 black, basal segments membraneous and whitish-yellow, posterior two visible tergites brownish. Genitalia (Figs. 591-593) elongated, polished blackish-brown.

Length: body 1.6 - 2.0 mm, wing 2.0 - 2.3 mm.

♀. Resembling male; basal three abdominal tergites and venter mainly membraneous and consequently pale yellow, tergite 4 and following tergites including ovipositor subshining blackish-brown.

Length: body 1.7 - 2.3 mm, wing 2.0 - 2.4 mm.

Common in Denmark and southern Fennoscandia to approximately 60°N, in Finland along the Baltic coast. - England, E to NW of European USSR (Leningrad region), common in central parts of Europe, but the records from S. Europe and N. Africa need verification. - June - August. On ground-vegetation and on leaves of bushes and trees.

Subgenus Drapetis s. str.

Shining black species (Fig. 583), hind tibiae without anterodorsal bristles. Antennae (Figs. 585-590) usually directed upwards, segment 3 short and rather ovate with lower margin convex. Anterior pair of ocellar bristles small, posterior pair much longer and distinct (as in Crossopalpus). Thorax broader when viewed from above, mesopleura with small hairs or bristles at least in upper hind corner. No distinct bristles on legs, at most femora with tiny preapical bristles, anterior tibiae without a trace of apical spurs. Wings (cf. Fig. 28) shorter and broader with well-developed axillary angle. Male genitalia (Fig. 33) small and short, left lamella very small.

The adults are only found individually, searching for prey in the bark of tree-trunks, less frequently on coniferous trees, but also often on windows in houses or in cowsheds, stables and green-houses; some species are common on ground-vegetation, particularly in fields. No hibernation in the adult stage has been observed. Larvae have been reared from debris in hollow trees and from rotten tree-stumps.

The subgenus is almost worldwide in distribution but is probably more frequent in temperate zones. 16 species are known from North America (Stone et al., 1965), and about 20 species from the Palaearctic region (but many others await discovery), one of which, assimilis, is recorded as Holarctic in distribution. Smith (1967, 1969) has recorded 8 Neotropical and 2 Ethiopian species respectively, but other records need verification.

Collin's (1961) excellent study of the British Drapetis species, together with the most recent revision of the European USSR fauna (Kovalev, 1972), can also be used for Fennoscandia; other records from Europe published before 1961 cannot be trusted. 9 species are recorded in this work from Fennoscandia but the occurrence of at least 3 further species seems to be likely. The species may easily be separated into two natural groups (assimilis- and exilis-groups), and their characteristics are given in the key above.

I. assimilis-group

104. DRAPETIS (DRAPETIS) ASSIMILIS (Fallén, 1815)
Figs. 9, 28, 583, 585, 594, 595, 766.

Tachydromia assimilis Fallén, 1815: 8.

Crossveins almost as widely separated as the length of last section of vein Cu. Polished black species with legs extensively darkened, prothoracic episterna bare and hind femora in ♂ with long pale pv bristly-hairs.

♂. Head (Fig. 9) shining black even on narrow face and clypeus, occiput above neck at middle thinly dull, all bristles black. Antennae (Fig. 585) blackish, segment 2 with 1 or 2 longer bristles at tip beneath, segment 3 short-ovate with long terminal arista; palpi blackish. Thorax shining black including pleura, prothorax polished except for a small greyish patch on bare episternum above. Short mesonotal pubescence pale, large bristles black.

Legs usually extensively dark brown to blackish, hind femora always almost black, fore coxae, base of femora and tibiae sometimes yellowish or at least lighter brown. Mid femora with a row of short black bristles posteroventrally; hind femora with 2 longer black preapical anterior bristles (much longer than other preapical bristles on anterior pairs), several long pale hairs above at base, and posteroventrally on basal two-thirds fringed with long pale hairs quite as long as femur is deep.

Wings (Figs. 28, 766) clear with brown veins, vein R4+5 considerably downcurved towards tip, ending parallel with vein M. Crossveins widely separated, squamae and halteres dark brown. Abdomen subshining blackish-brown; genitalia (Figs. 594-595) not broader than end of abdomen, all appendages quite free.

Length: body 1.2 - 1.4 mm, wing 1.5 - 1.6 mm.

♀. Mid tibiae with only fine hairs beneath, and no fringe of long pale hairs posteroventrally on hind tibiae.

Length: body 1.4 - 1.6 mm, wing 1.6 - 1.7 mm.

For the present known from Finland, where it is rather common along the Baltic coast in Ab and N. - England, NW of European USSR east to its central parts, C. Europe; ? Crimea, ? N. Africa; in North America (Stone et al., 1965) from Canada to S. Dak. and Ill.- May-September but mainly in August and September. On tree-trunks, less often on coniferous trees; also on flowers (Tuomikoski, 1952).

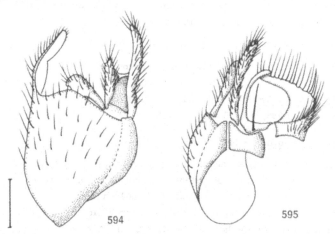

Figs. 594-595. Male genitalia of Drapetis (Drapetis) assimilis (Fall.). - 594: right periandrial lamella with cerci; 595: periandrium with cerci (left lamella in anterior view). Scale: 0.1 mm.

105. DRAPETIS (DRAPETIS) INGRICA Kovalev, 1972
Figs. 586, 596-598, 767.

Drapetis (s. str.) ingrica Kovalev, 1972: 181.

Resembling assimilis but legs extensively yellowish, face and clypeus dull grey, and veins R4+5 and M more evenly diverging.

♂. Head as in assimilis but face and clypeus dull greyish dusted, antennae (Fig. 586) rather brownish and basal segments sometimes paler. Palpi brownish in ground-colour. Thorax as in assimilis, without the upturned pale hair on prothoracic episterna. Legs yellowish, tarsi and hind femora brownish, or all legs rather uniformly brownish. Mid femora with the usual black posteroventral bristles brownish and more adpressed, indistinct on apical half of femur; pale posteroventral ciliation on hind femora long but scarcely as long as femur is deep.

Wings (Fig. 767) clear with yellowish-brown veins; vein R2+3 more evenly bowed towards costa, vein R4+5 almost straight, this and vein M somewhat diverging unlike assimilis; stalks of halteres paler. Slight differences also present in genitalia (Figs. 596-598).

Length: body 1.1 - 1.4 mm, wing 1.5 - 1.7 mm.

♀. Antennal segment 3 considerably shorter and rather triangular in shape, arista longer, at least 6 times as long as segment 3. Posterior four femora with only minute pale hairs beneath. Abdomen brown to yellowish-brown, all tergites almost equal in length, last segment polished blackish-brown.

Length: body 1.4 - 1.8 mm, wing 1.6 - 1.9 mm.

Rather rare along the Baltic coast of Finland, north to the south of Russian Carelia (Suistamo). - NW of European USSR (Leningrad region). - July-August.

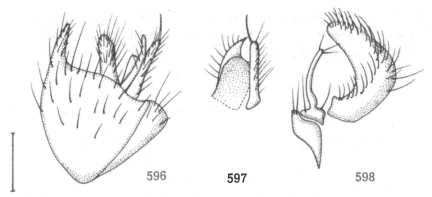

Figs. 596-598. Male genitalia of Drapetis (Drapetis) ingrica Kov. - 596: right periandrial lamella with cerci; 597: cerci in anterior view; 598: left periandrial lamella. Scale: 0.1 mm.

106. DRAPETIS (DRAPETIS) ARCUATA Loew, 1859

   Figs. 587, 599, 600, 768.

Drapetis exilis Meigen; Zetterstedt, 1838: 554.
Drapetis minima Meigen; Zetterstedt, 1842: 327.
Drapetis arcuata Loew, 1859: 40.
Drapetis assimilis Fallén; Lundbeck, 1910: 256.

Resembling assimilis but generally larger, antennae longer, prothoracic episterna with a long pale upturned hair, and wings with different venation; costal cell broad, veins R4+5 and M evenly diverging.

♂. Head with antennae larger than in assimilis, segment 3 (Fig. 587) distinctly longer than deep, broad, pointed at tip. Thorax polished black even on prothorax, episterna of latter with a long pale hair directed upwards. Legs often light brown to yellowish on front pair and sometimes also on mid femora, hind legs almost always blackish-brown; tarsi darkened. Hind femora with the pale posteroventral ciliation slightly shorter.

Wings (Fig. 768) with vein R1 distinctly bowed and consequently costal cell very broad, at least as deep as 2nd basal cell; veins R4+5 and M evenly diverging, vein R4+5 not downcurved before tip. Genitalia (Figs. 599-600) with quite different cerci.

Length: body 1.5 - 1.7 mm, wing 1.6 - 1.7 mm.

♀. Differing from male only by short hairs beneath posterior four femora and equally long abdominal segments, last segment dull brownish.

Length: body 1.7 - 2.0 mm, wing 1.7 - 1.9 mm.

Uncommon in Denmark and S.Sweden north to Nrk., in Finland to Om. - England, C.Europe (Germany, Czechoslovakia, Austria). - June-August. On tree-trunks but also often on windows.

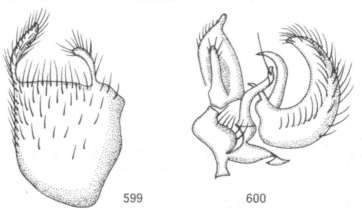

599 600

Figs. 599-600. Male genitalia of Drapetis (Drapetis) arcuata Loew. - 599: right periandrial lamella; 600: left periandrial lamella with cerci; (after Collin, 1961).

107. DRAPETIS (DRAPETIS) SIMULANS Collin, 1961
Figs. 33, 588, 601, 602, 769.

Drapetis simulans Collin, 1961: 34.

Prothoracic episternum with a long upturned hair as in arcuata but antennae smaller, vein R4+5 slightly downcurved before tip, and hind femora in ♂ without a ciliation of long hairs beneath.

256

♂. Head with small antennae (Fig. 588), uniformly blackish or with basal segments paler, segment 3 only slightly longer than deep and considerably smaller than in arcuata. Prothoracic episternum with a long pale upturned hair above fore coxae, otherwise with the same characters as in assimilis. Legs often uniformly yellowish to yellowish-brown but hind legs sometimes extensively blackish; hind femora with only minute pale hairs beneath.

Wings (Fig. 769) with dark veins, vein R2+3 considerably bowed and veins R4+5 and M with a tendency to be somewhat diverging towards tip, vein R4+5 not so much downcurved before tip as in assimilis; costal cell narrower than in arcuata, narrower or scarcely as deep as 2nd basal cell. Abdomen generally resembling that of assimilis but genitalia (Figs. 601, 602) with different appendages and cerci.

Length: body 1.3 - 1.7 mm, wing 1.6 - 1.8 mm.

♀. Resembling male except for sexual differences.

Length: body 1.4 - 1.6 mm, wing 1.6 - 1.8 mm.

Common in Denmark and southern Fennoscandia, in Sweden north to Nrk., in Finland to Sa; there is a ♀ labelled "Norway" in Coll. Becker. - England, NW and C parts of European USSR, C.Europe (Germany, Czechoslovakia, Austria). - June-September. On tree-trunks but also often on windows.

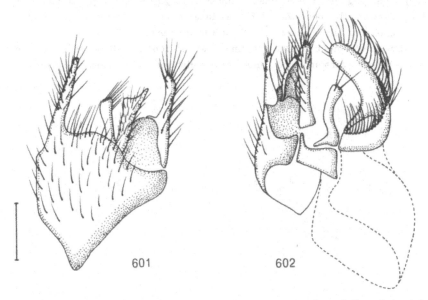

Figs. 601-602. Male genitalia of Drapetis (Drapetis) simulans Coll. - 601: right periandrial lamella with cerci; 602: periandrium with cerci (shape of hypandrium shown by broken line). Scale: 0.1 mm.

Note. D.convergens Collin, 1926, known from England and C.Europe, also belongs to this group of species, but it has silvery dusted prothorax including episternum, uniformly shorter bristles beneath antennal segment 2, yellowish legs and halteres, very small and deep black antennae, and veins R4+5 and M slightly undulating but almost parallel except at base. D.flavipes Macquart, 1834 is closely related to convergens.

II. exilis-group

108. DRAPETIS (DRAPETIS) PUSILLA Loew, 1859
Figs. 589, 603, 604, 770.

Drapetis exilis Meigen; Zetterstedt, 1842: 328.
Drapetis pusilla Loew, 1859: 42.
Drapetis nigripes Zetterstedt, 1859: 4997.

Wings clear with crossveins close together, cell R1 narrower and vein M slightly undulating. Legs extensively black even on tarsi, fore femora in ♂ with a recurved bristle at tip and dull tergite 4, hind tibiae in ♀ with short hairs.

♂. Head shining black but occiput mostly thinly dull grey. Antennae (Fig. 589) black, segment 3 very small, almost as long as deep; segment 2 with very short dark bristles beneath, not half as long as length of the segment. Prothoracic episterna silvery dusted, without a long upturned hair.

Legs uniformly blackish including tarsi, all femora with several dark preapical bristly-hairs and fore femora with a distinct recurved preapical bristle

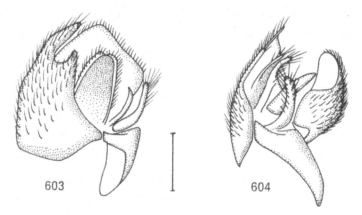

603

604

Figs. 603-604. Male genitalia of Drapetis (Drapetis) pusilla Loew. - 603: periandrium with cerci (cerci in anterior view); 604: the same (left lamella in anterior view). Scale: 0.1 mm.

in front, crossing the usual preapical bristles. Mid femora and tibiae practically bare beneath. Wings (Fig. 770) clear with crossveins closer together than in the assimilis-group of species; vein R2+3 evenly bowed and gradually curved towards costa, consequently cell R1 rather narrow. Vein R4+5 only slightly downcurved before tip, vein M somewhat undulating. Abdomen subshining but the enlarged tergite 4 somewhat dull; genitalia (Figs. 603, 604) very different from those of exilis.

Length: body 1.2 - 1.4 mm, wing 1.3 - 1.5 mm.

♀. Hind tibiae with only short hairs as in infitialis, but the small antennae, blackish legs and different wing-venation easily distinguish it from both exilis and infitialis.

Length: body 1.2 - 1.5 mm, wing 1.3 - 1.5 mm.

Commoner in Denmark but rather rare in S. Sweden (Sk.) and S. Finland (Helsinki). - England, C. Europe and C parts of European USSR; the record from Spain (Strobl, 1909: 179) is an error. - June-October. On tree-trunks, also on coniferous trees, on grave-stones and often on windows; Collin (1961) took it when sweeping the branches of oak trees.

109. DRAPETIS (DRAPETIS) EXILIS Meigen, 1822.
    Figs. 605, 606, 771.

Drapetis exilis Meigen, 1822: 91.
Drapetis pusilla Loew; Lundbeck, 1910: 259 (p.p.).

Resembling pusilla but fore tibiae and all tarsi yellowish, vein R2+3 more strongly curved towards costa and cell R1 broader; fore femora in ♂ without a recurved bristle at tip and tergite 4 shining; hind tibiae in ♀ with long pale hairs behind.

♂. Head as in pusilla but antennal segment 3 larger, slightly longer than deep. Legs paler, at least brownish on fore tibiae, and tarsi yellowish; fore femora without a recurved bristle at tip anteriorly. Wings (Fig. 771) clear with vein R2+3 more strongly bowed towards costa and cell R1 broader, vein R4+5 more decidedly downcurved before tip and vein M almost straight on apical two-thirds. Abdomen shining black on enlarged tergite 4; genitalia (Figs. 605, 606) superficially resembling those of infitialis.

Length: body 1.2 - 1.4 mm, wing 1.3 - 1.5 mm.

♀. Hind tibiae with a row of outstanding long whitish hairs posteriorly, abdomen somewhat brownish in ground-colour with all segments equally long.

Length: body 1.4 - 1.6 mm, wing 1.4 - 1.6 mm.

Rather common in Denmark, S.Sweden (Sk.) and in S.Finland north to Ta.
- England, NW and C parts of European USSR, C.Europe, S to Romania and
Spain (material examined). - May-September; Collin (1961) records April-
October in England. On tree-trunks and very often on windows; Kovalev (1972)
took it when sweeping bushes and the tops of small trees.

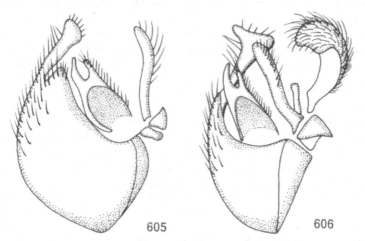

Figs. 605-606. Male genitalia of Drapetis (Drapetis) exilis Meig. - 605: periandrium with cerci (cerci in anterior view, appendage of left periandrial lamella removed); 606: periandrium with cerci (left lamella in anterior view); (after
Collin, 1961).

110. DRAPETIS (DRAPETIS) INFITIALIS Collin, 1961
     Figs. 607, 772.

Drapetis exilis subsp. infitialis Collin, 1961: 37.

Resembling exilis very closely, but there are slight differences in male genitalia and hind tibiae in ♀ without long pale hairs posteriorly.

♂. Antennal segment 3 somewhat longer than in exilis; legs usually a little
darker on front tibiae and basal segments of tarsi, particularly on hind legs.
Wing-venation (Fig. 772) as in exilis but crossveins obviously slightly wider
separated. Genitalia (Fig. 607) with a very close resemblance to that of exilis
in general but with considerable differences in detail: dorsal process of right
lamella broadened apically, left appendage straighter at tip, and right cercus
much longer, uniformly slender and not bent at tip.

Length: body 1.2 - 1.4 mm, wing 1.4 mm.

♀. Hind tibiae without a long ciliation of pale hairs posteriorly, only short hairs present as in pusilla, but the other main characters as in exilis.

Length: body 1.3 - 1.5 mm, wing 1.4 - 1.5 mm.

Rather rare, along the Baltic coast of Finland; also in the south of Russian Carelia. - England, C.Europe (Czechoslovakia) and C parts of European USSR (Rostov region, Kovalev, 1972). - July-August. Together with pusilla and exilis, but everywhere rather rare.

Note. Collin (1961) described infitialis as a subspecies of exilis, but in view of the sympatric distribution of both these forms subspecific status is not acceptable. The hopefully constant differential features, including the most recent records of distribution, prove that infitialis should be separated as a distinct species.

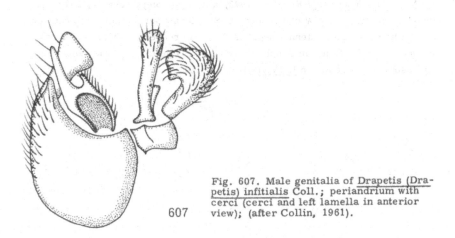

Fig. 607. Male genitalia of Drapetis (Drapetis) infitialis Coll.; periandrium with cerci (cerci and left lamella in anterior view); (after Collin, 1961).

607

111. DRAPETIS (DRAPETIS) PARILIS Collin, 1926.
Figs. 590, 608-610, 773.

Drapetis pusilla Loew; Lundbeck, 1910: 259 (p.p.).
Drapetis parilis Collin, 1926: 148 .

Venation as in pusilla but wings faintly brown clouded, legs paler and antennal segment 2 beneath with several longer bristly-hairs.

♂. Head as in pusilla but antennae (Fig. 590) with larger segments 2 and 3, very long arista (almost 3 times as long as antenna), and segment 2 beneath with longer dark bristly-hairs, distinctly more than half as long as the segment. Legs paler than in both pusilla and exilis; fore coxae, fore tibiae and all tarsi yellowish, sometimes also femora light brownish except for dorsum, and femora with very short hairs posteroventrally. Wings (Fig. 773) with venation as in pusilla, with narrower cell R1 and vein M undulating, but wings uniformly brown clouded. Abdominal tergite 4 subshining; genitalia (Figs. 608-610) with a very broad and blunt-tipped process on left lamella.

Length: body 1.2 - 1.4 mm, wing 1.3 - 1.6 mm.

♀. With the main diagnostic characters as in the male; hind tibiae without long ciliation posteriorly as in pusilla or infitialis. Abdomen subshining brown.

Length: body 1.3 - 1.7 mm, wing 1.4 - 1.6 mm.

Rather common in Denmark and S.Sweden north to Nrk. and Sdm., in Finland recorded only from Nylandia. - Great Britain, NW and C parts of European USSR, C.Europe (Czechoslovakia). - End June-July; in Great Britain (Collin,1961) to October. On leaves of bushes and trees.

Note. D.stackelbergi Kovalev, 1972, and D.completa Kovalev, 1972, described recently from NW of European USSR (Leningrad region), both have only short hairs beneath antennal segment 2, but in respect of their very small size (body 1.0 - 1.3 mm in length) and extensively yellow legs they superficially resemble the southern D.laevis Becker, 1913.

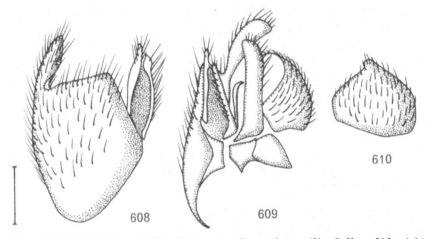

Figs. 608-610. Male genitalia of Drapetis (Drapetis) parilis Coll. - 608: right periandrial lamella with cerci; 609: periandrium with cerci; 610: appendage of left periandrial lamella. Scale: 0.1 mm.

112. DRAPETIS (DRAPETIS) INCOMPLETA Collin, 1926

Figs. 18, 611-613, 774.

Drapetis exilis Meigen; Lundbeck, 1910: 258; Engel, 1939: 113.

<u>Drapetis incompleta</u> Collin, 1926: 148.

Larger species with distinctly dark clouded wings, vein R4+5 abbreviated at tip. Antennae with short bristles beneath segment 2, segment 3 longer than usual.

♂. Head with vertex and occiput mostly thinly dark grey dusted; antennae (Fig. 18) black, segment 2 beneath with very short bristly-hairs, segment 3 enlarged and conical, at least 1.5 times as long as deep. Thorax silvery-grey dusted on prothorax and on a narrow anterior stripe on sternopleura. Legs yellow on fore coxae, rest of fore legs rather yellowish-brown, posterior four legs including tarsi blackish; hind femora and tibiae with short hairs.

Wings (Fig. 774) somewhat narrower, distinctly dark clouded (darker than in parilis) with thick blackish veins. Vein R4+5 and M almost parallel or very indistinctly undulating, vein R4+5 slightly upturned and more or less abbreviat-

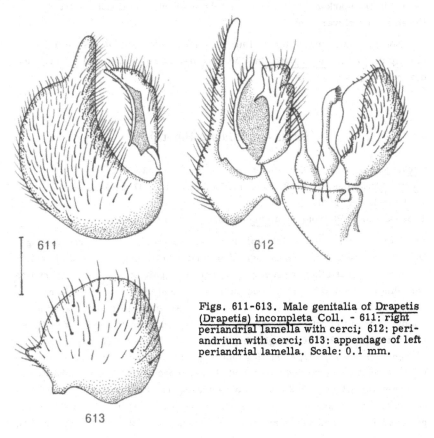

611

612

613

Figs. 611-613. Male genitalia of <u>Drapetis</u> (Drapetis) incompleta Coll. - 611: right periandrial lamella with cerci; 612: periandrium with cerci; 613: appendage of left periandrial lamella. Scale: 0.1 mm.

ed just before tip. Crossveins close together as in other species of the exilis-group. Abdomen with tergite 4 dull; genitalia (Figs. 611-613) larger with lamellae polished blackish-brown.

Length: body 1.5 - 1.7 mm, wing 1.6 - 1.7 mm.

♀. Resembling male; fore legs often darker but fore coxae always yellow, hind tibiae with short but rather closely-set hairs posteroventrally and antero-dorsally.

Length: body 1.6 - 2.0 mm, wing 1.6 - 1.9 mm.

Common in Denmark and S. Sweden in Scania, north to Nrk.; not yet recorded from Finland and Norway. - C. Europe (Czechoslovakia, Austria) and C parts of European USSR; not found in Great Britain and in NW parts of European USSR; Collin (1960) recorded it from Palestine (ssp. or sp. ?). - End May-early August, but mainly in June. On ground-vegetation and leaves of trees; in Denmark very common in fields of wheat, and Lundbeck (1910) collected it on a clover-field.

Note. D. fumipennis Strobl, 1906 (described as a var. of pusilla) is very closely related to incompleta, but the legs are entirely black and the male has different larger genitalia.

Genus *Crossopalpus* Bigot, 1857

Crossopalpus Bigot, 1857, Annls Soc. ent. Fr., 5: 563.
Eudrapetis Melander, 1918, Ann. ent. Soc. Am., 11: 187 (as subgenus of
    Drapetis Meigen).
Type-species: Platypalpus ambiguus Macquart, 1827 (mon.).

Small to medium-sized mainly polished black species (Fig. 614) resembling Drapetis but head (Fig. 10) deeper, produced below owing to deep jowls below eyes, and occiput slightly concave below neck. A pair of long vertical bristles very distinct from other short occipital hairs, posterior pair of ocellar bristles almost as long but anterior pair quite absent. Antennae (Fig. 19) placed above middle of head in profile, segment 3 with lower edge convex as in Drapetis s. str. but segment 2 with a single very long bristle beneath. Thorax broader when viewed from above, large bristles distinct, and sometimes with numerous similar long bristly-hairs evenly distributed over mesonotum; mesopleura bare. Anterior four tibiae with 2 distinct preapical spurs, hind tibiae (Figs. 617, 618) besides the usual preapical bristles sometimes with long bristles antero- and posterodorsally, and with a more or less developed tooth-like projection at tip behind. Wings (Fig. 29) clear, the common stem of veins R2+3 and R4+5 very short. Abdomen resembling that of <u>Drapetis</u> but male genitalia consisting of a large right periandrial lamella, left lamella absent but its dorsal appendage present and of very complicated structure.

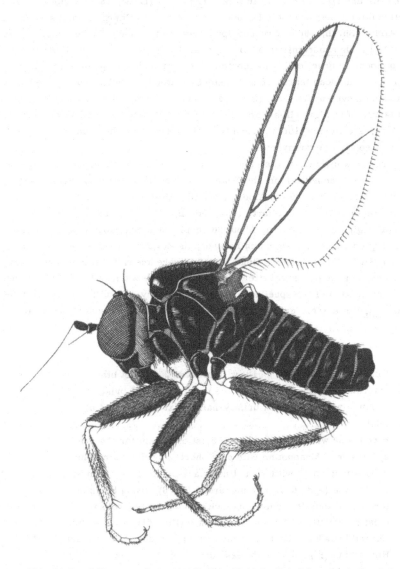

Fig. 614. Male of <u>Crossopalpus nigritellus</u> (Zett.). Total length: 1.6 - 2.0 mm.

The adults are sometimes very common; some species are typically terrestrial in habit, searching for their prey in heaps of cut sedge and litter, under dead leaves, in grass tufts, on human or animal dung, but also on trees and bushes; some occur in large numbers on the sandy or swampy banks of lakes and rivers, or are strictly coastal in habit; C.pilipes (Loew, 1859) occurs quite frequently on the sea-shore of S.Europe in company with Chersodromia species, and in S.Africa (Smith, 1969) Crossopalpus appears to occupy the sea-shore niche occupied by Chersodromia in other regions. The adults very often hibernate and may be found in practically any month of the year under heaps of vegetable-matter, under or inside dead leaves or under stones, often fully active during the winter months, and even under snow (Broen and Mohrig, 1965). Laurence (1953) has bred larvae of C.nigritellus from old cow pats, probably as feeders on fungi in cow dung, and Buxton (1961) reared two specimens from the young plasmodium of Fuligo septica (Myxomycetes).

The genus is worldwide in distribution but nowhere very numerous in species. 15 species are recorded from the Nearctic region (Stone et al., 1965), almost the same number from South America (Smith, 1967), and not very many more species occur in Africa (Collart, 1934; Smith, 1969); the records from the Oriental region need revision. About 20 species are described from the Palaearctic region, none of them being recorded as Holarctic in distribution, but this number is far from being final. 7 species are known from the colder parts of Europe, but the genus is better represented in the warmer southern regions.

Identification of the species does not present any difficulties as in the subgenus Drapetis s.str., and no differentiation of species-groups is necessary.

Key to species of Crossopalpus

1   Hind tibiae (Figs. 617, 618) with at least 2 distinct black anterodorsal bristles besides the usual preapical bristles. Mesonotum with long black bristly-hairs mixed with short brownish hairs 2

-    Hind tibiae without anterodorsal bristles except for the preapical ones. Mesonotum with only short brownish hairs (numerous acr and dc), without distinct black bristles except at sides 3

2(1) Hind tibiae (Fig. 617) with numerous (5 - 6) strong black anterodorsal bristles, mid tibiae with a large anterodorsal bristle near base. Antennal segment 3 (Fig. 616) rather shorter and broader, about 1.5 times as long as deep ...113. setiger (Loew)

-    Hind tibiae (Fig. 618) with less numerous (2 - 4, exceptionally 5) and thinner anterodorsal bristles, mid tibiae without an an-

terodorsal bristle near base. Antennal segment 3 (Fig. 615)
longer and rather slender, about twice as long as deep ....
.... 114. curvipes (Meig.)
3(1)   Thoracic pleura entirely shining black. Hind tibiae with a
       more or less distinct apical tooth posteriorly 4
-      Thoracic pleura partly dusted. Hind tibia with only a very small
       indistinct apical tooth posteriorly 6
4(3)   Fore coxae black. Last section of vein M slightly undulating
       but not curved upwards until extreme tip (Figs. 29, 777). Palpi
       and halteres darkened; hind femora with 3 long black antero-

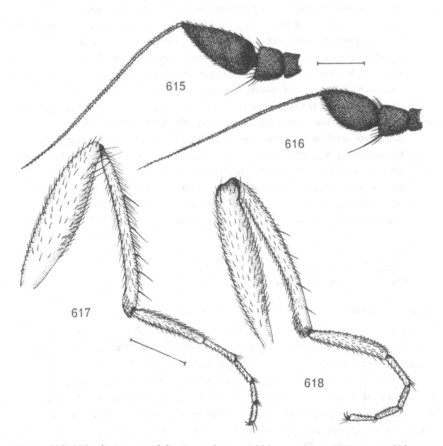

Figs. 615-616. Antennae of Crossopalpus. - 615: curvipes (Meig.), ♂; 616: setiger (Loew), ♂. Scale: 0.1 mm.
Figs. 617-618. Hind legs in anterior view of Crossopalpus. - 617: setiger (Loew),
♂; 618: curvipes (Meig.), ♂. Scale: 0.3 mm.

ventral bristles before tip, and hind tibiae with a rather large
and rounded apical tooth posteriorly. Larger, body about 1.6 -
2.2 mm in length.............................115. nigritellus (Zett.)
- Fore coxae yellowish. Last section of vein M more distinctly
  curved up before tip (cf. Fig. 778). Hind femora with 2 weaker
  anteroventral bristles before tip, and hind tibiae with a smaller
  and rather pointed tooth posteriorly. Smaller, body 1.2 - 1.5 mm
  in length.. 5
5(4) Legs mostly blackish. Arista shorter, less than 4 times as long
     as antenna. Hind femora thickened towards tip and somewhat
     curved; vein M more evenly bowed (Fig. 778). Halteres brown.
     Rather smaller species ....116. humilis (Frey)
- Legs yellowish, only hind coxae and four posterior femora at
  tip above blackish. Arista longer, more than 4 times as long as
  antenna. Hind femora more equally thickened but similarly curved;
  vein M slightly curved upwards on apical quarter. Halteres pale
  yellow. Slightly larger species......................minimus (Meig.)
6(3) Palpi pale yellow. Genitalia in ♂ (Figs. 631-633) and sternite
     8 in ♀ conspicuously large. Veins R4+5 and M slightly diverging
     towards tip (Fig. 779).......................117. curvinervis (Zett.)
- Palpi blackish. Genitalia in ♂ (Figs. 634-636) and sternite 8
  in ♀ of usual size, not conspicuously enlarged. Veins R4+5 and
  M converging towards tip, vein R4+5 with a downward curve
  before tip ....118. abditus Kov.

113. CROSSOPALPUS SETIGER (Loew, 1859)
     Figs. 616, 617, 619-621, 775.

Drapetis setigera Loew, 1859: 39.

Larger polished black species, with long black bristly-hairs on mesonotum
mixed with small brownish hairs; hind tibiae with a row of at least 5 strong
black anterodorsal bristles; antennal segment 3 short.

    ♂. Head shining black on frons, occiput thinly grey dusted; large bristles
(a pair of posterior ocellar and vertical bristles) long, black. Antennae (Fig.
616) black, segment 3 rather ovate, pointed at tip, about 1.5 times as long as
deep; ventral bristle at tip of segment 2 rather short, shorter than segment 3.
Palpi large, blackish. Thorax polished black, including most of pleura, fine
brownish hairs on mesonotum mixed with long black bristly-hairs, large bristles
very long, black.

Legs extensively blackish; besides the usual tibial and femoral preapical bristles with a row of 5 - 6 distinct strong black anterodorsal bristles on hind tibiae (Fig. 617), and a long black anterodorsal bristle on basal quarter of mid tibia. Hind femora (Fig. 617) usually with 3 long black preapical anteroventral bristles, and 1 or 2 longer erect hairs anterodorsally before tip. Hind tibia (Fig. 617) at tip behind rounded and usually yellowish, but not produced into a tooth-like projection.

Wings (Fig. 775) clear or indefinitely yellowish anteriorly, veins light brown; veins R4+5 and M very indistinctly diverging and scarcely undulating; anal fold distinct. Halteres with whitish-yellow knobs. Abdomen mainly shining black but basal and apical segments often dull brownish. Genitalia (Figs. 619-621) small, mostly brownish, with free appendages apically.

Length: body 1.6 - 2.1 mm, wing 2.0 - 2.4 mm.

♀. Resembling male except for sexual differences.

Length: body 1.7 - 2.5 mm, wing 2.3 - 2.5 mm.

Rather uncommon in Denmark and S.Sweden, and along the Baltic coast of Finland. - Widespread in Europe; England, NW and C parts of European USSR, and through C.Europe S to France and Spain. - May-September. Mainly a coastal species, or on the sandy banks of lakes and rivers and on adjacent ground-vegetation; rarely on tree-trunks (Kovalev, 1972).

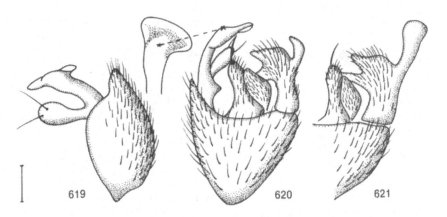

Figs. 619-621. Male genitalia of Crossopalpus setiger (Loew). - 619: periandrial lamella (lateral view, right side); 620: the same with cerci (dorsal view); 621: the same (lateral view, left side). Scale: 0.1 mm.

114. CROSSOPALPUS CURVIPES (Meigen, 1822)
      Figs. 615, 618, 622-624, 776.

Tachydromia nigra Meigen; Fallén, 1815: 8; Zetterstedt, 1842: 297.
Tachydromia curvipes Meigen, 1822: 75 (nom.n. for nigra Meigen sensu
      Fallén, nec Meigen, 1804).
Drapetis aterrima Haliday, in Curtis, 1834: 397 (Haliday, 1833: 149).
Tachydromia moriella Zetterstedt, 1838: 552.
Tachydromia picipes Zetterstedt, 1842: 298.

Closely resembling setiger but hind tibiae with less numerous and thinner anterodorsal bristles, no distinct anterodorsal bristle on mid tibia near base, and
antennae with longer and more slender segment 3.

    ♂. Head as in setiger but antennae (Fig. 615) distinctly longer, segment 3
rather slender and more conical, usually more than twice as long as deep; a
bristle on segment 2 at tip beneath not longer. Legs similarly black, but anterodorsal bristles on hind tibiae (Fig. 618) distinctly shorter and thinner, usually only 2 (near base and before middle of tibia), sometimes up to 4 and very exceptionally 5 in number (only seen in 2 ♀). Hind femora (Fig. 618) with thinner
anteroventral preapical bristles, and dorsum of femur with only short adpressed
hairs right up to tip. Anterodorsal bristle on mid tibia near base absent. The
different bristling on legs in setiger and curvipes seems to be quite constant
so far as the Fennoscandian fauna is concerned. Genitalia (Figs. 622-624)
larger than in setiger and with distinct differences in dorsal appendages.

    Length: body 1.9 - 2.3 mm, wing 2.2 - 2.3 mm.

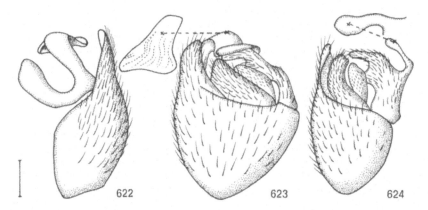

Figs. 622-624. Male genitalia of Crossopalpus curvipes (Meig.). - 622: periandrial lamella (lateral view, right side; 623: the same with cerci (dorsal
view); 624: the same (lateral view, left side). Scale 0.1 mm.

♀. Resembling male and with the same diagnostic features; antennal segment 3 sometimes slightly shorter, but still more slender and longer than in setiger.

Length: body 2.0 - 2.5 mm, wing 2.3 - 2.5 mm.

Very common and widespread in Denmark and Fennoscandia; in Norway north to TRy and Fi, in Sweden to P.Lpm. and Nb., in Finland to LkW; also from the Kola Peninsula. - Great Britain including Scotland, northern coast of Germany. No further records available to me; the specimens of curvipes from the central parts of European USSR (Kovalev, 1972: 192) very probably represent a distinct species, as has already been pointed out by Kovalev himself (in litt.). - May-September. Mainly a coastal species, and on the banks of lakes and rivers.

115. CROSSOPALPUS NIGRITELLUS (Zetterstedt, 1842)
Figs. 10, 19, 29, 614, 625-627, 777.

Tachydromia nigritella Zetterstedt, 1842: 298.
Drapetis nervosa Loew, 1859: 37.
Tachydromia parvicornis Zetterstedt, 1859: 4992.
Drapetis aterrima Curtis; Lundbeck, 1910: 254; Frey, 1913: 68; Engel, 1939: 111.

Larger polished black species with legs mostly black, including fore coxae, hind tibiae without long anterodorsal bristles but with a large apical tooth posteriorly. Mesonotum with only short brownish hairs besides the usual bristles at sides.

♂. Head greyish dusted, more thinly so on frons and vertex, clypeus and lower part of deep jowls below eyes polished black. Antennae small with short segment 3 and a very long bristle beneath segment 2 at tip. Palpi blackish. Thorax polished black even on mesopleura, mesonotal hairs short, brownish, large marginal bristles black.

Legs extensively black including fore coxae, anterior four tibiae and tarsi brownish; preapical tibial and femoral bristles long, hind femora slightly and evenly thickened, anteroventrally before tip with 3 long black bristles. Hind tibiae with only preapical bristles and a large brownish apical tooth behind. Wings (Fig. 777) clear with dark veins, vein M very indistinctly undulating, almost straight. Halteres dark. Abdomen mainly polished black except for grey dusted narrow sides; genitalia (Figs. 625-627) large and closed, polished black.

Length: body 1.6 - 2.0 mm, wing 1.7 - 2.1 mm.

♀. Resembling male except for sexual differences.

Length: body 1.7 - 2.2 mm, wing 1.8 - 2.2 mm.

A common species in Denmark and southern parts of Fennoscandia to approximately 62°N, in Sweden to Upl., in Finland to Ta; not yet recorded from Norway. - Widespread and common in Europe, from England and NW of European USSR through C.Europe S to Italy and Spain; Palestine. - The adults hibernate, and are found commonly in any month of the year. In grass, under dead leaves, in heaps of cut sedge, but also on bushes and trees, or on sandy coasts and the sandy banks of lakes and rivers.

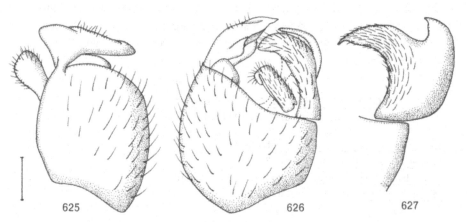

Figs. 625-627. Male genitalia of Crossopalpus nigritellus (Zett.). - 625: periandrial lamella (lateral view, right side); 626: the same with cerci (dorsal view); 627: left dorsal appendage. Scale: 0.1 mm.

116. CROSSOPALPUS HUMILIS (Frey, 1933)
    Figs. 628-630, 778.

Drapetis humilis Frey, 1913: 69.

Resembling nigritellus but smaller, fore coxae yellowish, hind femora apically thickened and slightly curved, vein M distinctly evenly bowed.

♂. Head as in nigritellus but jowls below eyes rather subshining, antennae with a conspicuously long thin apical bristle beneath segment 2, and somewhat shorter arista. Palpi smaller, dark. Legs yellowish on fore coxae, posterior four coxae yellowish-brown, sometimes front legs somewhat brownish. Hind femora distinctly thickened on apical third and slightly curved when viewed from above, anteroventrally before tip with only 2 thinner bristly-hairs. Hind tibiae with a smaller and apically more pointed tooth behind.

Wings (Fig. 778) often yellowish-brown tinged anteriorly, veins brown;

vein M distinctly evenly curved and consequently cell R5 broadest in apical third. Halteres dark brown. Genitalia (**Figs.** 628-630) not very large and with a very distinctive dorsal appendage on the left.

Length: body 1.2 - 1.4 mm, wing 1.4 - 1.6 mm.

♀. Resembling male except for sexual differences.

Length: body 1.3 - 1.5 mm, wing 1.5 - 1.6 mm.

Rather common in Denmark and southern parts of Norway and Sweden (north to Upl.), in Finland north to Om (a single ♂ also from LkW, Muonio); also in the south of Russian Carelia. - Great Britain, NW and C parts of European USSR, C.Europe (W.Germany, Czechoslovakia, Austria) and S.Europe (Bulgaria). - The adults hibernate, but are more often found in April-September. On horse and other animal dung, in heaps of vegetable matter, but often on windows, particularly of stables, together with Drapetis species.

Note. The closely related C.minimus (Meigen, 1838), known from England, C.Europe and Estonia has extensively yellowish legs, conspicuously longer arista, pale yellow halteres, and different wing-venation.

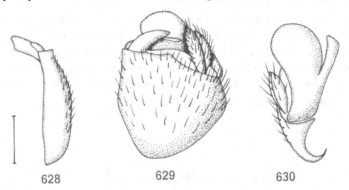

Figs. 628-630. Male genitalia of Crossopalpus humilis (Frey). - 628: periandrial lamella (lateral view, right side); 629: the same with cerci (dorsal view); 630: the same (lateral view, left side). Scale: 0.1 mm.

117. CROSSOPALPUS CURVINERVIS (Zetterstedt, 1842)
Figs. 631-633, 779.

Tachydromia curvinervis Zetterstedt, 1842: 301.

Larger species with thoracic pleura silvery-grey dusted above, palpi pale yellow; ♂ genitalia and sternite 8 in ♀ very enlarged.

♂. Head rather densely grey dusted, leaving lower margin of deep jowls polished. Antennae with segment 3 only slightly longer than deep, the usual bristle beneath segment 2 at tip rather short, as long as segment 3. Palpi pale yellow. Thorax with very dense pale hairing on shining mesonotum, but pleura densely silvery-grey dusted on the upper part, leaving most of sterno- and hypopleura polished black.

Legs black on coxae and femora except for tip, tibiae and tarsi brownish; sometimes also fore coxae and fore femora towards tip brown. Preapical fe- moral bristles rather small and thin; hind tibiae somewhat rounded at tip be- hind, apical tooth practically absent. Wings (Fig. 779) clear with dark veins, veins R4+5 and M undulating (vein M more conspicuously), slightly diverging towards tip. Halteres whitish-yellow. Abdomen rather shining black, basal two segments membraneous; genitalia (Figs. 631-633) very large, distinctly broader than apical abdominal segments and the large (right) periandrial la- mella almost as long as rest of abdomen.

Length: body 2.0 - 2.3 mm, wing 2.4 - 2.6 mm.

♀. Closely resembling male, abdominal sternite 8 conspicuously large and mainly shining, with distinct rounded tubercles at sides.

Length: body 1.8 - 2.2 mm, wing 2.3 - 2.6 mm.

Uncommon in southern Sweden north to Og. and in Finland to Ta; also in the south of Russian Carelia, but not yet recorded from Denmark and Norway. - Widespread; NW and C parts of European USSR, C.Europe, E as far as E. Asia (Ussuri, Amur). - May-September but mainly in August and September. On the banks of lakes and rivers and on bushes; in autumn inside the twisted dead leaves of oak-trees (Kovalev, 1972).

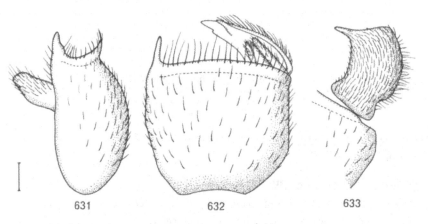

631   632   633

Figs. 631-633. Male genitalia of Crossopalpus curvinervis (Zett.). - 631: pe- riandrial lamella (lateral view, right side); 632: the same with cerci (dorsal view); 633: left dorsal appendage. Scale: 0.1 mm.

118. CROSSOPALPUS ABDITUS Kovalev, 1972
   Figs. 634-636, 780.

Crossopalpus abditus Kovalev, 1972: 194.

Resembling curvinervis but palpi blackish, ♂ genitalia and sternite 8 in ♀ small, and legs extensively darkened.

♂. Head as in curvinervis but palpi black in ground-colour and covered with dark brown pile. Thorax with dense silvery-grey pile on upper part of pleura as in curvinervis; legs more extensively blackish, tibiae and tarsi often dark brown, brownish in lighter specimens, but hind tibiae almost always black. Apical tooth at tip of hind tibiae behind practically absent.

Wings (Fig. 780) clear with dark veins, vein M distinctly undulating (but not so abruptly bent on last section near base in Fennoscandian specimens as is illustrated by Kovalev (1972, Fig. 33) in his original description); vein R4+5 decidedly downcurved before tip, anal fold distinct. Abdomen with much smaller genitalia (Figs. 634-636), which are at most as deep as last abdominal segment, the single periandrial lamella scarcely half as long as rest of abdomen.

   Length: body 1.8 - 2.4 mm, wing 2.3 - 2.7 mm.

♀. Resembling male, abdominal sternite 8 small and of normal shape, subshining.

   Length: body 1.8 - 2.5 mm, wing 2.3 - 3.1 mm.

   Rare. Sweden: Nb., (B.S.), ♂ (Boheman); Ås.Lpm., ♀ (Boheman); Fin-

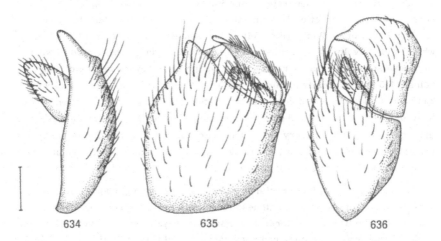

Figs. 634-636. Male genitalia of Crossopalpus abditus Kov. - 634: periandrial lamella (lateral view, right side); 635: the same with cerci (dorsal view); 636: the same (lateral view, left side). Scale: 0.1 mm.

land: Ks, Paanajärvi, ♀ (Frey) - on the Russian Carelia border. - NW and C
parts of European USSR, W.Ukraine. - June-early September. On the tops
of small trees, bushes and tree-trunks, and also on moorland (all from Kovalev,
1972).

Genus *Chersodromia* Walker, 1849

Chersodromia Walker, 1849, List Dipt.Brit.Mus., 4: 1157.
Coloboneura Melander, 1902, Trans.ent.Soc.Am., 28: 229.
Halsanalotes Becker, 1902, Mitt.zool.Mus.Berl., 2(2): 41.
Thinodromia Melander, 1906, Ent.News, 17: 370.
Type-species: Tachypeza brevipennis Zetterstedt, 1838 (design. by Rondani,
    1856: 147) = arenaria (Haliday).

Very small to rather large species (Fig. 637), ranging from less than 1 mm to
4 mm in body-length, black to blackish-brown in colour and often densely grey
dusted, legs sometimes yellow. Large bristles on head, thorax and legs well-
developed or practically absent. Head (Fig. 11) deep with strong heavily sclerotised proboscis, eyes microscopically pubescent and well separated both on
frons and face, jowls below eyes usually deep. 2 pairs of ocellar bristles, vertex with 2 pairs, or rarely with 1 pair, of vertical bristles. Thorax robust,
without differentiated humeri, large thoracic bristles usually in full number;
acr biserial, narrowly separated from multiserial dc. Legs simple in both
sexes, all femora practically of the same thickness; more or less bristled, at
least on hind tibiae. Wings (Fig. 30) clear, sometimes brown clouded or milky-
white, well-developed even if narrowed, at most slightly abbreviated in North
European species. Vein R1 rather long, ending at or beyond middle of wing,
veins R4+5 and M usually almost parallel; 1st basal cell very long, as long as
2nd basal cell, vein A and anal cell completely absent, but former often in the
form of a very faint fold. Abdomen consisting of eight fully sclerotised segments,
male genitalia (Fig. 34) very distinctive with left periandrial lamella partly fused
with complicated hypandrium, but unlike Stilpon with small cerci; female abdomen telescopic with cerci small, ovate.

The adults are typical coastal species, usually running about in large numbers on wet or dry sand or on stones close to water, often resting under sea-
weed thrown up on the seashore. The small, light-grey species can be seen on
sand only when moving; they run very rapidly, and if they fly then it is only for
short distances. The adults are highly predaceous, even in captivity. The occurrence of Chersodromia species is restricted to the seashore or to the shores

of salt lakes close to the sea, and only rarely do they penetrate far inland along rivers and lakes (cursitans). Nothing is known of the immature stages.

The genus is mainly Holarctic in distribution, extending to Central America (Panama), Hawaii, and the east African coast (Aldabra Is.). Further species certainly await discovery, but 24 of the **approximately** 35 known species occur in the Palaearctic region, and only 7 of these have been found in North Europe; all seven species occur in Denmark. A revision of all the Palaearctic species is also ready for press.

Chersodromia species form four clearly distinct natural groups (hirta-, cursitans-, speculifera-, and incana-groups) which are all represented in the North European fauna; this subdivision is also used in the following pages.

Fig. 637. Male of Chersodromia hirta (Walk.). Total length: 2.3 - 3.2 mm.

Key to species of Chersodromia

1   Hind femora with strong black anterodorsal and anteroventral
bristles. Jowls below eyes very deep, about one-third as deep
as the eye-height (Fig. 11). Large, blackish coloured and
black bristled species, body about 3 mm in length (I. hirta-
group) .... 119. hirta (Walk.)

\-   Hind femora with short hairs except for preapical bristles.
Jowls not so deep. Smaller, black to light grey species with
less distinct bristles everywhere 2

2(1)   Halteres dark, legs brown to blackish-brown, wings somewhat brownish. Jowls moderately deep, about one-fifth to
one-quarter as deep as the eye-height; palpi dark. Medium-
sized or smaller, blackish coloured and black bristled species (II. cursitans-group) 3

\-   Halteres pale with knobs often dusky, legs yellow (except
incana), wings clear or milky-white. Jowls narrow, at most
one-sixth as deep as the eye-height; palpi pale. Smaller,
grey to brownish-grey species.................................. 5

3(2)   1 pair of vt bristles, wings (Fig. 783) broad. Legs rather
shining, anterior tarsal segments shortened and slightly
dilated. No posthumeral bristle. Medium-sized, body 1.6 -
2.3 mm in length, blackish species ....121. difficilis Lundb.

\-   2 pairs of vt bristles, wings narrowed or abbreviated 4

4(3)   Wings (Fig. 782) narrowed but long, much longer than abdomen. Legs dull grey, anterior four tarsi long and slender.
Posthumeral bristle present. Medium-sized, body 1.4 - 2.4
mm in length, blackish species................120. cursitans (Zett.)

\-   Wings (Fig. 784) narrow and distinctly abbreviated, scarcely longer than abdomen. Legs rather shining, anterior four
tarsi shortened and dilated. No posthumeral bristle. Generally smaller, body 1.2 - 2.2 mm in length, rather dark
grey species .... 122. arenaria (Hal.)

5(2)   Face broad, as broad as frons in front, palpi small. Sternopleura largely polished. Legs yellow, bristles on head
and thorax distinct, wings clear or sligthly yellowish tinged
(III. speculifera-group)..6

\-   Face narrow, much narrower than frons, palpi large. Sternopleura with a very small polished patch. Legs extensively

dark, bristles on head and thorax practically absent. A small
body 1.1 - 1.7 mm in length, silvery-grey dusted species with
milky-white wings (IV. incana-group) .... 125. incana Walk.

6(5)   Antennae (Fig. 640) yellow with segment 3 darkened, halteres with knobs darker. Bristles on head and thorax black.
Wings (Fig. 785) tinged yellowish, rather broad. Medium-
sized, body 1.8 - 2.3 mm in length, greyish species ....
.... 123. speculifera Walk.

-   Antennae (Fig. 641) and halteres unicolourous yellowish.
Bristles on head and thorax brownish to brassy-yellow in ♀.
Wings (Fig. 786) clear, rather smaller and narrow. Smaller,
body 1.2 - 1.6 mm in length, rather greyish-brown species.....
.... 124. beckeri Mel.

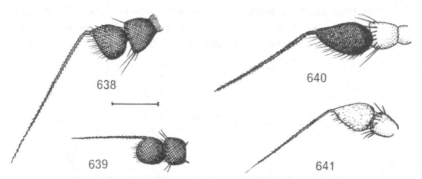

Figs. 638-641. Antennae of Chersodromia. - 638: arenaria (Hal.), ♂; 639:
incana Walk., ♂; 640: speculifera Walk., ♂; 641: beckeri Mel., ♂. Scale:
0.1 mm.

I. hirta-group

119. CHERSODROMIA HIRTA (Walker, 1836)
Figs. 11, 20, 30, 637, 642-643, 781.

Tachypeza ? hirta Walker, 1836: 180; 1851: 137.

A large blackish coloured and conspicuously black bristled species with deep
jowls below eyes and distinct black bristles on hind femora anterodorsally and
anteroventrally. Antennae, palpi and legs black, halteres pale.

♂. Head (Fig. 11) black with slight greyish dusting, face at least as deep
as frons in front, widening and very prominent below, and confluent with very
deep jowls. 2 pairs of ocellar and 2 pairs of vt bristles black, strong. Anten-

279

nae (Fig. 20) black, segment 3 short but pointed apically, arista slightly longer than antenna. Palpi blackish, ovate and flat, covered with short black bristles and with a long apical bristle. Thorax black, slightly grey dusted especially on pleura, sternopleura largely polished. Large bristles black, in full number : a humeral, a posthumeral, 4 - 5 notopleural, a supra-alar, a postalar, 5 large dc and a pair of apical scutellars. Distinct tufts of strong bristles between posterior four coxae.

Legs black, with short black hairs and conspicuous bristles on anterior four femora antero- and posteroventrally, and anterodorsally on apical half; hind femora with large bristles anteroventrally and anterodorsally; fore tibiae spindle-shaped dilated, armed dorsally with 3 distinct bristles, which are present also on the slender mid tibiae. Hind tibiae with 3 rows (dorsal, anterior and anteroventral) of several strong black bristles which are longer than tibia is deep.

Wings (Figs 30, 781) large, faintly brownish clouded especially on costal half, veins dark. Squamae dusky with dark fringes, halteres yellowish. Abdomen black, thinly greyish dusted and mainly with fine dark hairs, segment 8

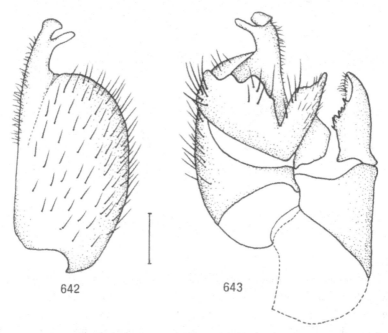

642   643

Figs. 642-643. Male genitalia of Chersodromia hirta (Walk.). - 642: right periandrial lamella; 643: periandrium with cerci (shape of hypandrium shown by broken line). Scale: 0.2 mm.

with long bristles posteriorly. Genitalia (Figs. 642-643) large, consisting mainly of the very convex right lamella, which is grey dusted and with short black
bristles; cerci conspicuously broad.

Length: body 2.3 - 3.2 mm, wing 2.3 - 3.0 mm.

♀. Resembling male except for sexual differences, abdomen long-pointed
with enlarged blunt-ended cerci.

Length: body 2.6 - 3.6 mm, wing 2.8 - 3.4 mm.

Common in Denmark and in S.Sweden, Sk. - Great Britain incl. Scotland, E through N.Germany to Sweden. - May-August. On sandy coasts close
to the sea, often in large numbers, more rarely on high dunes; adults can fly
quite well for short distances.

II. cursitans - group

120. CHERSODROMIA CURSITANS (Zetterstedt, 1819)
Figs. 644-646, 782.

Empis cursitans Zetterstedt, 1819: 82.
Chersodromia difficilis Lundbeck; Frey, 1913: 75, 1941: 7.

Medium-sized blackish species with 2 pairs of vt bristles, antennae and halteres dark. Legs dull brown, fore tarsi slender, hind femora with only preapical
bristles. Posthumeral bristle present.

♂. Head black with frons somewhat brownish, face almost as deep as
frons in front, produced below and confluent with moderately broad jowls; 2
pairs of ocellar and 2 pairs of vt bristles black, subequal. Antennae black, segment 3 almost spherical, with slightly supra-apical arista longer than antenna.
Palpi blackish-brown with silver dust, small. Thorax black, more greyish on
pleura, sternopleura largely polished. All hairs and bristles black, large bristles in full number including a posthumeral bristle.

Legs blackish-brown to brownish, dulled by greyish dust; femora and
tibiae with only preapical bristles but anterior four femora with a double row
of shorter bristly-hairs beneath, hind femora with dark hairs beneath, mid tibiae with a posteroventral comb of black spiny bristles on apical third, and hind
tibiae with 3 rows of about 2 or 3 black bristles on apical two-thirds. Fore tibiae
slightly spindle-shaped, all tarsi long and slender, longer than corresponding
tibiae.

Wings (Fig. 782) long but rather narrow, faintly brownish on costal half,
veins dark. Veins R4+5 and M almost parallel. Squamae blackish with brownish
fringes, halteres dark. Abdomen black to blackish-brown, greyish dusted and

with rather dense but fine dark hairs. Genitalia (Figs. 644-646) moderately large, right lamella thinly grey dusted with rather short but broad and apically pointed dorsal process.

Length: body 1.4 - 1.8 mm, wing 1.6 - 2.0 mm.

♀. Generally larger; no spine-like bristles in comb on mid tibiae, but with 2 distinct anterodorsal bristles and a similar bristle also on fore tibia. Abdomen long-telescopic with small cerci.

Length: body 1.6 - 2.4 mm, wing 1.6 - 2.0 mm.

Very common in Denmark, including Bornholm, and in southern parts of Fennoscandia, along Gulf of Bothnia in Sweden north to Nb. and in Finland to ObN, east to Russian Carelia. - Very rare in England, only 2 ♀ known (Smith, 1964), E to European USSR and along rivers far inland to C. and S. Europe (Czechoslovakia, Jugoslavia, Bulgaria). End May-August. Common on sandy coasts with seaweed close to the water, and on the sandy shores of freshwater inland.

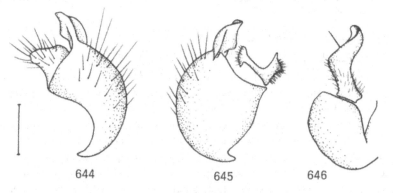

644   645   646

Figs. 644-646. Male genitalia of Chersodromia cursitans (Zett.). - 644: right periandrial lamella; 645: right periandrial lamella with cerci; 646: left periandrial lamella. Scale: 0.2 mm.

121. CHERSODROMIA DIFFICILIS Lundbeck, 1910
    Figs. 647-649, 783.

Tachypeza arenaria var.? b alata Walker, 1836: 180.
Chersodromia difficilis Lundbeck, 1910: 280.

Medium-sized blackish species resembling cursitans, but only 1 pair of vt bristles, wings broader, and anterior tarsi shortened and slightly dilated. No posthumeral bristle.

♂. Head as in <u>cursitans</u> but vertex with only 1 pair of long black vt bristles, about as long as the 2 pairs of ocellar bristles. Antennae black with segment 3 almost spherical, slightly supra-apical arista longer than antenna. Palpi dark brown with silver dust, covered with dark hairs and a black terminal bristle. Thorax black with greyish dusting, sternopleura largely polished. Small mesonotal hairs rather brownish, large bristles long, black, posthumeral bristle and the bristle on prothoracic episterna absent.

Legs dark brown to somewhat reddish-brown, rather shining. Dark femoral and tibial preapical bristles as in <u>cursitans</u> but with 1 or 2 additional bristly-hairs on fore tibiae above, 2 longer bristly-hairs on mid femora at base beneath, and mid tibiae with 2 black anterior bristles in addition to a ventral comb of short spine-like bristles. Hind tibiae with 3 rows of rather long bristles on apical two-thirds. Fore tarsi shortened and slightly dilated.

Wings (Fig. 783) broader than in <u>cursitans</u>, brownish on costal half, and vein R2+3 slightly longer, ending in costa halfway between veins R1 and R4+5; vein M slightly undulating near base. Squamae dark, halteres dark brown, stalk sometimes translucent yellowish. Abdomen blackish-brown, thinly greyish dusted and with fine brownish hairs. Genitalia (Figs. 647-649) rather large with right lamella silvery-grey dusted and with short black hairs, dorsal process conspicuously long, spoon-like.

Length: body 1.6 - 2.0 mm, wing 1.7 - 2.0 mm.

♀. Resembling male, but the black spine-like bristles in anteroventral comb on mid tibiae absent; cerci small.

Length: body 1.7 - 2.3 mm, wing 1.7 - 2.2 mm.

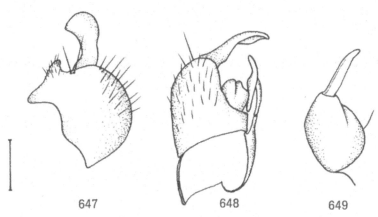

647   648   649

Figs. 647-649. Male genitalia of <u>Chersodromia difficilis</u> Lundb. - 647: right periandrial lamella; 648: periandrium with cerci and hypandrium; 649: left periandrial lamella. Scale: 0.2 mm.

Rather common in Denmark and in Sk., S.Sweden. - Within a small area from Great Britain including Scotland east to S.Sweden. - June-October. On sandy seashores not far from water, locally very common.

122. CHERSODROMIA ARENARIA (Haliday, 1833)
Figs. 638, 650-652, 784.

Tachypeza arenaria Haliday, 1833: 161.
Tachypeza brevipennis Zetterstedt, 1838: 548.
Tachypeza cursitans var. b Zetterstedt, 1842: 322.

A smaller dark greyish-black species with 2 pairs of vt bristles, resembling cursitans but wings brownish and abbreviated. Legs brownish, rather shining, anterior tarsi shortened with dilated apical segments.

♂. Head dull greyish-black to blackish-brown, frons narrowed at middle, face greyish and broad, broader than frons in front, widening below and confluent with rather deep and somewhat shining jowls. 2 pairs of ocellar and 2 pairs of vt bristles black but rather short. Antennae (Fig.638) brownish, segment 3 scarcely as long as deep, long arista slightly supra-apical. Palpi dark brown, grey dusted, small; terminal bristle long, black. Thorax blackish, thinly grey dusted, with sternopleura largely polished. Large bristles black with a very prominent humeral bristle and apical pair of scutellars, other bristles smaller, posthumeral bristle absent.

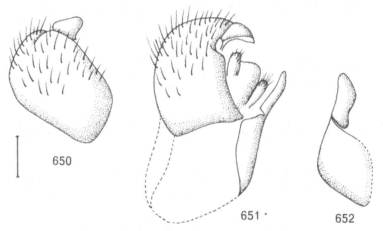

Figs. 650-652. Male genitalia of Chersodromia arenaria (Hal.). - 650: right periandrial lamella; 651: periandrium with cerci (shape of hypandrium shown by broken line); 652: left periandrial lamella. Scale: 0.1 mm.

Legs brown, somewhat shining, rather short and stout, obviously less bristled than in cursitans and difficilis. Preapical bristles on femora and tibiae black but rather weak, mid tibiae with a row of black bristly-hairs beneath and hind tibiae with long black bristles only on apical third (1 dorsal, 2-3 antero-dorsal, 2-3 anteroventral). Fore tibiae spindle-shaped dilated. Tarsi shortened, equal in length to their corresponding tibiae, apical segments on anterior two pairs short and dilated.

Wings (Fig. 784) somewhat brownish, narrow and abbreviated, scarcely longer than abdomen; veins brown, veins R4+5 and M parallel. Squamae dark brown, halteres blackish with stalk somewhat dirty yellow. Abdomen dark brown, greyish dusted and with fine pale hairs above, blackish on venter. Genitalia (Figs. 650-652) rather small, right lamella brown with grey dusting and black hairs, shining and darker apically; dorsal process small, rounded apically.

Length: body 1.2 - 1.6 mm, wing 0.9 - 1.1 mm.

♀. Resembling male but larger, abdomen usually more brownish, long-pointed apically, with small but slender cerci.

Length: body 1.5 - 2.2 mm, wing 1.0 - 1.3 mm.

Very common in Denmark and in western parts of Fennoscandia, in Nor-way north to Fn, and northern coast of Russian Lapland; less common in S. Sweden, penetrating north to Gtl and east only to Finnish N. - Widespread in North Europe: Iceland, Great Britain, but not recorded from European USSR except for extreme NW. - May-September. Common on coasts with gravel sand, on stones and rocks, and also on dirty beaches near towns and ports with decay-ing seaweed; the adults do not fly but jump very characteristically.

III. speculifera - group

123. CHERSODROMIA SPECULIFERA Walker, 1851
     Figs. 640, 653-655, 785.

Chersodromia speculifera Walker, 1851: 138.

Medium-sized greyish dusted species with black bristles and yellow legs. An-tennae yellow on basal segments, segment 3 dark, twice as long as deep, pointed, with terminal arista. Palpi and proboscis yellow.

♂. Head greyish with rather broad frons; face as broad as frons in front, silvery-grey dusted, widening below and confluent with rather narrow jowls. 2 pairs of ocellar and 2 pairs of vt bristles black, long. Basal antennal segments (Fig. 640) yellow, segment 3 blackish-brown, twice as long as deep and distinct-ly conical, with shorter concolourous terminal arista which is scarcely longer than antenna. Palpi yellow, small-ovate, with silvery and dark hairs, terminal bristle black. Proboscis yellowish. Thorax black, thinly grey dusted on mesonotum, more densely on pleura, sternopleura largely polished. Small thoracic hairs and large bristles black, posthumeral bristle present.

Legs yellow, apical four tarsal segments brownish; all hairs and bristles black. In addition to distinct preapical bristles with a double row of longer black hairs on anterior four femora beneath, similar hairs in anteroventral row on hind femora, fore tibiae rather slender with a tiny dark anterior bristle on basal third, mid tibiae with black ciliation anteroventrally on apical two-thirds, and hind tibiae with long black bristles (at least as long as tibia is deep) in 3 rows on apical two-thirds. Apical tarsal segments on anterior legs slightly dilated.

Wings (Fig. 785) rather broader, usually slightly yellowish tinged, veins pale; veins R4+5 and M slightly diverging. Squamae yellow, halteres pale yellow with knobs usually dusky. Abdomen blackish to blackish-brown, thinly grey dusted and with fine pale hairs, longer and darker posteriorly. Genitalia (Figs. 653-655) rather small, black, right lamella greyish dusted and with a tuft of long black bristles on a small hook on the left, smaller hairs black; dorsal process blunt apically but narrow.

Length: body 1.8 - 2.2 mm, wing 2.2 - 2.5 mm.

♀. Proboscis and squamae rather brownish; mid tibiae without the black

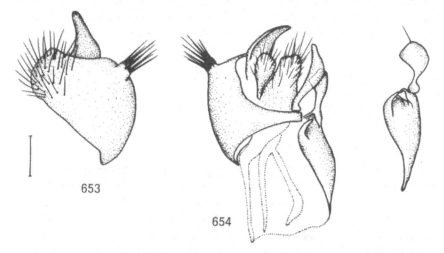

Figs. 653-655. Male genitalia of Chersodromia speculifera Walk. - 653: right periandrial lamella; 654: periandrium with cerci (shape of hypandrium shown by broken line); 655: left periandrial lamella. Scale: 0.1 mm.

anteroventral ciliation but on basal third with a distinct black **anterior bristle**. Abdomen subshining brown, cerci small, brownish.

Length: body 2.0 - 2.3 mm, wing 2.3 - 2.6 mm.

Rare in Denmark: F, Langeland, Lohals, ♀ (Lundbeck); NWZ, Røsnæs Fyr, 2♀ (Chvála). - England; N.Germany (Kröber, 1958), but can easily be confused with beckeri. - May-July. On sandy coasts close to water, and in England also in salt marshes close to the sea: on vegetation when the roots are covered at high tide (Collin, 1961).

124. CHERSODROMIA BECKERI Melander, 1928
Figs. 641, 656-658, 786.

Chersodromia beckeri Melander, 1928: 296.

Smaller, rather brownish species with yellow legs, resembling speculifera, but all bristles on head and thorax brownish or brassy-yellow in ♀, antennae uniformly yellowish except darker arista, and narrower clear wings.

♂. Head brownish, silvery-grey dusted, with broad frons and equally broad face which is produced below and confluent with rather narrow jowls. 2 pairs of ocellar and 2 pairs of vt bristles long, brownish. Antennae (Fig. 641) yellow to yellowish-brown, segment 3 shorter than in speculifera, terminal arista darker, long. Palpi small-ovate, pale yellow, covered with fine pale hairs and a dark terminal bristle. Proboscis yellowish-brown. Thorax brown to blackish-brown, densely silvery-grey dusted, sternopleura largely polished. Small mesonotal hairs pale, rather longer, as long as antennal segment 2. Large bristles as in speculifera but brownish in colour, at most dark brown.

Legs yellow except for the dark last tarsal segment, femora and tibiae with tiny dark preapical bristles. Fore femora with a row of about 10 dark bristly-hairs beneath, mid femora with 5 similar bristly-hairs on basal third, hind femora with only fine dark hairs beneath. Fore tibiae slender with a tiny dark anterior bristle on basal third, mid tibiae with a row of short, closely-set black bristles on apical third beneath, and hind tibiae with long black bristles at least on apical two-thirds, all bristles (2-3 dorsal, 3 anterior, 2 anteroventral) longer than tibia is deep. Tarsi long and slender.

Wings (Fig. 786) clear with pale yellow veins, **longer and narrower than** in speculifera. Veins R4+5 and M scarcely diverging. Squamae pale, halteres yellow to yellowish-brown. Abdomen brown to blackish-brown, thinly grey dusted and with fine pale hairs, more densely at sides and on venter. Genitalia (Figs. 656-658) rather small, brownish, right lamella greyish dusted and with sparse pale hairs, but the left (longer) dorsal process pointed apically and distinctly polished black.

Length: body 1.2 - 1.4 mm, wing 1.3 - 1.5 mm.

♀. Larger, and all bristles on head and thorax considerably paler, rather golden- to brassy-yellow in colour. Mid tibiae without the row of closely-set small bristles apically beneath but with 1 or 2 small black anterior bristles in basal and apical thirds; hind tibiae with more conspicuous bristles along the entire length. Cerci small, brownish.

Length: body 1.4 - 1.6 mm, wing 1.5 - 1.7 mm.

Rare. Denmark: NWJ, Agger, 2♂ 2♀ (Mortensen); also in Russian Ib, Terijoki, Rajajoki, ♂ (Krogerus). - Rather a common species along the German and Polish Baltic coasts. - June-August. On sandy coasts of the Baltic Sea.

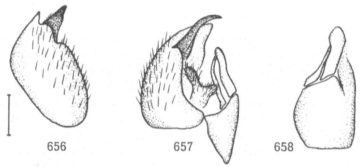

Figs. 656-658. Male genitalia of <u>Chersodromia beckeri</u> Mel. - 656: right periandrial lamella; 657: periandrium with cerci; 658: left periandrial lamella. Scale: 0.1 mm.

IV. <u>incana</u> - group

125. <u>CHERSODROMIA INCANA</u> Walker, 1851
    Figs. 639, 659-661, 787.

<u>Chersodromia incana</u> Walker, 1851: 138.

A small, light silvery-grey species with milky-white wings and only whitish points on head and thorax instead of the usual bristles. Palpi large, whitish; antennae and legs brownish-black.

♂. Head dark brown in ground-colour, densely silvery-grey dusted. Frons broad in front and widening above, face narrow below antennae, widening towards mouth and confluent with narrow jowls. 2 pairs of ocellar and 2 pairs of vt bristles hardly visible, in the form of small whitish points. Antennae (Fig. 639)

288

brownish-black, segment 3 short and ovate, arista supra-apical, bare, slightly longer than antenna. Palpi whitish-yellow, very large and egg-shaped, almost as long as polished brown proboscis. Thorax very densely light grey dusted, sternopleura with only an indication of a small polished dark brown patch. All thoracic hairs and bristles in the form of small whitish punctures.

Legs rather brownish-black, thinly grey dusted, but knees and posterior four metatarsi yellowish-brown. Pubescence indistinct, whitish; femora without preapical bristles but fore femora with a double row of longer pale hairs beneath. Fore tibiae spindle-shaped with 2 longer dark preapical bristles; mid tibiae with 2 small additional black bristles on apical quarter in front, and with a row of short black bristles on apical third beneath. Hind tibiae with several tiny dark bristles on apical third, tarsi rather slender.

Wings (Fig. 787) blunt-tipped, milky-white with indistinct whitish veins, veins R4+5 and M almost parallel. Squamae dark brown with paler margin and fine whitish fringes, halteres yellowish with knobs slightly dusky. Abdomen dark brown, thinly silvery-grey dusted and with fine pale hairs, especially on venter. Genitalia (Figs. 659-661) small, greyish dusted except for three darker somewhat shining strip-like dorsal processes on right lamella.

Length: body 1.1 - 1.4 mm, wing 1.1 - 1.3 mm.

♀. Resembling male, mid tibiae without short dark bristles beneath.

Length: body 1.2 - 1.7 mm, wing 1.2 - 1.4 mm.

Common in Denmark incl. Bornholm, east to SW.Sweden (Sk., Hall.). - England including Ireland, the Netherlands, German North Sea islands and ? Polish Pomerania. - End May-August. Locally very common, running around

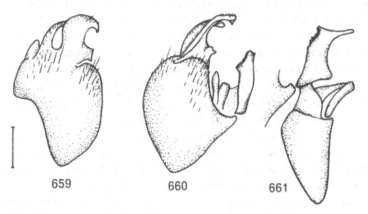

659   660   661

Figs. 659-661. Male genitalia of Chersodromia incana Walk. - 659: right periandrial lamella; 660: periandrium with cerci; 661: left periandrial lamella. Scale: 0.1 mm.

rapidly on coasts of fine dry sand not far from the sea; almost invisible when stationary.

Genus *Stilpon* Loew, 1859

Stilpon Loew, 1859, Neue Beitr., 6: 34 (as subgenus of Drapetis Meigen).
? Pseudostilpon Séguy, 1950, Vie et Milieu, 1: 83.
Type-species: Tachydromia graminum Fallén, 1815 (design. by Loew, 1864: 5).

Very small species (Fig. 662) with yellowish legs and very thickened fore femora, wings often narrowed and pointed with characteristic pattern. Head (Fig. 12) very deep even though jowls below eyes indistinct, slightly concave in front in profile, but facial part below antennae distinctly convex. Eyes microscopically pubescent, separated above antennae by long parallel frons, meeting on a shorter distance below antennae and facets enlarged here. 2 pairs of very small ocellar bristles and a pair of rather short crossed vertical bristles. Antennae (Fig. 21) placed at or below middle of head in profile, basal segment very small; segment 2 cup-shaped and very large, beneath at tip with 2 long diverging bristles; segment 3 small, shorter than deep and somewhat kidney-shaped; arista very long, distinctly supra-apical or almost dorsal. Palpi long-ovate with short apical bristle, proboscis somewhat recurved. Thorax slightly longer than deep with well developed metathorax as in Drapetis, humeri not differentiated. Mesonotum uniformly covered with numerous short hairs, all bristles (a humeral, 2 notopleural, a postalar and 2 scutellars) small. Legs short and strong with very thickened fore femora, mid femora slender; no distinct preapical bristles except on mid femora. Wings (Figs. 31, 788-790) with specifically distinct dark pattern or at least slightly clouded, rather broad and blunt-tipped or very narrow and pointed (both types may even be present in one species and are not determined by the season), or wings very abbreviated. Vein R2+3 very short, ending before middle of wing, sometimes abbreviated; vein R4+5 long, ending at wing-tip and practically parallel with vein M, even if evenly bowed. 1st basal cell shorter than 2nd basal cell, anal cell and vein A completely absent. Abdomen consisting of eight weakly sclerotised but equally long segments as in Chersodromia, segment 8 in ♂ short and concealed beneath segment 7. Male genitalia of very complicated structure, particularly the cerci; right periandrial lamella (on the left because of twisting) very large, left lamella partly fused with hypandrium as in Chersodromia. Female ovipositor short, cerci ovate.

The adults are exclusively terrestrial; they are often found in grass tufts or sphagnum moss, or in heaps of cut sedge, and are sometimes obtained when sieving soil or flood refuse in early spring. Hibernation of the adults has been proved in S.graminum. Nothing is known of the immature stages and life-cycle.

The genus is mainly Holarctic in distribution and is poor in species: 5 species are recorded from North America (Stone et al., 1965), 7 species from the Palaearctic region, including a species from the Canary Is. (Frey, 1936) and another from Japan (Saigusa, 1964), 1 species from Nepal (Smith, 1965) and 2 species from the Ethiopian region from Rhodesia and S.Africa (Smith, 1969).

The genus Pseudostilpon Séguy, 1950, with the species P.paludosa (Perris, 1852) and P.delamarei Séguy, 1950 from the Pyrenees, was erected on the ba-

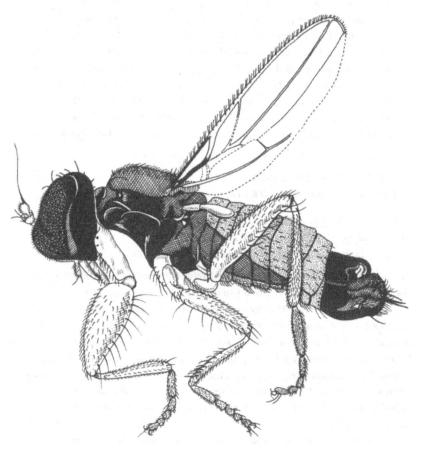

Fig. 662. Male of Stilpon graminum (Fall.). Total length: 1.1 - 1.3 mm.

sis of an abbreviated vein R2+3 but its systematic position remains unclear.

5 species of Stilpon are known from Europe; 3 have also been found in Fennoscandia, and the occurrence there of the fourth British species (sublunata Collin, 1961) is also possible; S.machadoi Smith, 1965a is known only from Portugal.

Key to species of Stilpon

1 Wings (Figs. 788, 789) uniformly faintly yellowish-brown tinged, without a distinct dark pattern. Preapical bristle beneath fore tibiae indistinct .... 126. graminum (Fall.)

- Wings (Fig. 790) with distinct dark brown clouding at middle, base and tip of wing almost clear. Fore tibiae with distinct even if small dark preapical bristle beneath 2

2(1) Wings with a large dark brown cloud at middle, cell R3 without a clear patch. Male genitalia with long black terminal spines on both cerci (Fig. 667)..................... 127. nubila Coll.

- Dark cloud on wing with a clear ovate patch at middle of cell R3. Right cercus of male hypopygium without terminal black spines .... 128. lunata (Walk.)

126. STILPON GRAMINUM (Fallén, 1815)
    Figs. 12, 31, 662-664, 788, 789.

Tachydromia graminum Fallén, 1815: 15 (p.p., nec var.b).
Tachydromia celeripes Meigen, 1830: 343.
Agatachys flavipes Winthem, in Meigen, 1830: 343 (MS).

Wings uniformly yellowish to yellowish-brown clouded, no distinct dark pattern; costa uniformly dark brown. Fore tibiae with minute preapical bristle beneath.

♂. Head (Fig. 12) dull grey on parallel frons and occiput, lower postocular margins silvery dusted. 2 pairs of ocellar and a pair of vt bristles small, dark. Antennae yellow on basal segments, segment 3 and the long supra-apical arista dark brown; segment 2 with 2 long blackish bristles beneath. Palpi very pale, small-ovate, terminal bristle dark. Thorax dull grey on mesonotum, upper part of pleura and whole of prothorax: lower part of pleura (sterno- and hypopleura) polished black. Mesonotum with minute brownish hairs posteriorly, the usual bristles small, black.

Legs yellow to somewhat dirty yellow, tarsi slightly darkened; fore femora conspicuously thickened. Anterior four femora with several long pale bristly-hairs beneath, mid femora with a long dark preapical bristle anteriorly; hind femora with several longer dark bristly-hairs above near base, and with similar anteroventral bristles becoming longer and darker towards tip. Fore tibiae with an indistinct minute anterior preapical bristle, mid tibiae fringed with several strong black anteroventral bristles on apical half.

Wings (Figs. 31, 788, 789) faintly yellowish-brown tinged, slightly darkened along veins R2+3 and Cu, costa uniformly dark brown, almost blackish towards tip of vein M. Wings rather broad and blunt-tipped, or narrowed (reduced posteriorly) and pointed; in latter case veins R4+5 and M less diverging. Abdomen dull grey, large genitalia (Figs. 663, 664) partly polished black above; left cercus armed with 3 strong black apical spines.

Length: body 1.1 - 1.3 mm, wing 0.8 - 1.2 mm.

♀. Resembling male, but wings usually broader and blunter at tip, seldom narrowed; mid tibiae with only brownish ventral bristly-hairs placed more anteriorly. Abdomen dull brownish with small apical segment, and cerci blackish.

Length: body 1.2 - 1.6 mm, wing 1.1 - 1.3 mm.

Common in Denmark and southern parts of Fennoscandia, in Norway north to AK, in Sweden to Sdm., in Finland along the Baltic coast north to Ta and Sa, and along the Gulf of Bothnia to ObS; also in the south of Russian Carelia. -

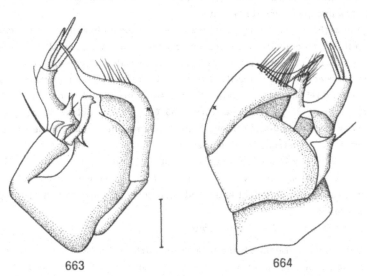

663   664

Figs. 663-664. Male genitalia of _Stilpon graminum_ (Fall.). 663: hypopygium with cerci on the left, right periandrial lamella (marked x) on the right, left lamella, partly fused with hypandrium, on the left; 664: the same, cerci on the right. Scale: 0.1 mm.

Widespread and common in Europe; England, NW and C parts of European USSR, C. and W. Europe, S to Spain and the Azores. - The adults hibernate and are found commonly in any month of the year; often in grass tufts, in heaps of cut sedge or in moss; Lundbeck (1910) collected it by sieving flood refuse.

127. STILPON NUBILA Collin, 1926
    Figs. 21, 665-667, 790.

Tachydromia graminum var. b Fallén, 1815: 15.
Drapetis lunata Walker, 1851: 136 (p.p., figs.).
Stilpon nubila Collin, 1926: 149.

Wings with a distinct dark brown pattern in the form of a large ovate patch anteriorly, base and tip of wing almost clear; costa black on median half. Fore tibiae with a tiny dark preapical bristle.

♂. Head including antennae as in **graminum**; thorax similarly grey dusted on mesonotum and the upper part of pleura, but the small brownish hairs, acr and dc, perhaps more distinct even anteriorly. Legs paler yellow with anterior tibiae and hind metatarsi often slightly brownish, and last segment on tarsi, particularly on fore legs, distinctly dark. Long bristly-hairs on anterior four femora beneath and on hind femora at base above paler, and fore tibiae with a distinct dark preapical bristle beneath.

Wings (Fig. 790) with a conspicuous dark brown cloud anteriorly, occupying cell R3 and cell R5 except for lower part, and leaving basal and apical quarters of wing almost hyaline; similar clouding also along vein Cu. Costa blackish between tips of veins R1 and R2+3, basal section and apical part between veins R4+5 and M very pale. Wings rather broad and blunt-tipped. Abdomen dull grey, large genitalia (Figs. 665-667) polished black above, dusted grey below, and very distinct from those of graminum; right cercus with 1, left cercus with 2, strong black spines at tip.

Length: body 1.2 - 1.4 mm, wing 1.0 - 1.1 mm.

♀. Tarsi often extensively brownish and mid tibiae without a comb of small black ventral spines on apical half. Abdomen dull greyish with last segment and cerci darkened.

Length: body 1.2 - 1.4 mm, wing 1.1 mm.

Uncommon in Denmark and S. Sweden, north to Ög. and Gtl.; not yet recorded from Finland and Norway. - Great Britain, NW of European USSR (Estonia), Jugoslavia, the Azores; widespread but everywhere uncommon. - June-early October. In grasses but mainly on sandy coasts; M. Ackland collected adults on a stone path in his garden at Oxford, England.

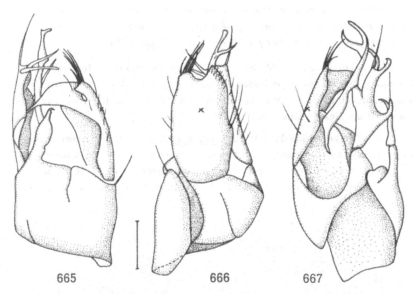

Figs. 665-667. Male genitalia of Stilpon nubila Coll. - 665: hypopygium with cerci on the left; 666: the same with right periandrial lamella (marked x) in front; 667: the same with cerci in front, left periandrial lamella, partly fused with hypandrium, on the right. Scale: 0.1 mm.

128. STILPON LUNATA (Walker, 1851)

Drapetis lunata Walker, 1851: 136 (p.p.; descr., not figs.).
Hemerodromia femorata Heeger, 1852: 779.

Wings rather large and broad with a large dark brown cloud anteriorly as in nubila, but with a clear ovate patch in cell R3, and male genitalia with black spines only on left cercus.

    ♂. Head, thorax and legs as in nubila, fore tibiae with a distinct dark preapical bristle beneath. Wings with a similar dark brown pattern as in nubila but cell R3 with a distinct clear patch at middle, and veins R4+5 and M more decidedly though evenly upcurved; vein R2+3 often incomplete. Costa mainly pale, blackish for a short distance beyond tip of vein R1 and again towards tip of vein M.

    Abdomen dull grey with somewhat smaller genitalia of different structure from those of nubila; right cercus without a black spine at tip, left cercus with 2 longer and 1 shorter black spines apically, and 2 further shorter black spines near base.

♀. Resembling male except for sexual differences; abdomen dull grey, brownish in ground-colour, last segment and cerci blackish.

Length (both sexes): body 1.25 - 1.5 mm (Collin, 1961).

Very rare. Denmark: SJ, Sønder-Bork, 2♀ (Wüstnei). - Great Britain, N.Germany, NW of European USSR; ? C.Europe, ? Azores - can easily be mistaken for sublunata. - May-September, in Denmark in June. Appears to be a sea-coast species (Collin, 1961).

Note. The very closely related S.sublunata Collin, 1961, known with certainity only from England, has the wing-pattern as in lunata but the dark clouding is more intense, wings shorter and narrower with vein M not curved upwards, and there are also slight differences in male genitalia (Collin, 1961, Fig. 26).

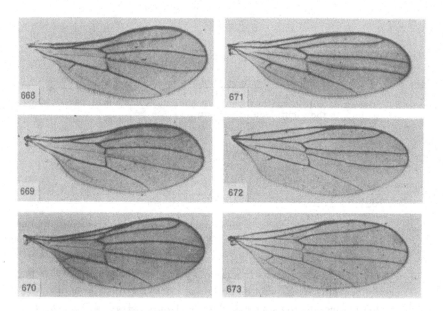

Figs. 668-673. Wings of Symballophthalmus (668-670) and Platypalpus (671-673). - 668: dissimilis (Fall.), x 17; 669: fuscitarsis (Zett.), x 17; 670: pictipes (Beck.), x 16; 671: ciliaris (Fall.), x 16; 672: confiformis Chv., x 16; 673: confinis (Zett.), x 16.

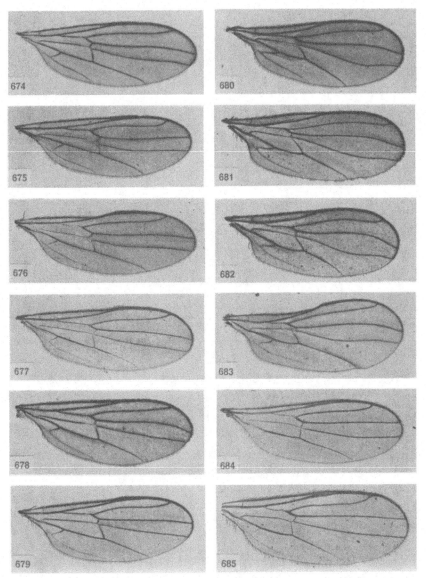

Figs. 674-685. Wings of Platypalpus. - 674: stigmatellus (Zett.), x 13; 675: pectoralis (Fall.), x 15; 676: mikii (Beck.), x 15; 677: nonstriatus Strobl, x 14; 678: maculus (Zett.), x 13; 679: pallipes (Fall.), x 13; 680: albiseta (Panz.), x 17; 681: albisetoides Chv., x 17; 682: albocapillatus (Fall.), x 18; 683: niveiseta (Zett.), x 18; 684: unguiculatus (Zett.), x 21; 685: zetterstedti Chv., x 26.

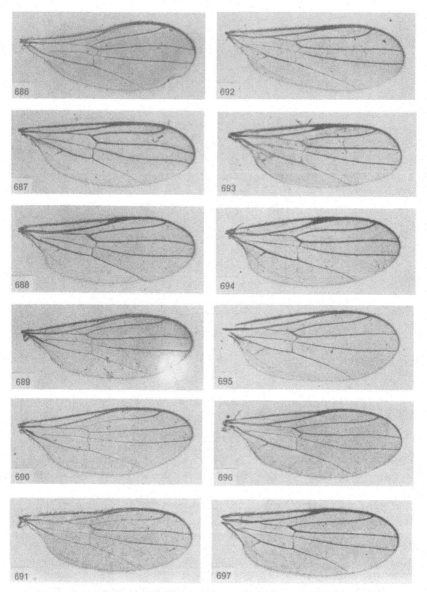

Figs. 686-697. Wings of Platypalpus. - 686: alter (Coll.), x 24; 687: laestadianorum (Frey), x 16; 688: lapponicus Frey, x 17; 689: sahlbergi (Frey), x 18;
690: boreoalpinus Frey, x 18; 691: alpinus Chv., x 18; 692: commutatus (Strobl),
x 16; 693: longicornis (Meig.), x 20; 694: brunneitibia (Strobl), x 16; 695: difficilis (Frey), x 20; 696: scandinavicus Chv., x 17; 697: tuomikoskii Chv., x 19.

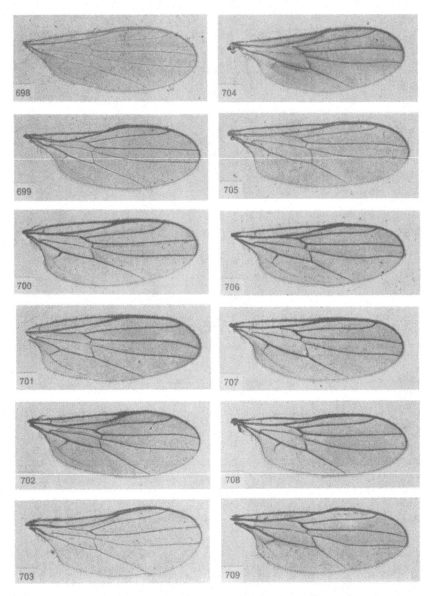

Figs. 698-709. Wings of Platypalpus. - 698: exilis (Meig.), x 17; 699: pulicarius (Meig.), x 23; 700: nigricoxa (Mik), x 16; 701: luteus (Meig.), x 13;
702: nigritarsis (Fall.), x 17; 703: sylvicola (Coll.), x 27; 704: fuscicornis
(Zett.), x 15; 705: ruficornis (v. Roser), x 20; 706: ater (Wahlbg.), x 17; 707:
minutus (Meig.), x 18; 708: niger (Meig.), x 22; 709: aeneus (Macq.), x 18.

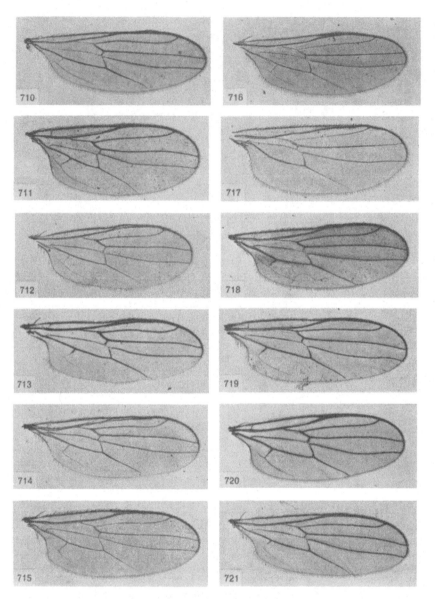

Figs. 710-721. Wings of Platypalpus. - 710: maculipes (Meig.), x 15; 711: rapidus (Meig.), x 18; 712: pallidicoxa (Frey), x 19; 713: agilis (Meig.), x 18; 714: pseudorapidus Kᴏᴠ., x 20; 715: nigrosetosus (Strobl), x 16; 716: cothurnatus Macq., x 21; 717: cryptospina (Frey), x 28; 718: optivus (Coll.), x 16; 719: annulatus (Fall.), x 14; 720: melancholicus (Coll.), x 12; 721: notatus (Meig.), x 17.

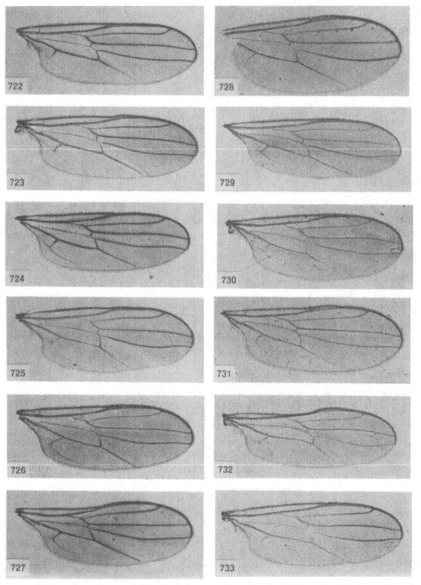

Figs. 722-733. Wings of Platypalpus. - 722: strigifrons (Zett.), x 14; 723: infectus (Coll.), x 17; 724: interstinctus (Coll.), x 11; 725: coarctatus (Coll.), x 15; 726: clarandus (Coll.), x 17; 727: articulatus Macq., x 24; 728: articulatoides (Frey), x 26; 729: annulipes (Meig.), x 15; 730: ecalceatus (Zett.), x 18; 731: calceatus (Meig.), x 21; 732: stabilis (Coll.), x 18; 733: pallidiventris (Meig.), x 17.

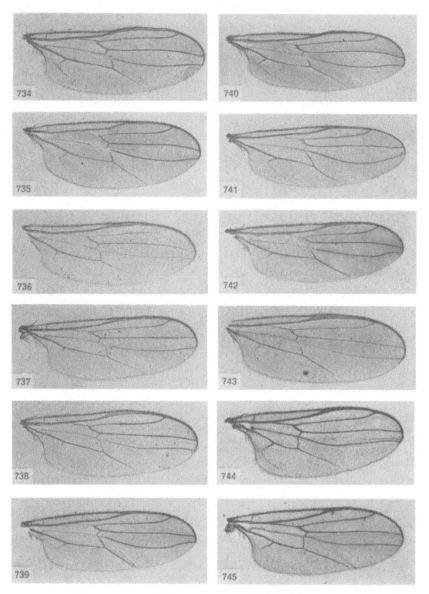

Figs. 734-745. Wings of Platypalpus. - 734: longiseta (Zett.), x 17; 735: laticinctus Walk., x 10; 736: albicornis (Zett.), x 15; 737: flavicornis (Meig.), x 19; 738: pallidicornis (Coll.), x 21; 739: major (Zett.), x 9; 740: analis (Meig.), x 12; 741: candicans (Fall.), x 11; 742: cursitans (Fabr.), x 10; 743: verralli (Coll.), x 20; brevicornis (Zett.), x 20; 745: sordidus (Zett.), x 16.

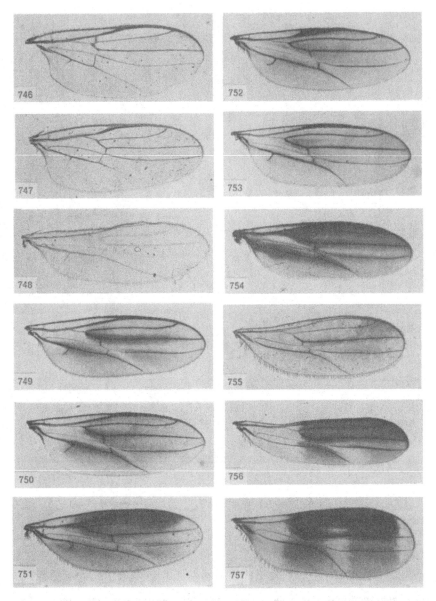

Figs. 746-757. Wings of Platypalpus (746-747), Dysaletria (748), Tachypeza (749-754) and Tachydromia (755-757). - 746: subbrevis (Frey), x 20; 747: hackmani Chv., x 20; 748: atriceps (Boh.), x 27; 749: nubila (Meig.), x 16; 750: truncorum (Fall.), x 13; 751: fuscipennis (Fall.), x 19; 752: heeri Zett., x 12; 753: fennica Tuomik., x 14; 754: winthemi Zett., x 16; 755: terricola Zett., x 22; 756: sabulosa Meig., x 25; 757: connexa Meig., x 24.

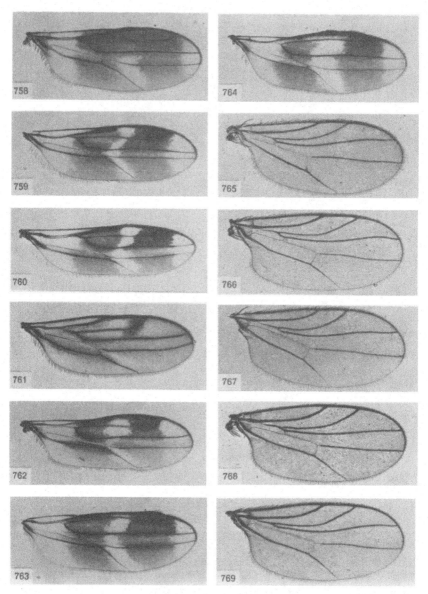

Figs. 758-769. Wings of Tachydromia (758-764) and Drapetis (765-769). - 758: morio (Zett.), x 22; 759: arrogans (L.), x 23; 760: aemula (Loew), x 26; 761: punctifera (Beck.), x 20; 762: incompleta (Beck.), x 19; 763: umbrarum Hal., x 18; 764: woodi (Coll.), x 22; 765: ephippiata (Fall.), x 21; 766: assimilis (Fall.), x 26; 767: ingrica Kov., x 32; 768: arcuata Loew, x 25; 769: simulans Coll., x 26.

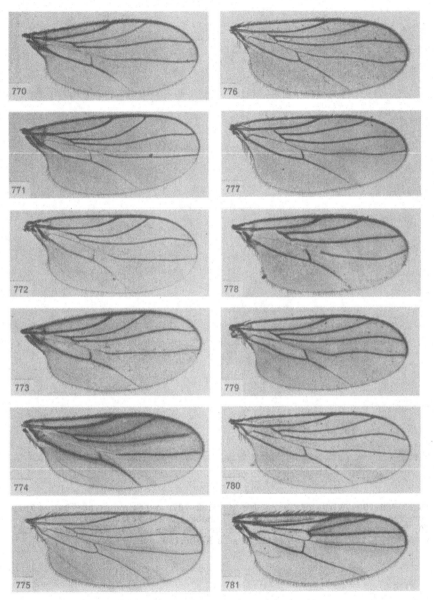

Figs. 770-781. Wings of Drapetis (770-774), Crossopalpus (775-780) and Chersodromia (781). - 770: pusilla Loew, x 36; 771: exilis Meig., x 32; 772: infitialis Coll., x 26; 773: parilis Coll., x 30; 774: incompleta Coll., x 26; 775: setiger (Loew), x 24; 776: curvipes (Meig.), x 20; 777: nigritellus (Zett.), x 23; 778: humilis (Frey), x 32; 779: curvinervis (Zett.), x 19; 780: abditus Kov., x 21; 781: hirta (Walk.), x 17.

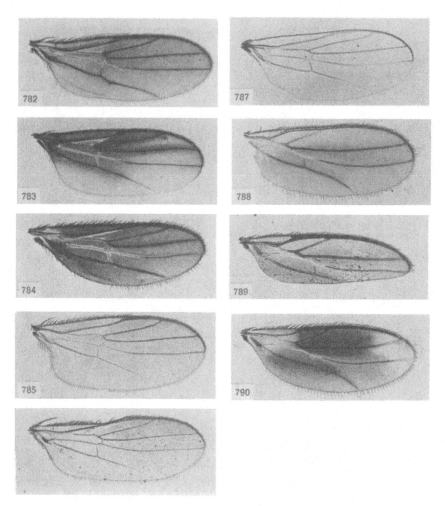

Figs. 782-790. Wings of Chersodromia (782-787) and Stilpon (788-790). - 782: cursitans (Zett.), x 27; 783: difficilis Lundbk., x 23; 784: arenaria (Hal.), x 50; 785: speculifera Walk., x 20; 786: beckeri Mel., x 28; 787: incana Walk., x 40; 788 & 789: graminum (Fall.), x 38 & 50; 790: nubila Coll., x 46.

Literature

Andrewes, C.H., 1968: Some records of local Empididae (Dipt.). - Entomologist's mon. Mag., 104: 249.

Ardö, P., 1957: Studies in the marine shore dune ecosystem with special reference to the dipterous fauna. - Opusc. ent., Suppl. 14: 1-255.

Backlund, H., 1945: Wrack fauna of Sweden and Finland. Ecology and chorology. - Opusc. ent., Suppl. 5: 1-237.

Bährmann, R., 1960: Vergleichend-morphologische Untersuchungen der männlichen Kopulationsorgane bei Empididen. - Beitr. Ent., 10: 485-540.

Becker, P., 1958: Some parasites and predators of biting midges, Culicoides Latreille (Dipt., Ceratopogonidae). - Entomologist's mon. Mag., 94: 186-189.

Becker, Th., 1887: Beiträge zur Kenntnis der Dipteren-Fauna von St. Moritz. - Berl. ent. Z., 31: 93-141.

- 1889: Neue Dipteren aus Dalmatien, gesammelt auf einer Reise im Mai 1889. - Ibid., 33: 335-346.

- 1889a: Altes und Neues aus der Schweiz. - Wien. ent. Ztg, 8: 73-8

- 1889b: Berichtigung. - Ibid., 8: 285.

- 1890: Altes und Neues aus Tirol und Salzburg. - Ibid., 9: 65-70.

- 1900: Beiträge zur Dipteren-Fauna Sibiriens. - Acta Soc. Sci. fenn., 26 (9): 1-66.

- 1902: Aegyptische Dipteren. - Mitt. zool. Mus. Berl., 2(2): 1-195.

- 1907: Die Ergebnisse meiner dipterologischen Frühjahrsreise nach Algier und Tunis. 1906. - Z. syst. Hymenopt. Dipterol., 7: 97-128.

- 1908: Dipteren der Kanarischen Inseln. - Mitt. zool. Mus. Berl., 4 (1): 1-180.

- 1913: Dipteren aus Marokko. - Ezheg. zool. Muz., 18: 84-85.

Beling, T., 1882: Beitrag zur Metamorphose zweiflügeliger Insekten aus den Familien Tabanidae, Leptidae, Asilidae, Empididae, Dolichopodidae und Syrphidae. - Arch. Naturgesch., 44: 187-240.

- 1888: Beitrag zur Metamorphose einiger zweiflügeliger Insekten aus den Familien Tabanidae, Empididae und Syrphidae. - Verh. zool.-bot. Ges. Wien, 38: 1-4.

Bezzi, M., 1899: Contribuzioni alla fauna ditterologica Italiana. - Boll. Soc. ent. ital., 30: 121-164.

- 1903: Katalog der paläarktischen Dipteren, 2, 396 pp., Budapest.

- 1904: Empididae Indo-australiani raccolti dal Signor L. Biró. - Annls Mus. Nat. hung., 2: 320-361.

Boheman, C.H., 1852: Entomologiska Anteckningar under en resa i Södra Sverige 1851. - K. svenska Vetensk-Akad. Handl., 1851 (1852): 53-210.

Brauns, A., 1949: In Deutschland und Schleswig-Holstein neuaufgefundene Zweiflüglerarten (Diptera). - Entomon, 1: 155-161.

- 1959: Autökologische Untersuchungen über die thalassicolen Zweiflügler (Diptera) im schleswig-holsteinischen Bereich der Nord- und Ostsee. - Arch. Hydrobiol., 55: 453-594.

Broen, B. & W.Mohrig, 1965: Zur Frage der Winteraktivität von Dipteren in der Bodenstreu. - Dt.ent.Z., 12: 4-5.

Brunetti, E., 1920: The fauna of British India including Ceylon and Burma. Diptera, Brachycera, I, IX + 401 pp., London.

Buxton, P.A., 1960: British Diptera associated with fungi. III. Flies of all families reared from about 150 species of fungi. - Entomologist's mon.Mag., 96: 61-94.

Chillcott, J.G., 1958: A new Nearctic species of Symballophthalmus Becker (Diptera: Empididae). - Can.Ent., 90: 647-649.

- 1961: A revision of the genus Roederioides Coquillett (Diptera: Empididae). - Ibid., 93: 419-428.

- 1962: A revision of the Platypalpus juvenis complex in North America (Diptera: Empididae). - Ibid., 94: 113-143.

Chvála, M., 1966: Notes on the genus Tachydromia Meig. and the annulimana-group (Diptera, Empididae). - Acta ent.bohemoslov., 63: 464-477.

- 1970: Revision of Palaearctic species of the genus Tachydromia Meig. (= Tachista Loew) (Diptera, Empididae). - Acta ent.Mus.Nat.Pragae, (1969) 38: 431-540.

- 1970a: Descriptions of nine new species of Palaearctic Chersodromia Walk. (Diptera, Empididae), with notes on the genus. - Acta ent.bohemoslov., 67: 384-407.

- 1970b: Some extracts from the correspondence between J.E.Collin and W. Lundbeck. - Entomologist, 1970: 144-146.

- 1971: A revision of the Scandinavian Tachydromiinae (Dipt., Empididae) described by J.W.Zetterstedt. - Ent.Scand., 2: 1-28.

- 1971a: A third Palaearctic species of Drapetis subgenus Elaphropeza, from Sicily (Insecta, Diptera, Empididae). - Steenstrupia, 1: 127-130.

- 1972: Notes on Scandinavian Platypalpus (Dipt., Empididae) with description of four new species and new synonymies. - Ent.Scand., 3: 1-11.

- 1973: European species of the Platypalpus albiseta-group (Diptera, Empididae). - Acta ent.bohemoslov., 70: 117-136.

- 1973a: Notes on British Tachydromia (Dipt., Empididae), with description of a new species from Inverness-shire. - Entomologist's mon.Mag., 108 (1972): 214-218.

Coe, R.L., 1962: A further collection of Diptera from Jugoslavia, with localities and notes. - Bull.Mus.Hist.nat.Belgrade, 1962, Sér.B, Livre 18: 95-136.

Collart, A., 1934: Description de deux Drapetis nouveaux du Congo Belge (Diptera: Empididae). - Bull.Annls Soc.r.ent.Belg., 74: 61-67.

Collin, J.E., 1926: Notes on the Empididae (Diptera) with additions and corrections to the British List. - Entomologist's mon.Mag., 62: 146-159, 185-190.

- 1950: A new Palaearctic species of Chersodromia with very short wings (Diptera: Empididae). - Proc.R.ent.Soc.Lond., (B), 19: 78-79.

- 1960: Some Empididae from Palestine. - Ann.Mag.nat.Hist., 1959, (13) 2: 385-420.

- 1961: British Flies 6. Empididae, VIII + 782 pp., Cambridge.

- 1969: Some new species of Empididae from Central Europe collected by Dr. O. Ringdahl. - Opusc. ent., 34: 150-157.

Coquillett, D. W., 1900: Papers from the Harriman Alaska Expedition. IX. Entomological results (3): Diptera. - Proc. Wash. Acad. Sci., 2: 389-464.

- 1903: The genera of the dipterous family Empididae, with notes and new species. - Proc. ent. Soc. Wash., 5: 245-272.

Corti, E., 1907: Eine neue Art der Dipterengattung Tachydromia (Mg.) Lw. - Wien. ent. Ztg, 26: 101-102.

Curtis, J., 1833: British Entomology: Being illustrations and descriptions of the genera of insects found in Great Britain and Ireland, 10: 434-481 pls., 11: 482-529 pls., London.

Czerny, L. & G. Strobl, 1909: Spanische Dipteren. III. Beitrag. - Verh. zool. -bot. Ges. Wien, 59: 121-301.

Dyte, C. E., 1967: Some distinctions between the larvae and pupae of the Empididae and Dolichopodidae (Diptera). - Proc. R. ent. Soc. Lond., (A), 42: 119-128.

Emeis, W., 1970: Zur Verbreitung und Ökologie der Empididen (Ins. Dipt.) in Schleswig-Holstein. - Schr. naturw. Ver. Schlesw.-Holst., 40: 79-96.

Engel, E. O., 1938-

Engel, E. O., 1938-1939: Empididae, Tachydromiinae. In: Lindner, E.: Die Fliegen der palaearktischen Region, 4, 4: 1-119, Stuttgart.

Fabricius, J. C., 1775: Systema entomologiae, 832 pp., Flensburgi et Lipsiae.

- 1781: Species insectorum, 2, 517 pp., Hamburgi et Kilonii.

- 1794: Entomologia systematica emendata et aucta, 4, 472 pp., Hafniae.

Fallén, C. F., 1815-1816: Empididae Sveciae, 34 pp., Lundae.

- 1823: Supplementum Dipterorum Sveciae, 16 pp., Lundae.

Fleschner, C. A. & D. W. Ricker, 1953: An Empidid fly predaceous on citrus red mites. - J. econ. Ent., 46: 155.

Frey, R., 1907: Uebersicht der finnischen Arten der Gattung Tachydromia Meig. (= Platypalpus Macq.) (Dipt.). - Z. syst. Hymenopt. Dipterol., 7: 407-413.

- 1908: Finlands Tachydromia-arter. - Meddn Soc. Fauna Flora fenn., (1907) 34: 20-21.

- 1909: Mitteilungen über finnländische Dipteren. - Acta Soc. Fauna Flora fenn., 31 (9): 1-24.

- 1913: Zur Kenntnis der Dipterenfauna Finlands. II. Empididae. - Ibid., 37 (3): 1-89.

- 1913a: In Lundström C & R. Frey, Beitrag zur Kenntnis der Dipterenfauna des nördl. europäischen Russlands. - Ibid., 37 (10): 1-20.

- 1915: Résultats scientifiques de l'expédition polaire Russe en 1900-1903, sous la direction du Baron E. Toll. Section E: Zoologie. II, 10. Diptera Brachycera aus den arktischen Kunstengegenden Sibiriens. - Acad. Sci. Russ. Mém., Cl. Phys.-Math. ser. 8, 29 (10): 1-35.

- 1918: Beitrag zur Kenntnis der Dipterenfauna des nördl. europäischen Russlands. II. Dipteren aus Archangelsk. - Acta Soc. Fauna Flora fenn., 46 (2): 1-32.

- Några nykomlingar för Finlands empidid- och dolichopodidfauna. - Memoranda Soc. Fauna Flora fenn., (1928-1929) 5: 111-113.
- 1936: Die Dipterenfauna der Kanarischen Inseln und ihre Probleme. - Commentat. biol., 6: 1-237.
- 1941: Diptera Brachycera (excl. Muscidae, Tachinidae). - Enumeratio Ins. fenn., VI. Diptera, 31 pp., Helsingfors.
- 1943: Übersicht der paläarktischen Arten der Gattung Platypalpus Macq. (= Coryneta Meig.). (Dipt. Empididae). - Notul. ent., 23: 1-19.
- 1950: Dipterenfaunan vid Tana älv i Utsjoki sommaren 1949. Mit einem Anhang: Synonymische Bemerkungen einigen neuen Diptera brachycera aus Utsjoki in Finnisch-Lappland. - Ibid., 30: 5-18.

Griffiths, G.C.D., 1972: The phylogenetic classification of Diptera Cyclorrhapha with special reference to the structure of the male postabdomen. - Series Entomologica, 8, 340 pp., The Hague.

Hackman, W., 1963: Studies on the dipterous fauna in burrows of voles (Microtus, Clethrionomys) in Finland. - Acta zool. fenn., 102: 1-64.

Haliday, A.H., 1833: Catalogue of Diptera occuring about Holywood in Downshire. - Entomologist's mon. Mag., 1: 147-180.

Heeger, E., 1852: Beiträge zur Naturgeschichte der Insecten. - Sber. Akad. Wiss. Wien, 9: 123-144, 263-286, 473-490, 774-781.

Hennig, W., 1952: Die Larvenformen der Dipteren, 3, VII + 628 pp., Berlin.
- 1970: Insektenfossilien aus der unteren Kreide. II. Empididae (Diptera, Brachycera). - Stuttg. Beitr. Naturk., Nr. 214: 1-12.
- 1971: Insektenfossilien aus der unteren Kreide. III. Empidiformia ("Microphorinae") aus der unteren Kreide und aus dem Baltischen Bernstein; ein Vertreter der Cyclorrhapha aus der unteren Kreide. - Ibid., Nr. 232: 1-28.
- 1972: Eine neue Art der Rhagionidengattung Litoleptis aus Chile, mit Bemerkungen über Fühlerbildung und Verwandtschaftsbeziehungen einiger Brachycerenfamilien (Diptera: Brachycera). - Ibid., Nr. 242: 1-18.

I.C.Z.N., 1963: The suppression under the Plenary Powers of the pamphlet published by Meigen, 1800. - Bull. zool. Nom., 20: 339-342.

Kovalev, V.G., 1966: On the fauna and ecology of Tachydromiinae (Diptera, Empididae) in the middle belt of the European part of the USSR. - Ent. Obozr., 45: 774-778 (In Russian).
- 1968: The use of taxonomic analysis in systematics of Diptera. - Zool. Zh., 47: 720-731 (In Russian).
- 1969: Empididae, Tachydromiinae. In Bej-Bienko, G.J.: Opredelitel nasekomych evropejskoj casti SSSR, 5 (1): 578-607, Leningrad (In Russian).
- 1969a: Systematic review of the subfamily Tachydromiinae (Diptera, Empididae). (Thesis of dissertation) Avtoreferat, 098 entomologia, Izd. Mosk. Univ., 20 pp., Moscow (In Russian).
- 1970: A redescription of the type specimen of Drapetis (s. str.) laevis Becker (Diptera, Empididae) from Tangier. - Ent. Obozr., 49: 687-690 (In Russian).
- 1971: New European species of Diptera of the Genus Platypalpus Macq. (Diptera, Empididae). - Ibid., 50: 200-213 (In Russian).
- 1972: Diptera of the genera Drapetis Mg. and Crossopalpus Bigot (Diptera, Empididae) from the European part of the USSR. - Ibid., 51: 173-196 (In Russian).

Kröber, O., 1958: Nachträge zur Dipteren-Fauna Schleswig-Holsteins und Niedersachsen (1933-35). - Verh. Ver. naturw. Heimatforsch. 33: 39-96.

Krogerus, R., 1932: Uber die Okologie und Verbreitung der Arthropoden der Triebsandgebiete an den Küsten Finnlands. - Acta zool. fenn., 12: 1-308.

Krystoph, H., 1961: Vergleichend-morphologische Untersuchungen an den Mundteilen bei Empididen. - Beitr. Ent., 11: 824-872.

Latreille, P.A., 1796: Précis des caractères génériques des Insectes, disposés dans un ordre naturel, 179 pp., Paris.

Laurence, B.R., 1951: The prey of some tree trunk frequenting Empididae and Dolichopodidae (Dipt.). - Entomologist's mon. Mag., 87: 166-169.

- 1953: Some Diptera bred from cow dung. - Ibid., 89: 281-283.

Lindroth, C.H., 1931: Die Insektenfauna Islands und ihre Probleme. - Zool. Bidr. Upps., 13: 105-599.

Lindroth, C.H., H. Andersson, H. Bödvarsson & S. G. Richter, 1970: Preliminary report on the Surtsey investigation in 1968. Terrestrial Invertebrates. - Surtsey Res. Prog. Rep., 5: 1-7.

Linné, C., 1761: Fauna svecica sistens animalia Sveciae regni, 2, 578 pp., Stockholmiae.

Lioy, P., 1863: I ditteri distribuiti secondo un nuovo metodo di classificazione naturale. - I. R. Ist. Veneto di Sci., Let. ed Arti, Atti ser. 3, 9: 187-236.

Loew, H., 1859: Neue Beiträge zur Kenntnis der Dipteren. Sechster Beitrag. - K. Realschule zu Meseritz, Programm 1859: 1-50.

- 1863: Diptera Americae septentrionalis indigena. Centuria tertia. - Berl. ent. Z., 7: 1-55.

- 1864: Ueber die schlesischen Arten der Gattungen Tachypeza Meig. (Tachypeza, Tachista, Dysaletria) und Microphorus Macq. (Trichina und Microphorus). - Z. Ent., (1860) 14: 1-50.

Lundbeck, W., 1910: Diptera Danica, genera and species of flies hitherto found in Denmark. Part 3, Empididae, 324 pp., Copenhagen.

Lyneborg, L., 1965: The entomology of the Hansted Reservation, North Jutland, Denmark. 9. Diptera, Brachycera & Cyclorrhapha - Fluer. - Ent. Meddr, 30: 201-262.

Macquart, J., 1823: Monographie des insectes Diptères de la famille des Empides, observés dans le nord-ouest de la France. - Soc. d'Amateurs des Sci., de l'Agr. et des Arts, Lille, Rec. des Trav., 1819/1822: 137-165.

- 1827: Insectes Diptères du Nord de la France. Vol. 3: Platypezines, Dolichopodes, Empides, Hybotides, 159 pp., Lille.

- 1834: Histoire naturelle des Insectes. Diptères, Vol. 1, 578 pp., Paris.

Malloch, J.R., 1918: A preliminary classification of Diptera, exclusive of Pupipara, based upon larval and pupal characters, with keys to imagines in certain families. 1. - Bull. Ill. St. Lab. nat. Hist., 12: 161-409.

Meigen, J.W., 1800: Nouvelle classification des mouches à deux ailes (Diptera L.) d'après un plan tout nouveau, 40 pp., Paris.

- 1804: Klassifikazion und Beschreibung der europäischen zweiflügeligen Insecten (Diptera Linn.). 1, 1: XXVIII + 152, 2: VI + 153-314 pp., Braunschweig.

- 1822-1838: Systematische Beschreibung der bekannten europäischen zweiflügeligen Insekten. 3 (1822): X + 416 pp., 6 (1830): IV + 401 pp., 7 (1838): XII + 434 pp., Hamm.

Meijere, J.C.H. de, 1907: Eerste supplement op de nieuwe naamlijst van nederlandsche Diptera. - Tijdschr.Ent., 50: 151-195.

Melander, A.L., 1902: American Diptera. A monograph of the North American Empididae. I. - Trans.Am.ent.Soc., 28: 195-367.

- 1906: Some new or little-known genera of Empididae. - Ent.News, 17: 370-379.

- 1918: The Dipterous genus Drapetis. - Ann.ent.Soc.Am., 11: 183-221.

- 1928: Diptera, fam.Empididae. - Genera Insect., (1927) 185: 434 pp., Bruxelles.

- 1965: Family Empididae, in: Stone, A. et al., A catalog of the Diptera of America North of Mexico, pp. 446-481, Washington.

Mik, J., 1894: Dipterologische Miscellen. - Wien.ent.Ztg, 13: 164-168.

Nielsen, P., O.Ringdahl & S.L.Tuxen, 1954: Diptera 1 (exclusive of Ceratopogonidae and Chironomidae). - Zoology Iceland, 3 (48a), 189 pp., Copenhagen and Reykjavík.

Oldenberg, L., 1924: Die Empididen v.Rosers in Stuttgart. (Dipt.). - Dt.ent. Z., 1924: 226-236.

Oldroyd, H., 1949: A wingless Empid (Diptera) from Tasmania. - Entomologist's mon.Mag., (1948) 84: 278-279.

Olivier, A.G., 1791: Encyclopédie méthodique. Dictionnaire des Insectes, 6: 704 pp., Paris.

Panzer, G.W.F., 1806: Faunae insectorum germanicae. Heft 103, 24 Tfl., Nürnberg.

Parmenter, L., 1952: Diptera in grass tussocks. - Entomologist's mon.Mag., 88: 13.

Perris, E., 1852: Seconde excursion dans les Grandes-Landes. - Annls Soc. linn.Lyon, 1850/1852: 145-216.

Peterson, B.V., 1960: Notes on some natural enemies of Utah black flies (Diptera: Simuliidae). - Can.Ent., 92: 266-274.

Peterson, B.V. & D.M.Davies, 1960: Observations on some insect predators of black flies (Diptera: Simuliidae) of Algonquin Park, Ontario. - Can.J. Zool., 38: 9-18.

Poulton, E.B., 1913: Empidae and their prey in relation to courtship. - Entomologist's mon.Mag., 49: 177-180.

Quate, L.W., 1960: Diptera: Empididae. - Insects Micronesia, 13: 55-73.

Ringdahl, O., 1921: Bidrag till kännedomen om de skånska stranddynernas insektfauna. - Ent.Tidskr., 42: 21-92.

- 1939: Diptera Brachycera i Regio alpina. - Ibid., 60: 37-50.

- 1941: Bidrag till kännedomen om flugfaunan (Diptera Brachycera) på Hallands Väderö. - Ibid., 62: 1-23.

- 1945: För svenska faunan nya Diptera. - Ibid., 66: 1-6.

- 1947: Förteckning över flugor från Ölands alvar. - Ibid., 68: 21-28.

- 1950: Dipterologiska antekningar från sydsvenska mossar. - Ibid., 71: 111-119.

- 1951: Flugor från Lapplands, Jämtlands och Härjedalens fjälltrakter (Diptera Brachycera). - Opusc.ent., 16: 113-186.

- 1954: Några dipterologiska antekningar från Råå kärr och vassar. - Ent. Tidskr., 75: 223-234.

- 1957: Fliegenfunde aus den Alpen. - Ibid., 78: 115-134.

Rondani, C., 1856: Dipterologiae Italicae prodromus. Vol.1. Genera Italica ordinis dipterorum ordinatim disposita et distincta et in familias et stirpes aggregata, 228 pp., Parmae.

Roser, C.von, 1840: Erster Nachtrag zu dem in Jahre 1834 bekannt gemachten Verzeichnisse in Württemberg vorkommender zweiflügliger Insekten. - KorrespBl. Württ. landw. Ver., 1: 49-64.

Saigusa, T., 1963: A new Japanese species of the genus Symballophthalmus with bicolorous thorax (Diptera, Empididae). - Sieboldia, 3: 183-185.

- 1964: Empididae of the Yaeyama Group (Diptera, Brachycera). Reports of the Committee of Foreign Sci. Res., Kyushu Univ., No.2 (2nd Report of the Kyushu Univ. Exp. to the Yaeyama Group, Ryukyus), pp. 173-177 (In Japanese).

Schiner, J.R., 1862: Fauna Austriaca. Die Fliegen (Diptera), 1, LXXX + 672 pp., Wien.

Scopoli, J.A., 1763: Entomologia carniolica exhibens insecta carnioliae indigene et distributa in ordines, genera, species, varietates methodo Linnaeana, 421 pp., Vindobonae.

Séguy, E., 1950: Un nouveau genre de Corynétine du Midi de la France (Dipt., Empididae). - Vie Milieu, 1: 83-87.

Service, M.W., 1969: Tachydromia spp. (Dipt., Empididae) as predators of adult Anopheline mosquitoes. - Entomologist's mon. Mag., (1968) 104: 250-251.

Siebke, H., 1877: Enumeratio Insectorum Norvegicorum. Fasc.IV, Catalogus Dipterorum Continentem, XIV + 255 pp., Christianiae.

Smith, K.G.V., 1962: Studies on Brazilian Empididae. - Trans. R. ent. Soc. Lond., 114: 195-266.

- 1964: Chersodromia cursitans Zetterstedt (Dipt., Empididae) reinstated as a British species. - Entomologist's mon. Mag., (1963) 99: 127-128.

- 1964a: A remarkable new genus and two new species of Empididae (Tachydromiinae, Drapetini) from the Cook Islands. - Pacif. Insects, 6 (2): 247-251.

- 1965: Diptera from Nepal, Empididae. - Bull. Br. Mus. nat. Hist. Entomology, 17 (2): 61-112.

- 1965a: A new species of Stilpon Loew, 1859 (Diptera: Empididae) from Portugal. - Proc. R. ent. Soc. Lond. (B), 34: 48-50.

- 1967: Family Empididae, in: A catalogue of the Diptera of the Americas south of the United States, Part 39: 1-67, Sao Paulo.

- 1967a: A second Palaearctic species of Drapetis subgenus Elaphropeza Macquart from Spain. - Proc. R. ent. Soc. Lond. (B), 36: 153-155.

- 1967b: Afrikanische Empididae (Dipt.). - Stuttg. Beitr. Naturk. Nr.179: 1-16.

- 1969: The Empididae of Southern Africa (Diptera). - Ann. Natal Mus., 19: 1-347.

- 1969a: Platypalpus (Cleptodromia) longimana Corti, new to Britain and the male of P. altera (Collin) (Dipt., Empididae). - Entomologist's mon. Mag., 105: 108-110.

Sommerman, K.M., 1962: Notes on two species of Oreogeton, predaceous on black fly larvae (Diptera: Empididae and Simuliidae). - Proc. ent. Soc. Wash., 54: 123-129.

Stone, A. et al., 1965: A Catalog of the Diptera of America north of Mexico. - Agric. Handb. Forest Serv. U.S. No. 276, IV + 1696 pp., Agric. Res. Serv., Washington.

Strobl, G., 1880: Dipterologische Funde um Seitenstetten. Ein Beitrag zur Fauna Nieder-Österreichs. - Gymn.-Progr. Seitenstetten, 14: 1-65.

- 1893: Die Dipteren von Steiermark. I. - Mitt. naturw. Ver. Steierm., (1892) 29: 1-199.

- 1898: Die Dipteren von Steiermark. IV., Nachträge. - Ibid., (1897) 34: 192-298.

- 1899: Spanische Dipteren. - Wien. ent. Ztg, 18: 12-83.

- 1906: Spanische Dipteren. II. Beitrag. - Mems. R. Soc. esp. Hist. nat., (1905) 3: 271-422.

- 1909: Empididae, in : Czerny, L. & G. Strobl, Spanische Dipteren. III. Beitrag. - Verh. zool.-bot. Ges. Wien, 59: 121-301.

- 1910: Die Dipteren von Steiermark. V., II. Nachtrag. - Mitt. naturw. Ver. Steierm., (1909) 46: 45-293.

Teschner, D., 1961: Beiträge zur Kenntnis der Fauna eines Müllplatzes in Hamburg. 6. Die Fliegen eines Hamburger Müllplatzes. - Ent. Mitt. Zool. Staatsinst. Zool. Mus. Hamburg, Nr. 35 (1961), 2: 189-204.

- 1962: Fliegen einer Hamburger Wohnung und in Hamburg neuaufgefundene Fliegenarten (Diptera). - Ibid., Nr. 37 (1962), 2: 221-232.

Theowald, B., 1962: Some remarks on Empididae (Diptera, Brachycera). - Ent. Ber., 22: 192.

Trehen, P., 1962: Contribution à l'étude de l'anatomie de l'hypopygium dans la familie des Empidinae (Diptera - Empididae). - Bull. Soc. zool. Fr., 87: 498-508.

Tuomikoski, R., 1932: Zwei neue Empididen aus Finnland. - Notul. ent., 12: 46-50.

- 1935: Ein vermutlicher Fall von geographischer Parthenogenesis bei der Gattung Tachydromia (Dipt., Empididae). - Annls ent. fenn., 1: 38-43.

- 1938: Phänologische Beobachtungen über die Empididen (Dipt.) Süd- und Mittelfinnlands. - Ibid., 4: 213-247.

- 1939: Beobachtungen über das Schwärmen und die Kopulation einiger Empididen (Dipt.). - Ibid., 5: 1-30.

- 1952: Über die Nahrung der Empididen-Imagines (Dipt.) in Finnland. - Ibid., 18: 170-181.

- 1966: The Ocydromiinae group of subfamilies (Diptera, Empididae). - Ibid., 32: 282-294.

Ulrich, H., 1972: Zur Anatomie des Empididen-Hypopygiums (Diptera). - Veröff. zool. StSamml. Münch., 16: 1-28.

Vaillant, F., 1952: Quelques Empididae nouveaux pour l'Algérie (Diptera). - Revue fr. Ent., 19: 64-67.

Wahlgren, E., 1910: Diptera 1. Orthorrhapha, Brachycera. 24. Fam. Dansflugor. Empididae. - Svensk Insektfauna, 11, 1 (2): 41-95.

Walker, F., 1836: Notes on Diptera. - Entomologist's mon. Mag., 3: 178-182.

- 1837: Notes on Diptera. - Ibid., 4: 226-230.

- 1851: Insecta Britannica. Diptera, 1, 313 pp., London.

Wahlberg, P.F., 1844: Nya Diptera från Norrbotten och Luleå Lappmark. - Öfvers. K. VetenskAkad. Förh., 1: 106-110.

Wéber, M., 1972: New Platypalpus Macquart species from Hungary (Diptera: Empididae). - Annls Mus. nat. Hung., 64: 305-313.

Westwood, J.O., 1840: An introduction to the modern classification of Insects. Synopsis of the genera of British Insects. 2, XI + 587 + 158 pp., London.

Whitfield, F.G.S., 1925: The natural control of the leaf-miner Phytomyza aconiti Hendel (Diptera) by Tachydromia minuta Meigen (Diptera). - Bull. ent. Res., 16: 95-97.

Zetterstedt, J.W., 1819: Några nya Svenska Insect-arter. - K. svenska Vetensk -Akad. Handl., 1819: 69-86.

- 1838: Insecta Lapponica descripta. Diptera, pp. 477-868, Lipsiae.

- 1842-1859: Diptera Scandinaviae. Disposita et descripta, 1 (1842): 1-440, 8 (1849): 2935-3366, 11 (1852): 4091-4545, 12 (1855): 4547-4942, 13 (1859): 4943-6190, Lund.

Zusková, L., 1966: Czechoslovak species of the genus Platypalpus Macquart (Diptera, Empididae). - Acta faun. ent. Mus. Nat. Pragae, 11 (113): 331-372.

			DENMARK												
			N. Germany	G. Britain	SJ	EJ	WJ	NWJ	NEJ	F	LFM	SZ	NWZ	NEZ	B	Sk.	Bl.
Symballophthalmus dissimilis (Fall.)	1		●										●	●	●	
S. fuscitarsis (Zett.)	2	●	●						●				●	●	●	
S. pictipes (Beck.)	3		●													
Platypalpus ciliaris (Fall.)	4	●	●		●			●	●				●	●	●	
P. confiformis Chv.	5															
P. confinis (Zett.)	6		●													
P. stigmatellus (Zett.)	7		●													
P. pectoralis (Fall.)	8	●	●	●	●				●				●	●	●	
P. mikii (Beck.)	9														●	
P. nonstriatus Strobl	10															
P. maculus (Zett.)	11		●												●	
P. pallipes (Fall.)	12	●	●	●					●	●				●	●	
P. albiseta (Panz.)	13	●	●					●	●	●				●	●	
P. albisetoides Chv.	14															
P. albocapillatus (Fall.)	15	●	●					●	●					●	●	
P. niveiseta (Zett.)	16		●												●	
P. unguiculatus (Zett.)	17														●	
P. zetterstedti Chv.	18															
P. alter (Coll.)	19		●													
P. laestadianorum (Frey)	20															
P. lapponicus Frey	21															
P. sahlbergi (Frey)	22															
P. boreoalpinus Frey	23															
P. alpinus Chv.	24															
P. commutatus (Strobl)	25		●													
P. longicornis (Meig.)	26	●	●	●	●	●		●	●	●	●	●	●	●	●	
P. longicornioides Chv.	27														●	
P. brunneitibia (Strobl)	28														●	
P. difficilis (Frey)	29		●				●						●	●	●	
P. scandinavicus Chv.	30															
P. tuomikoskii Chv.	31															
P. exilis (Meig.)	32	●	●	●				●	●					●	●	
P. pulicarius (Meig.)	33		●										●			
P. nigricoxa (Mik)	34															
P. luteus (Meig.)	35	●	●	●	●				●	●	●			●	●	
P. nigritarsis (Fall.)	36	●	●	●	●	●	●	●	●	●	●		●	●	●	
P. excisus (Beck.)	37														●	
P. sylvicolus (Coll.)	38		●												●	

316

#	Hall.	Sm.	Öl.	Gtl.	G. Sand.	Ög.	Vg.	Boh.	Dlsl.	Nrk.	Sdm.	Upl.	Vstm.	Vrm.	Dlr.	Gstr.	Hls.	Med.	Hrj.	Jmt.	Ång.	Vb.	Nb.	Ås. Lpm.	Ly. Lpm.	P. Lpm.	Lu. Lpm.	T. Lpm.
1	●	●									●			●		●				●			●					
2	●	●	●			●		●			●									●								
3																												
4	●	●				●				●	●			●	●		●			●						●		
5																				●						●		●
6																				●	●		●		●		●	●
7																	●	●	●	●	●		●	●	●		●	
8	●	●		●	●	●				●	●	●																
9																												
10																												
11		●		●							●			●	●					●			●	●	●	●	●	
12	●			●							●			●						●	●							
13	●		●								●																	
14								●																				
15		●	●		●	●		●			●									●			●					
16																												
17		●													●					●			●	●	●		●	●
18																												●
19																												
20																							●	●		●		●
21																												●
22																												
23																				●			●	●	●			●
24																												
25																												
26		●	●	●		●					●	●			●					●	●		●	●	●		●	●
27																												
28								●																				
29	●	●				●					●																	
30																				●			●					
31																												
32			●		●																●							
33			●																									
34																												●
35	●	●				●	●	●			●	●								●								
36	●	●	●	●		●				●	●	●		●									●		●		●	●
37																											●	
38	●	●																										

		Ø+AK	HE (s+n)	O (s+n)	B (ø+v)	VE	TE (y+i)	AA (y+i)	VA (y+i)	R (y+i)	HO (y+i)	SF (y+i)	MR (y+i)	ST (y+i)	NT (y+i)	Ns (y+i)
Symballophthalmus dissimilis (Fall.)	1	●													●	
S. fuscitarsis (Zett.)	2															
S. pictipes (Beck.)	3										●					
Platypalpus ciliaris (Fall.)	4	●			●											
P. confiformis Chv.	5													●		
P. confinis (Zett.)	6	●		●	●						●					
P. stigmatellus (Zett.)	7	●		●	●						●				●	●
P. pectoralis (Fall.)	8															
P. mikii (Beck.)	9															
P. nonstriatus Strobl	10															
P. maculus (Zett.)	11	●		●							●		●			
P. pallipes (Fall.)	12	●	●	●												
P. albiseta (Panz.)	13										●	●				
P. albisetoides Chv.	14															
P. albocapillatus (Fall.)	15	●													●	
P. niveiseta (Zett.)	16															
P. unguiculatus (Zett.)	17			●							●				●	●
P. zetterstedti Chv.	18															
P. alter (Coll.)	19															
P. laestadianorum (Frey)	20															
P. lapponicus Frey	21															
P. sahlbergi (Frey)	22															
P. boreoalpinus Frey	23										●				●	●
P. alpinus Chv.	24															
P. commutatus (Strobl)	25															
P. longicornis (Meig.)	26	●	●								●					
P. longicornioides Chv.	27															
P. brunneitibia (Strobl)	28															
P. difficilis (Frey)	29															
P. scandinavicus Chv.	30															
P. tuomikoskii Chv.	31															
P. exilis (Meig.)	32															
P. pulicarius (Meig.)	33															
P. nigricoxa (Mik)	34															
P. luteus (Meig.)	35			●							●	●		●		
P. nigritarsis (Fall.)	36	●		●	●					●	●	●				●
P. excisus (Beck.)	37															
P. sylvicolus (Coll.)	38															

					FINLAND																				USSR		
	Nn (ø + v)	TR (y + i)	F (v + i)	F (n + ø)	Al	Ab	N	Ka	St	Ta	Sa	Oa	Tb	Sb	Kb	Om	Ok	Ob S	Ob N	Ks	LkW	LkE	Le	Li	Ib	Kr	Lr
1						●	●	●		●				●		●			●				●		●	●	
2						●	●													●							
3																											
4					●	●	●	●	●	●		●	●	●		●									●	●	
5																				●	●		●	●			●
6	◖			◖	●		●						●	●					●	●	●		●	●		●	●
7						●	●			●				●	●				●	●	●	●	●	●		●	●
8					●	●	●		●	●		●		●											●	●	
9																											
10						●								●	●										●	●	
11				●		●	●	●				●	●	●							●		●			●	●
12							●	●		●			●	●	●		●								●	●	
13						●	●	●		●	●														●		
14						●																					
15					●	●	●	●	●	●	●			●		●		●	●	●						●	●
16																											
17		●		●	●	●	●		●	●			●		●	●	●		●	●	●		●	●	●	●	●
18																					●						●
19		◖																		●							
20				◗																●	●		●	●			●
21			◗	◗																●	●		●	●			●
22				◖																●	●		●				●
23		●			●	●	●			●				●	●					●	●		●	●		●	●
24							●			●																	
25																											●
26	◖	◗	◗	●	●	●	●	●		●		●		●	●				●	●			●	●	●	●	●
27																											
28					●		●			●																●	
29					●	●	●			●		●				●										●	
30																										●	
31							●																				
32	◖				●	●	●			●				●												●	
33						●	●			●																	
34																				●			●	●			●
35						●	●	●		●	●		●	●	●										●	●	
36				◖	●	●	●	●	●	●		●	●	●	●	●		●	●	●	●		●		●	●	●
37																											
38						●				●																	

		N. Germany	G. Britain	SJ	EJ	WJ	NWJ	NEJ	F	LFM	SZ	NWZ	NEZ	B	Sk.	Bl.
P. fuscicornis (Zett.)	39												●		●	●
P. ruficornis (v. Roser)	40	●							●						●	●
P. fenestella Kov.	41															
P. ater (Wahlbg.)	42															
P. minutus (Meig.)	43	●	●	●	●	●	●	●	●	●	●	●	●	●	●	●
P. niger (Meig.)	44	●	●			●			●		●		●			
P. aeneus (Macq.)	45		●													
P. maculipes (Meig.)	46	●	●												●	
P. rapidus (Meig.)	47		●													
P. pallidicoxa (Frey)	48		●												●	
P. agilis (Meig.)	49	●	●												●	●
P. pseudorapidus Kov.	50															
P. nigrosetosus (Strobl)	51															
P. cothurnatus Macq.	52		●	●					●			●	●	●		
P. cryptospina (Frey)	53		●													
P. optivus (Coll.)	54		●							●						
P. annulatus (Fall.)	55	●	●			●	●		●			●	●	●	●	
P. melancholicus (Coll.)	56		●													
P. notatus (Meig.)	57	●	●	●	●	●	●	●	●	●			●		●	
P. strigifrons (Zett.)	58	●	●	●	●	●	●	●	●	●			●		●	
P. infectus (Coll.)	59	●	●	●	●	●	●	●	●		●		●		●	
P. interstinctus (Coll.)	60	●	●	●	●	●	●	●	●	●			●		●	
P. coarctatus (Coll.)	61	●	●	●	●	●	●	●	●				●		●	
P. clarandus (Coll.)	62		●													
P. articulatus Macq.	63	●	●						●				●		●	
P. articulatoides (Frey)	64												●			
P. annulipes (Meig.)	65	●	●	●	●				●	●	●		●	●	●	
P. ecalceatus (Zett.)	66		●						●	●			●			
P. calceatus (Meig.)	67	●	●	●												
P. stabilis (Coll.)	68		●	●						●	●		●	●	●	
P. pallidiventris (Meig.)	69	●	●	●	●	●		●	●	●	●		●	●	●	
P. longiseta (Zett.)	70	●	●	●				●	●	●	●	●	●	●	●	●
P. laticinctus Walk.	71		●	●					●	●			●			
P. albicornis (Zett.)	72	●	●	●						●			●			
P. flavicornis (Meig.)	73	●	●	●					●				●			
P. pallidicornis (Coll.)	74		●	●						●			●			
P. major (Zett.)	75	●	●	●	●		●	●	●	●			●	●	●	
P. analis (Meig.)	76		●	●									●	●		

320

SWEDEN

	Hall.	Sm.	Öl.	Gtl.	G. Sand.	Ög.	Vg.	Boh.	Dlsl.	Nrk.	Sdm.	Upl.	Vstm.	Vrm.	Dlr.	Gstr.	Hls.	Med.	Hrj.	Jmt.	Ång.	Vb.	Nb.	Ås. Lpm.	Ly. Lpm.	P. Lpm.	Lu. Lpm.	T. Lpm.
39						●																						
40																												
41																												
42																						●				●	●	●
43	●	●	●	●		●	●	●		●	●	●		●	●		●			●	●	●	●	●	●			
44																												
45				●																								
46		●	●	●		●		●				●		●								●		●				
47																												
48												●																
49		●					●	●	●		●	●		●														
50																												
51																												
52						●				●	●																	
53			●																									
54																												
55	●	●	●	●		●				●																		
56																												
57		●				●		●																				
58	●		●	●	●									●														
59																												
60	●	●	●	●							●				●								●					
61	●	●				●						●																
62		●																										
63											●												●					
64				●																								
65	●	●										●																
66						●		●						●	●	●	●	●	●	●		●		●	●	●	●	●
67		●		●		●	●				●																	
68	●	●	●	●																								
69	●	●	●	●	●	●		●			●										●							
70	●	●	●	●				●		●	●										●				●			
71																												
72			●			●																						
73			●			●									●													
74	●					●				●	●																	
75	●	●				●	●	●			●	●		●					●									
76																												

	No.	Ø+AK	HE (s+n)	O (s+n)	B (ø+v)	VE	TE (y+i)	AA (y+i)	VA (y+i)	R (y+i)	HO (y+i)	SF (y+i)	MR (y+i)	ST (y+i)	NT (y+i)	Ns (y+i)
P. fuscicornis (Zett.)	39															
P. ruficornis (v. Roser)	40															
P. fenestella Kov.	41															
P. ater (Wahlbg.)	42															
P. minutus (Meig.)	43	◗	◗	●	●						◗	◗	◗		◗	
P. niger (Meig.)	44															
P. aeneus (Macq.)	45															
P. maculipes (Meig.)	46										◗	◗				
P. rapidus (Meig.)	47															
P. pallidicoxa (Frey)	48															
P. agilis (Meig.)	49															
P. pseudorapidus Kov.	50															
P. nigrosetosus (Strobl)	51	◗														
P. cothurnatus Macq.	52	◗														
P. cryptospina (Frey)	53															
P. optivus (Coll.)	54															
P. annulatus (Fall.)	55				◗											◗
P. melancholicus (Coll.)	56															
P. notatus (Meig.)	57				◗					◖	◗	◖				
P. strigifrons (Zett.)	58	◖			●				◖							
P. infectus (Coll.)	59															
P. interstinctus (Coll.)	60	●	◖	◖	●				●							
P. coarctatus (Coll.)	61	◗	◗		●											
P. clarandus (Coll.)	62															
P. articulatus Macq.	63	◗												◗		
P. articulatoides (Frey)	64		◗													
P. annulipes (Meig.)	65											◗				
P. ecalceatus (Zett.)	66			◗	◗						●		◗	◗	◗	◗
P. calceatus (Meig.)	67															
P. stabilis (Coll.)	68															
P. pallidiventris (Meig.)	69	◗							◖	●						
P. longiseta (Zett.)	70	◗		●					◖				◗		◗	
P. laticinctus Walk.	71															
P. albicornis (Zett.)	72														◗	
P. flavicornis (Meig.)	73	◗														
P. pallidicornis (Coll.)	74															
P. major (Zett.)	75	◗		●	●						●	◗				
P. analis (Meig.)	76															

	Nn (ø+v)	TR (y+i)	F (v+i)	F (n+ø)	Al	Ab	N	Ka	St	Ta	Sa	Oa	Tb	Sb	Kb	Om	Ok	ObS	ObN	Ks	LkW	LkE	Le	Li	Ib	Kr	Lr
39																											
40																											
41									●																		
42				▶		●									●			●	●	●	●	●					●
43	◀	◀			●	●	●	●	●	●	●	●	●	●	●	●			●						●	●	
44																											
45																											
46						●	●	●		●	●				●	●									●	●	
47										●																	
48						●	●	●		●															●	●	
49					●					●																	
50					●			●		●	●	●	●	●	●		●			●					●	●	
51					●	●	●			●															●		
52					●	●	●			●																	
53					●																						
54																											
55						●			●	●	●					●	●	●		●					●		
56																	●								●		
57						●	●	●												●					●		
58						●	●																				
59					●	●	●			●																	
60					●	●	●	●		●	●		●		●	●		●		●					●		
61					●	●	●																				
62																											
63						●	●							●		●	●										
64																	●										
65																											
66				▶	●	●	●	●		●	●	●	●	●	●	●	●	●	●	●		●	●	●	●	●	●
67					●	●	●			●								●		●		●			●		
68						●	●	●																	●		
69					●	●	●	●		●	●					●				●					●		
70					●	●	●	●		●	●																
71																											
72					●	●																					
73										●																	
74					●					●																	
75					●	●	●	●		●																●	●
76																											

		N. Germany	G. Britain	SJ	EJ	WJ	NWJ	NEJ	F	LFM	SZ	NWZ	NEZ	B	Sk.	Bl.
P. candicans (Fall.)	77	●	●	●	●		●	●	●	●			●	●	●	
P. cursitans (Fabr.)	78	●	●	●				●	●	●		●	●		●	
P. verralli (Coll.)	79		●						●			●	●	●	●	
P. brevicornis (Zett.)	80														●	
P. sordidus (Zett.)	81														●	
P. subbrevis (Frey)	82															
P. hackmani Chv.	83														●	
Dysaletria atriceps (Boh.)	84															
Tachypeza nubila (Meig.)	85	●	●	●	●		●	●	●	●	●	●	●	●	●	
T. truncorum (Fall.)	86		●						●				●		●	●
T. fuscipennis (Fall.)	87		●													
T. heeri Zett.	88		●													
T. fennica Tuomik.	89														●	
T. winthemi Zett.	90															
T. sericeipalpis Frey	91															
Tachydromia terricola Zett.	92	●							●			●	●		●	
T. sabulosa Meig.	93	●			●			●							●	
T. connexa Meig.	94	●	●												●	
T. morio (Zett.)	95	●	●						●							
T. lundstroemi (Frey)	96															
T. arrogans (L.)	97	●	●	●				●	●	●		●			●	
T. aemula (Loew)	98	●	●	●				●	●	●		●	●		●	
T. punctifera (Beck.)	99															
T. incompleta (Beck.)	100															
T. umbrarum Hal.	101	●	●	●	●		●	●	●	●		●	●	●	●	
T. woodi (Coll.)	102		●					●								
Drapetis (E.) ephippiata (Fall.)	103	●	●					●	●	●					●	
D. (D.) assimilis (Fall.)	104		●													
D. (D.) ingrica Kov.	105															
D. (D.) arcuata Loew	106		●										●		●	
D. (D.) simulans Coll.	107		●						●	●			●		●	
D. (D.) pusilla Loew	108		●	●						●			●	●	●	
D. (D.) exilis Meig.	109	●	●										●		●	
D. (D.) infitialis Coll.	110															
D. (D.) parilis Coll.	111		●		●			●	●	●			●		●	
D. (D.) incompleta Coll.	112						●	●	●	●	●	●	●		●	
Crossopalpus setiger (Loew)	113		●			●								●		
C. curvipes (Meig.)	114	●	●		●	●		●	●				●		●	

	Hall.	Sm.	Öl.	Gtl.	G. Sand.	Ög.	Vg.	Boh.	Dlsl.	Nrk.	Sdm.	Upl.	Vstm.	Vrm.	Dlr.	Gstr.	Hls.	Med.	Hrj.	Jmt.	Ång.	Vb.	Nb.	Ås. Lpm.	Ly. Lpm.	P. Lpm.	Lu. Lpm.	T. Lpm.
77	●	●	●	●			●				●			●	●					●								
78	●	●	●			●					●	●		●	●					●	●			●	●	●	●	
79	●	●				●	●			●				●						●		●		●				
80						●																		●				●
81		●		●		●	●																	●	●			●
82																												
83																												
84			●			●					●																	
85	●	●	●		●	●	●	●			●	●		●	●		●			●	●		●	●	●			
86		●									●			●		●					●			●				
87		●				●					●			●					●	●	●	●	●	●	●	●	●	●
88								●						●						●	●			●			●	●
89	●					●		●			●	●			●					●	●		●	●	●		●	●
90														●						●	●		●	●	●		●	●
91																												
92		●		●																								
93	●																											
94											●																	
95						●		●			●																	
96						●					●																	
97	●					●					●	●																
98		●	●	●		●		●			●			●														
99																												●
100																								●				●
101	●	●	●	●		●	●	●			●	●		●		●				●	●	●	●	●			●	●
102																												
103		●	●	●		●		●		●	●																	
104																												
105																												
106		●				●				●																		
107		●				●				●	●																	
108																												
109																												
110																												
111		●					●			●	●																	
112						●				●																		
113			●	●																								
114		●	●	●		●	●	●			●	●					●					●				●		

325

		Ø+AK	HE (s+n)	O (s+n)	B (ø+v)	VE	TE (y+i)	AA (y+i)	VA (y+i)	R (y+i)	HO (y+i)	SF (y+i)	MR (y+i)	ST (y+i)	NT (y+i)	Ns (y+i)
P. candicans (Fall.)	77	◗	◗	◗	◗	●					◖				◗	
P. cursitans (Fabr.)	78	●	●	◗							●				◗	●
P. verralli (Coll.)	79	◗	◖		◗						◖				◗	
P. brevicornis (Zett.)	80															
P. sordidus (Zett.)	81															
P. subbrevis (Frey)	82															
P. hackmani Chv.	83															
Dysaletria atriceps (Boh.)	84															
Tachypeza nubila (Meig.)	85	◗			◖	●					●				◗	◗
T. truncorum (Fall.)	86		●	●	●						◖		◗		◗	
T. fuscipennis (Fall.)	87	◗														
T. heeri Zett.	88				◖						●					
T. fennica Tuomik.	89															
T. winthemi Zett.	90			◗												
T. sericeipalpis Frey	91															
Tachydromia terricola Zett.	92															
T. sabulosa Meig.	93				●											
T. connexa Meig.	94															
T. morio (Zett.)	95		◗												◗	
T. lundstroemi (Frey)	96															
T. arrogans (L.)	97															
T. aemula (Loew)	98								◖							
T. punctifera (Beck.)	99				◗											
T. incompleta (Beck.)	100				◗										◗	
T. umbrarum Hal.	101	◗			◖						●	◖			◗	
T. woodi (Coll.)	102															
Drapetis (E.) ephippiata (Fall.)	103	◖														
D. (D.) assimilis (Fall.)	104															
D. (D.) ingrica Kov.	105															
D. (D.) arcuata Loew	106															
D. (D.) simulans Coll.	107															
D. (D.) pusilla Loew	108															
D. (D.) exilis Meig.	109															
D. (D.) infitialis Coll.	110															
D. (D.) parilis Coll.	111															
D. (D.) incompleta Coll.	112															
Crossopalpus setiger (Loew)	113															
C. curvipes (Meig.)	114										◗				◗	

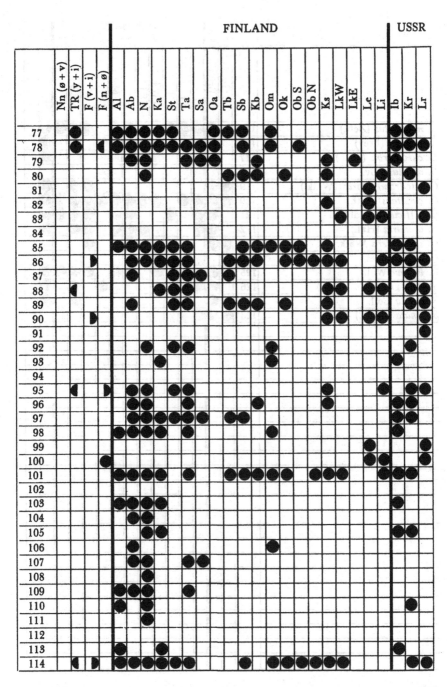

		N. Germany	G. Britain	SJ	EJ	WJ	NWJ	NEJ	F	LFM	SZ	NWZ	NEZ	B	Sk.	Bl.
C. nigritellus (Zett.)	115	●	●	●	●				●	●		●	●	●	●	
C. humilis (Frey)	116	●								●	●		●		●	
C. curvinervis (Zett.)	117														●	
C. abditus Kov.	118															
Chersodromia hirta (Walk.)	119	●	●										●	●	●	
C. cursitans (Zett.)	120	●	●	●	●	●		●	●	●	●	●	●	●	●	
C. difficilis Lundbk.	121		●	●			●		●	●				●	●	
C. arenaria (Hal.)	122		●	●			●		●	●				●	●	
C. speculifera Walk.	123	●	●						●				●			
C. beckeri Mel.	124	●					●									
C. incana Walk.	125	●	●			●		●		●	●		●	●	●	
Stilpon graminum (Fall.)	126	●	●	●							●		●		●	
S. nubila Coll.	127	●	●										●		●	
S. lunata (Walk.)	128	●	●	●												

SWEDEN

	Hall.	Sm.	Öl.	Gtl.	G. Sand.	Ög.	Vg.	Boh.	Dlsl.	Nrk.	Sdm.	Upl.	Vstm.	Vrm.	Dlr.	Gstr.	Hls.	Med.	Hrj.	Jmt.	Ång.	Vb.	Nb.	Ås. Lpm.	Ly. Lpm.	P. Lpm.	Lu. Lpm.	T. Lpm.
115	●							●			●	●																
116			●							●	●	●																
117		●				●																						
118																							●	●				
119																												
120	●	●	●	●			●																●					
121																												
122				●																								
123																												
124																												
125	●																											
126	●	●		●							●																	
127				●	●																							
128																												

		Ø+AK	HE (s+n)	O (s+n)	B (ø+v)	VE	TE (y+i)	AA (y+i)	VA (y+i)	R (y+i)	HO (y+i)	SF (y+i)	MR (y+i)	ST (y+i)	NT (y+i)	Ns (y+i)
C. nigritellus (Zett.)	115															
C. humilis (Frey)	116	▶														
C. curvinervis (Zett.)	117															
C. abditus Kov.	118															
Chersodromia hirta (Walk.)	119															
C. cursitans (Zett.)	120															
C. difficilis Lundbk.	121															
C. arenaria (Hal.)	122														▶	
C. speculifera Walk.	123															
C. beckeri Mel.	124															
C. incana Walk.	125															
Stilpon graminum (Fall.)	126	▶														
S. nubila Coll.	127															
S. lunata (Walk.)	128															

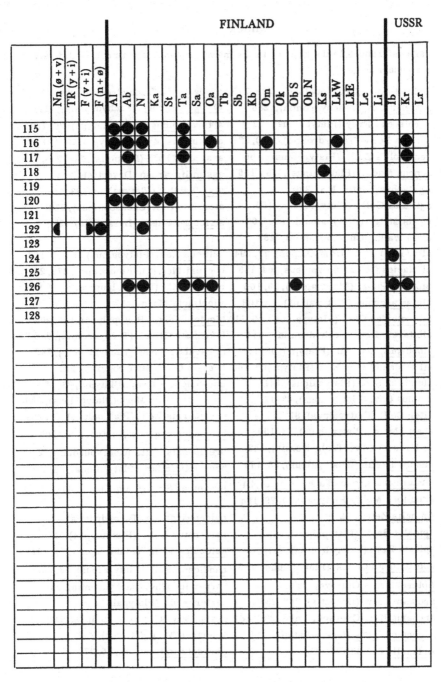

Index

Page references to the key and to the main taxonomic treatment. Valid names are underlined.

Author's address:
Katedra systematické Zoologie Přírodovědcké Faculty University Karlovy,
Viničná 7, 128 44 Praha 2, Czechoslovakia.